PRAISE FOR *Anger I*

"A companion to Lydon's 1994 memoir, *Rotte*
warrant another . . . and he's a gifted enough

—Billboard.com

"Vintage Johnny Rotten." —***Daily News*** **(New York)**

"A dishy chronicle." —***Details***

"Ridiculously entertaining. . . . His tales of a near-Dickensian life in Sixties working-class London suggest how an inchoate rebel found his purpose in punk." —***The Telegraph*** **(UK)**

"Lydon's storied career and pull-no-punches attitude is on full display in the memoir. . . . It's clear that he's a seeker of truth and perspective." —***Esquire***

"Fills in the gaps that his previous autobiography, *Rotten* . . . left wide open, notably his pre-Sex Pistols days, while also going over old ground with a fully-toothed rake. . . . Fascinating." —***Irish Times***

"Lydon is at his best when writing about his family—his parents were working-class Irish immigrants—and . . . is quite moving in his account of Vicious." —***Los Angeles Times***

"Features plenty of morbidly fascinating tidbits from one of England's least likely national treasures." —**RollingStone.com**

"It is clear that, though fond of zingers (he once called Ozzy Osbourne a 'senile delinquent') and political put-downs, Lydon is also a serious and thoughtful artist, bookish and unafraid of hard work, and thus serving as a model citizen in a more ideal republic than ours. Besides, he's a philosopher: We're capable of horrible evil, he writes, but 'because we are also capable of analyzing that, that is exactly why we're better.' A lucid, literate pleasure."

—***Kirkus Reviews*** **(starred review)**

"Coarse, plain-speaking and mischievous, pitched somewhere between Dennis the Menace and Diogenes the Cynic." —***Financial Times***

"Lydon's voice remains as acerbic as ever, idiosyncratic grammar and all, albeit humanized and warmed considerably in his maturity."

—***Austin Chronicle***

"Lydon is an unabashed grammatical scofflaw who can deploy an earthy colloquialism with the best of them. *Anger Is an Energy* is packed with this brand of vivid storytelling." **—SFGate.com**

"You don't pick up a John Lydon book expecting safe, careful prose. This punk rock icon possesses a vast, self-justifying ego—but you wouldn't want him any other way. His passion and his intellect remain an inspiration." ***—NME***

"Lydon's newly published second memoir *Anger Is an Energy* establishes that there's much more to the person than the public persona. Indeed, the book goes out of its way to present the hidden nooks and crannies in the architecture of Lydon's psyche." **—*Paste* magazine**

"Fascinating. . . . Both elegant and blunt." **—*The Guardian* (UK)**

"A ripe, breathless romp through an extraordinary life. . . . But this is a serious book too, about how poverty and illness can create pain that can be turned into something positive, presenting a man keen to fill out the nihilistic cartoon that has persisted in pop culture." **—*The Observer***

"An accurate reflection of the man it seeks to portray: unique, uncompromising, and . . . fascinating." **—*The Mail on Sunday* (UK)**

"The book is most fascinating about his childhood. I was gripped." **—*The Times* (UK)**

"Both thoughtful and irascible. . . . Throughout, Lydon's skills as a storyteller are in evidence. . . . Lydon brings a humour to his recollections and is at pains not to take himself, or the music business, too seriously." ***—Irish Independent***

"Rollicking [and] rambunctious." ***—Irish Examiner***

"A great autobiography, if you enjoyed *Rotten*, then you'll enjoy this too. . . . Lydon is always engaging, challenging and entertaining." **—*The Register* (UK)**

ANGER IS AN ENERGY

PUBLISHER'S NOTE

This autobiography is by John Lydon *in his own words*. Sometimes, the organization of those words does not conform to the traditional rules of grammar. In some cases, the reader will happen upon words not listed in the dictionary, or used in ways one might describe as "unorthodox." The publisher is aware of this—they are not typos and misspellings we have missed; they are part of Mr. Lydon's unique "lingo" and, as such, have been given (mostly) free rein. As John might say, "Don't let tiffles cause fraction."

ANGER IS AN ENERGY

MY LIFE UNCENSORED

JOHN LYDON

WITH ANDREW PERRY

DEY ST.

An Imprint of William Morrow*Publishers*

The Lydons. I can't thank my family for giving me a career, because I did that to myself, but I can thank them for standing by me. Thank you.

Nora. The love of my life. My best friend. The rows are beautiful but the making up is more so. You give me nothing but love and support. Which I hope I'm repaying. Thank you.

I dedicate this book to integrity.

© Joe Stevens

First published in Great Britain by Simon & Schuster UK Ltd., 2014

HarperCollins books may be purchased for educational, business, or sales promotional use. For information please e-mail the Special Markets Department at SPsales@harpercollins.com.

A hardcover edition of this book was published in 2015 by Dey Street Books, an imprint of William Morrow Publishers.

FIRST DEY STREET BOOKS PAPERBACK EDITION PUBLISHED 2016.

Library of Congress Cataloging-in-Publication Data has been applied for.

ISBN 978-0-06-240023-9

HB 01.03.2024

CONTENTS

INTRODUCTION

MAY THE ROAD RISE WITH YOU

Anger is an energy. It really bloody is. It's possibly the most powerful one-liner I've ever come up with. When I was writing the Public Image Ltd song "Rise," I didn't quite realize the emotional impact that it would have on me, or anyone who's ever heard it since.

I wrote it in an almost throwaway fashion, off the top of my head, pretty much when I was about to sing the whole song for the first time, at my then new home in Los Angeles. It's a tough, spontaneous idea.

"Rise" was looking at the context of South Africa under apartheid. I'd be watching these horrendous news reports on CNN, and so lines like "They put a hotwire to my head, because of the things I did and said," are a reference to the torture techniques that the apartheid government was using out there. Insufferable.

You'd see these reports on TV and in the papers, and feel that this was a reality that simply couldn't be changed. So, in the context of "Rise," "Anger is an energy" was an open statement, saying, "Don't view anger negatively, don't deny it—use it to be creative." I combined that with another refrain, "May the road rise with you." When I was growing up, that was a phrase my mum and dad—and half the surrounding neighborhood, who happened to be Irish also—used to say. "May the road rise, and your enemies always be behind you!"

So it's saying, "There's always hope," and that you don't always have to resort to violence to resolve an issue. Anger doesn't necessarily equate directly to violence. Violence very rarely resolves anything. In South Africa, they eventually found a relatively peaceful way out. Using that supposedly negative energy called anger, it can take just one positive move to change things for the better.

When I came to record the song properly, the producer and I were arguing all the time, as we always tend to do, but sometimes the arguing actually helps; it feeds in. When it was released in early 1986, "Rise" then became a total anthem, in a period when the press were saying that I was finished, and there was nowhere left for me to go. Well, there was, and I went there. Anger *is* an energy. Unstoppable.

When I sing it onstage nowadays, it's very emotional for me, because there's such a connection with the audience. I'll get these melodramatic responses, that people are bang in empathy with the actual statement, and the point and purpose of the song. They fully understand it and they share it back with me. Now, that takes your breath away. Often, I can forget my place in the song. I'm so impressed listening to the audience singing it, that they take over. For me, that's complete success: something really generous has been understood by everybody in the building.

Anger is the root core of why I write songs. Sometimes I barely think I'm in control of myself when I'm writing. If there's such things as guardian angels out there—well, mine's a real bleedin' piece of work. There's a great deal of forethought and experience that goes into these things, you see, in the preamble, in my life in general. Once I'm *on*, then the words just flow. And when I'm on, I'm *ON*.

Whatever that thing in me is, it keeps me going and being like this, and being relentless, and understanding things in my way—it's not so far-fetched, after all, from the rest of humanity. It really isn't. We all go through this, but I'm just the one who gets up and says it.

I come from the dustbin. I was born and raised in a piss-poor neighborhood in North London, which was pretty much what you'd imagine Russia to be today. It was very, very controlled. Everything. And the presumption of control too. And people were being born into this "shitstem," as the Jamaicans call it, of just believing that others had the right to dictate to them in that way. Like I said to the Royal Family, "You can ask for my allegiance, but you certainly can't demand it. I'm not anybody's cannon fodder."

I don't think that way of thinking had really come into the British psyche for many years. It had done in previous centuries but it had been nullified, shall we say, through the Victoriana approach. The British have a really delicious history of civil disorder, but by the time the Second World War was over it had all been mollycoddled under the carpet, and was not mentioned in history lessons—but for some of us out there who love to read, well, look what we found.

I could read and write at the age of four or five. My mum taught me, but after I got meningitis aged seven, I lost everything—all my memory, including who my mum and dad were. It took a long time to come back. I'd go to the library after school and just sit there and read until the place closed. Mum and Dad were very good, they trusted me that I'd find my way home, even though many a time I couldn't—I'd literally forgotten where I lived.

I loved getting back into reading, though—history, geology, or anything about wildlife, and then later I progressed into Dostoyevsky. By eleven, I was finding *Crime and Punishment* very insightful—very miserable but sometimes when you wallow in other people's misery and dourness, it's fulfilling and rewarding. Like, "Well, sod his luck, I'm a lot higher up the ladder of tragedy than him!" So books were incredibly important—my life preservers.

There have been conversations here in the United States about why every ex-President opens a library when politicians do not read the books. Hello, America! Kind of explains your politics. For me, reading saved me, it brought me back. And I found myself in there, so when the memories and bits came back, they kind of made sense to me and I realized I was the same person that I was before I lost everything—it's just I was ever so much better at it and able to look at myself and go outside of myself and ask, "Look, what do you think you're doing? Try getting it right instead of just bumping into situations without any forethought."

Maybe I was being hard on myself there—what am I expecting from myself, up to the age of seven? But I'm very, very demanding of me, and that's always gonna be the case. Nobody can write anything really that bad about me that I haven't already thought of, and half the time when they're really being hateful, I go, "Phew, they let me off lightly." As you will see in the pages ahead, I am my own hardest taskmaster, and this book is all part and parcel of me researching myself—a lifelong and ongoing process.

Back in my late teens, I was definitely ready for something. I was fully loaded, and it happened in a most amazing way, because it wasn't anything I was looking for. But as soon as "Would you want to sing in our band?" came in, it was "Wow, yeah! Cor, now all the pieces fit!" and I wasn't gonna give it up too damn easy. I was very resilient even with the others not turning up at the first rehearsals and all of the other negatives that befell the early Sex Pistols.

I didn't arrive with notebooks full of lyrics, they just came straight out. I use my brain as a library. I like to keep notes but usually I'm very dismissive once I've written things down. I can think quicker than I can write, so therefore I've got good storage space between the ears.

It felt bloody fantastic to be able to shout these things out. In all honesty, it was not in my imagination to foresee quite the huge numbers of people that ended up listening. I'd just seen the Pistols as a nightclub act, really. I didn't see much hope in it. Because, like everything else, the music business was well and truly sewn up by then. All of the free-loving bands from the '60s, they'd grabbed all the top-notch seats and they weren't making room on the bus for no one.

Within a year or two, however, a couple of the first things I wrote—"Anarchy In The UK" and "God Save The Queen"—really hit their target. I'd like to thank the British public library system: that was my training ground, that's where I learned to throw those verbal grenades. I wasn't just throwing bricks through shop windows as a voice of rebellion, I was throwing words where they really mattered. Words count.

I was discussed openly by councilors and parliamentarians, who angrily cited the Traitors and Treason Act. That was a deadly thing to be brought up against. It was a very old law, and actually from what my lawyer was telling me, it still carried the death penalty. Ouch! What? For using words? To dictate from a government point of view what you think your population should or should not be doing is absurd. We're the ones that vote them in—not for them to tell us in return what *we're* doing wrong. They should be emphasizing what we're doing right. Civil rights for us all, I say. Judge not, lest ye be judged.

The whole fiasco aroused that naggy little git in me, the idea that words are actually weapons, and are perceived as such by the powers-that-be. What a thrill that became. Absolutely—wow! It was justification for me. It was hardcore and serious, it wasn't done for a laugh. I utterly resent all forms of government. This one was telling me I wasn't allowed to say certain things—in other words, I wasn't allowed an opinion. And so I discovered that I really am toxic for the powers-that-be.

Not many "pop singers" push it that far. I mean, you've got Pussy Farts in Russia now, and I'm so much on their side. I do love bravery. But before them mine was the most extreme predicament I've heard of any pop star ever being in. It was the most political, and the most dangerous, and I laughed all the way through it. Our so-called manager, Malcolm McLaren, shit himself, as did the rest of the group. That's basically why we started falling apart: they were terrified of being dragged into what they viewed as scandals. For me, these were the questions that needed to be asked. It was absolute public research. What can you say, and what can you not say? Why on earth is "bollocks" a word you can't touch? Who's to tell me that? That's what set me off on the road I now follow. Tell it like it is. And never back down.

I saw a live video of Iggy Pop once, just one song, and he was doing "Down In The Street" and I was just so impressed with the bravery of the racket—in no way at all being weak, just FULL ON. There he was with his long, blond, *luxurious* hair, and mascara—Iggy! And it worked for me, because the man wasn't shying away from what his message was. I'm here, get used to it. The sheer relentless bravery of it.

You can't always expect to be accepted and sometimes it's equally beneficial that you're not, but either way, once you've had the bravado to stand on that stage, it's *yours*. Do not run from it. And I do *not* run from it.

I never allowed myself a big pat on the back for what I'd achieved, even though I'd come from nothing, because the next problem was already plonked on my lap, and then the one after that. This is not a trophy hunt for me. These were just things that I felt needed to be stated.

I'd said my bit. The political restraints and presumptions of being British—I thought I'd dealt with them in the Sexy Piss-ups, so then what you do is you move on to the next thing which was internal politics—sort *myself* out, and find out what's wrong with *me*. Before you make your career of pointing fingers at others, you've really got to sort out what might be going wrong inside your own self. So that's how I used my next band, Public Image

Limited—PiL, for short—to stop being a big head, with the complete faith that we would all go into this as equals.

That way, we managed to get some great work done. Really important stuff and thrilling to this day. I love my Pubic Hairs Limited. We totally challenged what everybody considered music to be at that time. It fundamentally changed the concept of music forever. In fact, I changed music twice.

It's difficult to remember the details, but somewhere back in the '80s or '90s, it was communicated to me, "Wouldn't it be a nice idea if you got an MBE?" I suppose they thought I was becoming tame, but you see, they weren't really listening to the content of *Metal Box* and *Album*. The vocals were presumed to be not as insightful, but they really were. The subject matter was internal rather than external, therefore it was presumed that somehow I could be cosseted into the shitstem and Johnny don't go that way. I am very wary of those self-aggrandizing titles; I don't find them necessary. I am actually rather fond of pomp and ceremony—I just don't want a place in it.

And yet, I recently had dealings with the American government while applying for U.S. citizenship, and they told me the British still keep an open file on me, to this day. Go figure!

All I want in life is clarity, transparency, so I know who is doing what, and to whom, at all times. My only real enemies in life are liars, and they'll do everything to stop me because they want the contamination to continue, because it's comfortable for them, or completely ignorant mindless fools who believe every word they read in a daily rag.

I know damn well that the people who will draw most entertainment from this book will be the haters, and practically every second line is going to be justification for their contempt. Well, that's fine. That is somewhat also part of the point. As long as they are thinking, even negatively, at least it's thought! Anger is an energy, remember?

• • •

So, here's *My Life Uncensored*. There should be a caveat to that—*Even Though They Try*. Censorship is something I've always been against. It's the kind of ordinance that comes down from people that don't like to think very hard and aren't prepared to analyze themselves, just judge others, and are scared of the future. The future's unknown, let's leap in, see where it takes us. There's an old quote but it's absolutely true: "The only thing we have to fear is fear itself."

This book is basically the life of a serious risk-taker. Risk-taking's in me. It's what gets the best out of me. In early 2014, I pitched myself forwards for one of my biggest risks ever—three months on the road in America playing King Herod in *Jesus Christ Superstar*. Yes, I know. I was well aware of the shock value in it, and the condemnation I would garner—I love it . . . *love it!*—but that don't matter tuppence compared to what I would get out of it as a human being. It was forcing myself to take orders, and follow a script. The final challenge! Then a week before the show was due to open, it got canceled without any real explanation.

But listen, I'll try to be as accurate as possible without causing too much personal damage, because everybody deserves a chance to get back and repair themselves no matter how many times they fall down. I've led a hard life here, and I don't want it to be dragged down as an unnecessary act of spite against lesser players in the bigger picture. I'll leave the spite to those dogs and rats.

I'll do my best to remember who the fuck I am. I may occasionally refuse to stick to my life's chronology, but I want this to be honest, and open, and the whole truth and nothing but . . . BUT! *I could be wrong, I could be right.*

Everything in life is interrelated. Unpredictability is the story of my life. I make things safe for other people to follow in my wake. I am the elephant in the room. I'm a stand-up-and-be-counted fella, the last man standing—but that's in a world where nobody seems to be able to count.

1
BORN FOR A PURPOSE

"Trials and tribulations!" I wrote in the early '80s, trying to come to grips with the chaos and confusion in which I entered the world. "When I was born, the doctor did not like me/He grabbed my ankles, held me like a turkey/Dear Mummy, why d'you let him hit me/This was wrong, I knew you did not love me." Three verses later you arrive at the conclusion that I was a very disgruntled baby.

That song, "Tie Me To The Length Of That," I'm really proud of. At the time on TV, there were a lot of medical programs where they were showing actual births. They were breaking new ground about what you could actually show, so watching all these babies popping out all over the shop, I was like, "Look at that, they're smacking that poor little babby's bottom as it comes out." They do it with good reason, but I was just thinking, how traumatic must that be, from the sanctity of the womb to "There's a good slap on your backside, fella!"

My father was furious when he heard the song, because there's a reference to him, the "stupid drunk—then the bastard dropped me." This was a story my aunt told me, and one my mum later reiterated, that he turned up drunk, the proud father. He'd taken a day off work and, in the panic of it all, one thing led to another. I was born in the early wee hours of the 31st, in the bitterly cold January of 1956, and he'd been "panicking" all night.

He was furious at his portrayal. "It whaddn't loike dat! *Well* . . . it moighta happened but not for da reasons yer t'ink!" Poor Daddy. I wasn't doing it to be spiteful or get back at him. As I say, I was just trying to translate into song the emotions I must've been going through as a newborn. That's why I love writing songs; it's absolutely researching myself to the nth degree.

There's a picture from my own parents' wedding which is of fabulous interest to me because there, in the far-right corner, is my Auntie Agnes holding a baby. The most likely explanation is that this baby must have been me. So: I'm a bastard! In recent years, I've even had to deal with other children apparently born to my mum out of wedlock. I never could get honest answers out of all of the relevant family members. None of them like talk, everything is hush-hush, and so everything is a mystery. Certainly, until I've sorted out the mystery in my own life and my own position, I find it very difficult to deal with other alleged family members.

I had no birth certificate, and I suspected I possibly wasn't born in London because maybe my father was worried about being drafted for National Service, so he had to duck and dive a bit. For obvious reasons, I have to be vague about it, as indeed my mother and father were with information about themselves, or anything at all. It was like trying to get blood out of a stone. "Hello, am I a member of this family?" "*Well*, ye know . . ." that would be my mother's sense of humor, which was very hard to grasp when you're young. It kept me in a constant state of alertness—to come back at things from another angle. So many games of noughts and crosses your parents can play, teasing their children. It all becomes very useful in adulthood.

It taught me to be sharp. Rather than them just ignore you, and tell you about the tooth fairy—it was a higher level than that. They're not plying you with fantasy. It was obvious in our house that if Santa Claus tried to come down the chimney, one, he'd be burnt, and two, he'd be beaten to a pulp as a very suspicious character—a priest of child-molestation quality!

In them days, it was a little bit different to now. You didn't trust no one. Mum and Dad were very backwards folk—not dumb, they were clever in their own way, because they were survivalist—but as to how situations in England worked, they always felt manipulated.

My dad, John Christopher Lydon, came from Galway, and he was used to working on all manner of heavy-duty equipment. He came over to London at fourteen, looking for work on the building sites, and he quickly got a license so he could drive cranes and things. He'd never seen himself as a shit-shoveler.

His father was a violent, brawling, *weird thing*. He came to England before my father, and he lived nearby, but the two never liked each other very much. My dad was always over there at his place, trying to connect somehow. It was very grim. We used to call him the "Owl Fella," as in the Old Fellow—he never looked much like an owl. He was a prolific smoker. He used to smell of cigarettes all the time, and he always had a fag butt sticking out at the corner of his mouth. He talked very guttural, and it was hard to make out what he was saying, because he was obviously a fully fledged alcoholic, and a definite playboy for the prostitutes. It was very odd watching their relationship.

My mum, Eileen, was very loving, but in a very quiet way. There wasn't much said. That's all you need when you're little—attention from adults, but the right kind of attention. Mum always had something wrong going on health-wise. They were only seventeen or eighteen when they married and started having us.

My mother's family, the Barrys, came from County Cork—a place called Carrigrohane. Apparently they met while he was working there. We'd have to go to their farm every summer holiday, all to please my mum, really. They could hardly bear us and, annoyingly, we had to tolerate them. They'd sit around not talking to each other. My granddad and grandmother from my mother's side weren't big talkers. In fact, the whole family would sit in silence for days, but for forcing words out of them. A very quiet way of being—very strange. That would drive my dad spare, because Dad was a talker, in his own way.

There was some kind of resentment buried about my father. They wouldn't talk to him, but he endured it. I think it was all to do with . . . he wasn't good enough for her, which was a very strange proposition, because again years later we find out that my grandmother from my mother's side was ostracized from her family for marrying Jack Barry, my mother's father, who was something of a war hero in the "fight for independence," ha ha.

Apparently her side "had money"—whatever that means. It's hard to explain outside of Ireland, but money meant you owned the farm. Jack built his own farm after the war, when the South won their rights. So he obviously did well for himself, but he was prejudged, and the Irish can be incredible snobs—much more so than anything in Britain, even with the class structure. It's always lurking there.

Life for us in London was very inner-city and deprived. Everyone around us was piss-poor. We had no concept of what money really was. We lived on Benwell Road, which is where Arsenal have now built their Emirates Stadium. It was right by the railway bridge, in a Guinness Trust block called Benwell Mansions. There was a shop out front that was occupied at the time we moved in by a tramp called Shitty Tom. You went down a hallway, and we lived around the backyard in two rooms—a kitchen and a bedroom, with an outdoor toilet, which was available to the public. You'd find drunks passed out in it at night, which meant we had to grow up very accustomed to using the pisspot. There was also a bomb shelter there, but because people used it to dump rubbish, it was full of rats.

In the bedroom was Mum, Dad, me, and then my younger brothers, as they arrived—Jimmy, Bobby, and finally Martin.

Then it was six—four kids, two parents. We weren't touchy-feely as a family; you really didn't need to be. You imagine—two double beds and a cot, in a tiny room with an oil heater, and you're touching each other all the time accidentally. The very last thing you want on top of that is huggy-poo. Because come winter you're all wrapped under your old coats anyway.

The rent was £6 a month, something like that. To this day, when I hear that racial slur, like, "Look at them Pakis, eight to a room"—I think, "Well, hello, not only are those the words of racist bigots, but I actually grew up like that." I know most people around me did too. We weren't thinking it had anything at all to do with the color of your skin. It's economic deprivation.

When Shitty Tom died, we moved into the front room. That man never ever threw anything out, so you can imagine the pile. And the smell didn't go away for a long time, because he was in there for a week before anyone found him. There always seems to have been stinky, smelly dead bodies around me.

I had to learn botty-wiping at a very early age for my younger brothers. It was through necessity, that's just how it was. My mum was very ill for much of the time, and somebody had to do it. I'm not at all disgusted by it now, that's humanity. I think it was a great thing that my mum asked me, would I? And I did. I liked the responsibility of it. I knew I could be up at the crack of dawn, and I didn't mind making porridge. I liked sorting things out.

Around our neighborhood I think there was a lot of that: people looking out for the younger ones. These are all community values that are sorely being dissipated. I don't mean that in a romantic delusional way, because I imagine things before the Second World War were, "I hate you more than you hate me." I don't imagine there was much of a sense of community other than the incredibly arrogant Victorian toffs and the incredibly starving-to-death others. But after the war I suppose community was a different thing; it had to be pulled together because that was the only way to survive.

Dad was away a lot of the time. Often we'd go with him, wherever his work was. When I was about four, we lived in Eastbourne. What a hellhole that was. My memory of it was terrifying, because our flat was right on the ocean, and listening to the sea at night absolutely scared the hell out of me. I just couldn't help but think a wave would come in and drown us.

For the vast majority of the time, it was Mum looking after us. With Dad not around, I absolutely didn't mind looking out for her. I liked the responsibility. It's instinctively in me to look out for people—that's what I do.

My mother was always very worried. In them days, it'd always be the players trying to pop round, thinking, "Hmmm, a woman unprotected." There'd be a knock on the door, and she'd say, "Close the curtains, be quiet, wait till he goes away." We grew up very wary of strangers in that respect. Of men. Don't trust them. I felt very, very protective of her. It's the one area where I go into overdrive, when I think my family or my very close friends are threatened. A different situation comes on. That's where Gandhi gets a bazooka.

My mum was always ill. Endless miscarriages didn't help her none. I don't suppose they knew much about safe-sex procedures in them days. Indeed, they would've viewed that as a mortal sin, as indoctrinated into them from on high—Catholic priests inflicting children upon you.

One time she had a miscarriage, and I was the only one with her in the flat. There were relatives all around, but sometimes you're on your own, Jack. It's quite a thing to carry a bucket of miscarriage—and you can see little fingers and things in it—and have to flush it all down the outdoor toilet. There wasn't a phone in the house, so I had to deal with all that first and then go to the doctor, which was a long walk.

There were various other family members on hand to help out. Auntie Agnes, who'd married my father's brother, lived in the same housing as us in Benwell Road. Then there was Auntie

Pauline, who first came to live with us when we still only had the two rooms at Benwell. Looking back on it now that I'm an adult, I can't conceive of how difficult that must've been for my dad and my mum in one bed, with my mum's sister with me and Jimmy in the other. That's up close and comfortable—not!

But I loved Auntie Pauline. She was like the big sister I never had—fantastically warm, but at the same time absolutely remote, in the Barry style. Once Shitty Tom died, we had an extra room for Auntie Pauline, which is where Uncle George came in. I loved that fella. He was so great.

At Christmas we had to go to church but Auntie Pauline refused. By the time we came back, she'd gnawed the heads off all the toy soldiers I'd just been given as presents. To this day, I don't know why. When George came back, he'd bought me a house-building kit on the principles of Lego, but obviously cheaper. He opened that up and stole away my tears. I played with him all afternoon and I'll never forget it, because he spent such a long time teaching me things and got me involved.

After a few years, he married Pauline and they moved to Canada. I was very impressed at the wedding for so many different reasons, chiefly for meeting George's brother. I can't remember his name, but he was an absolute Celtic hooligan with a 45-degree crevice across his face. He was like, "Aye reet. Ah goat hit wi' an axe!" Gosh, how impressive! That's a fucking street fighter, mate. *Wowzers!*

My mother was devoted to making me an intelligent human being. It was her who taught me to read and write at four, a long time before school. By the time I finally got to Eden Grove Primary School, a Catholic school, it was a very serious problem for the nuns, because I was left-handed, and fluent. It was like, sit in the corner and wait for the rest of the class to catch up. The indolence crept in, and—for whatever reason, even though I was very shy and quiet—resentment from the nuns. So they'd hit me with "Oh, you're left-handed, that's the sign of the devil." What kind of message is that, to give a five-year-old who can already read and write? What evil, spiteful nonsense is that?

That followed through bitterly, this absolute dislike of me, for being a smarty-pants or whatever. They'd beat you with the sharp edge of a ruler on your right hand but, because I wrote with my left hand, they hit me on the left . . . to make sure that I'd write with my right hand! But you can't do that. That's the way my brain's wired. And it was utterly ridiculous because I didn't need reading or writing lessons. I'd done that at home.

Eden Grove was a small school directly connected to a Catholic church—all the upstairs classes led in through a gangplank into the church, and the downstairs ones through a courtyard, so you really couldn't avoid it. Everything was holier-than-thou, and everything you did was wrong and God would punish you—such a peculiar attitude. It wasn't anything I'd been expecting, up to the age of five, just how wicked they were.

Priests always frightened me. Going to church was terrifying as a young kid. They just always struck me as being very similar to Dracula or characters in Hammer horror movies. Christopher Lee! They always came over in that dogmatic, dictatorship way, and that condescending judgment. The nuns were worse because they were smelly old women with a bitter hatred of mankind. Brides of Jesus? I'm sure that's not what He had in mind.

Many of the locals weren't too happy with Irish immigrants full stop, but they certainly weren't happy with a Catholic school, attached to a church, in the middle of these working-class council flats. They viewed that very much, I suppose, as people view a mosque today, as an alien agenda, and considered you an outsider for having anything to do with it.

I never felt Irish. I always felt, "I'm English, this is where I come from, and that's that." Because you'd be reminded of that when you went to Ireland: "Ye're not Oirish!" the locals would say. So it was like, "Bloody hell, shot by both sides here." I still love that Magazine song—so relevant to me, those lyrics.

My brothers and I talked the local lingo, but I'd really forgotten how broad my parents' accents were. My mother's in particular was very deep Cork, and very country. After Malcolm's passing, we were looking through Sex Pistols footage, and I found a tape of my mother being interviewed. It was all buried away in warehouses and, when I heard it back, I was shocked at how broad and hard to understand her accent was. It was almost unintelligible to me.

Mum and Dad tried to be religious, but obviously that didn't work too well. The Catholic Church is all about money, and we didn't have any. On Sundays we'd be dragged to church, but Mum and Dad were good in that it was never early-morning church when we were very young, it was always the 7 p.m. service, which was great because that meant we missed Jess Yates doing *Stars On Sunday* on the TV.

At school, I was working all this out for myself. Did I know there was sexual abuse going on there? Oh yeah, abso-fucking-lutely. It's institutionalized abuse, and covered up and condoned. Everybody knew to run when the priest came a-visiting, and by no means ever get yourself involved in the choir, or any altar-boy nonsense, because that was direct contact number one, so I learned how *not* to sing very successfully—deliberately—bum notes, because I knew that would be a really dangerous thing to be waltzing into. So the love of singing was kicked out of me because of bloody priests. Imagine the joy of eventually joining the Sex Pistols, and making the world a better place—in a very vengeful way.

But for all that I was a quiet but happy little bunny. There was dirt and poverty and England was just out of rationing, but a nice hot English summer's day seems to have mattered more to me. That's my fondest memories, moments like that. What they call salad days. I never understood what that term meant when I was young, because salad was something I dreaded. My mum's idea of a salad was Heinz Salad Cream, and awful pale-looking green leaf things. The only joy in it, of course, was the beetroot, because I love pickled beetroot. I can sit and eat a whole jar at a time. I love it! And I loved gooseberries too; my mum would buy them in the summer. Now, I can't bear them. They're vile. I don't know how on earth I could tolerate something so sour. It was punishing to eat them, but maybe it was scurvy or Vitamin C deficiency that made my body crave them.

I liked the clothes that my mum would put us in. I adored the tartan waistcoats, and the little checked suits with the jackets, shorts, and waistcoats. I liked all of that. She dressed us well, very matchy-matchy with Jimmy, but that was all right. It was kind of like, our gang wear this, and that's that. That wasn't what other kids were wearing, so maybe that somehow crept into me, as being important to be individual.

I appreciated it very much over time, because I know how poor we were. I know how much effort it took to dress us at all. It was always there, that we couldn't afford nothing. There's almost a fond memory, too, of near-starvation once—no money at all, so all there was for dinner was one can of Heinz Mulligatawny between all of us. It was Dad's homecoming present to us, so there we are, all sitting around the one can of Mulligatawny. I don't think they make it any longer, and with good reason. It was like a curried soup, and at the time for us the curry in it was inedible—burny-hot. And so, "I'd rather starve." "Well, *starve*, den!"

You'd see big houses and things, but you wouldn't have any relationship to it at all, didn't understand it. It didn't make sense to me that people could live in such large places. I always used to think, "What do they do with all them rooms? How do you sleep at night knowing there's so many windows to lock?"

I loved the summers, because it meant we could be out all day long, with no need to go home at all—in fact, even forget that was home. And be so bitterly upset when it got dark in the evening. You'd hear the yelling and the screaming, "Wherr *aaiir* ye?" There were bombsites from the war, and thousands of kids running rampant in them. They were absolutely like adventure playgrounds, thrilling. Amazing, a wonderful thing, a bombsite, to a kid. Never get bored, always something new to unravel and explore, and of course the factories too.

Bloody hell, at five, six, seven, trying to break into the factories was thrilling. The whole area around Benwell Road and Queensland Road was still all blown up from the war but they were putting factories in and around it. There'd be a whole bunch of us—everything you did in them days, there were twenty kids involved—and we'd build makeshift ladders out of bricks from the bombsite, to climb up the walls. Once you were on the roof, everything was easy, you'd just drop in. It was a challenge, and I liked that.

There was a Wall's Ice Cream factory at the top of Queensland Road, and that was a magnet to try and break in there, but it was impossible—it was too modern, and had iron shutters and grilles and padlocks. Instead, you'd wait for the vans when they were loading, and when the workers would go in to fill up the trolley and bring it back, you'd try to nick a lolly. Every and any way to nick a Raspberry Split—that was the lolly of the day. Wall's ice cream inside and raspberry ice on the outside—absolutely the most delicious lolly, and *anything* to get one for nothing.

The ice they used to pack the ice creams in between—it wasn't liquid nitrogen, but something like that; there's some chemical in it to keep them cold while transferring between the factory and the truck. One time, for a dare, I put my tongue on what I thought was an ice block, and it wasn't, and it took a layer off. "Go on, I dare you to lick it!" "Uuuurrhh, I'll do anything, I'm mad!" "Run, here they come!" "Ulluullullulleh!"

Another time, I got caught breaking in with my cousin Peter, Jimmy, and two other kids. These coppers dragged Jimmy and me back to the house, and they must have seen the anxiety on our faces. My dad answered the door, and they said, "Are these your kids? We caught them breaking in . . ." He went, "Therr not mine, nottin' to do wi me!" It was obvious they were nodding and winking at each other, and the police go, "Well, we don't know what to do with them, maybe we should take them up north and leave them there?" Oh, the sense of abandonment! I cried my eyes out. It sounded very real.

As adults, I suppose they were having a laugh about it, both sides. It was only an empty garage we got caught in, there was nothing in there. It was a smart way of telling you, "Stay out of what's not yours." And, "Don't get caught"—that was always my dad's bottom line. "If yer goanna do stupid t'ings, don' get caught—don' fockin' embarrass me!"

So we were eventually let in, but made to stand outside for a while, and think about what we were doing. It worked. It ended the "letting ourselves into other people's property" phase. Who knows where that would've led? It's a slippery slope, thievery and burglary and all of that, and presuming other people's things are your right.

But that's how London was. Not a lot of cars, empty streets, street lighting was poor, and there were just hundreds and hundreds of kids unsupervised, getting up to God knows what on bombsites. But not really unsupervised, it was, "Get oat an' lerrrrn, an' when ye com' home, don' bring da police wit' ye!"

Meningitis came from the rats. They were all over the place. They piss on the ground and, as rodents do, drag their bums leaving a urine trail. Meanwhile, I'd make paper boats and float them in the potholes in our backyard, so I'd touch the water, and then touch my mouth, and that's how I got infected.

It didn't come on overnight. I'd had very bad headaches, dizzy spells, fainting fits, and imagining things that I knew weren't there, like green dragons breathing fire. That was the awful thing about it, watching myself inside myself, panicking over something I knew wasn't there. But I could not stop my body doing that. Screaming fits of total fear.

The night before I went into hospital, I had a pork chop, and I've never been able to eat pork chops since. I absolutely can't go near 'em. Even the smell. I don't mind crispy bacon, but a pork chop—no! Because I blamed everything on that, for many a year, so I ended up convincing myself that it was the pork what did it! How very healthy of me.

The next morning, when my mother thought, "Oh gosh, this is getting bad," the doctor came and I blacked out while he was in the house. The next thing I knew I was in an ambulance, and I blacked out again, and then months later I woke up in a hospital. I was in a total coma for six or seven months. Once I went into that, that was it, there was nothing that went on at all.

When I came to, I remember them waving fingers in front of my eyes, going, "Follow my finger." I deliberately didn't, because even though I was really seriously ill, I thought I should feign illness on top. What on earth convinced me to do that? But I remember doing it at the time, so I was always a cheeky little sod, even to myself. Quietly malevolent, even in illness!

I was in the Whittington Hospital, which always made me think of Dick Whittington, a positive association. I was on a huge ward of forty kids, many worse off than me, so self-pity was not an option. There was a great library in the middle, loads of fascinating books, some way beyond my capabilities, but that just enticed me more. It's odd what goes and what doesn't. I hadn't forgotten how to read, yet I couldn't talk—language was gone. I'd be thinking I was formulating words, but they told me after I was just making noises.

Sometimes, as much as three times a day, they'd drain the fluid in my spinal column—the "lumbar punch." "This is gonna feel like a punch up your lumbar, John!" The needle was very painful as they'd insert it in the base of the spine, and then when they'd draw the fluid out, you could feel it all the way up your back and into your head. Absolutely nauseating. I have a complete fear of needles from that. I hate them. I recommend, before anyone becomes a heroin addict, they go get a lumbar punch—that'll change their mind about it. A most dreadful thing, and so embarrassing too, even at seven and a half, to have something like that prodded up your rear. I always felt my bottom was my own, and I don't like bottom-watchers. They'd quite literally pin me down, the nurses, while they did that. I'd be so screaming in fear of it, because I knew the pain that was about to come.

It definitely had a long-term effect on my posture. It curved my spine—if they drain too much fluid, it can do that. I was supposed to then walk around with a broom handle between my arms to arch my back and make me stand up straight but, to this day, if I try to stand up completely dead straight, I feel very dizzy; it cuts off the blood supply to the brain, so I'd rather walk around like Richard III there.

It also totally affected my eyesight. I had to wear glasses for a long time, but in the end I couldn't bear them. I've got very good distance eyesight; I can see far away very clearly, but up close it's a torment for me even to clip my nails, because it's all a blur, so I wear glasses for that. I have to glare to focus on people. Lucky me, huh? People think, "That scary cunt!" Ha ha.

After another four or five months recovering in hospital, I'd become totally institutionalized. I got comfortable not knowing anything. That's a condition that, thank God, the doctors, my parents, and the whole lot of them just wouldn't tolerate. My mum and dad dragged me out of there kicking and screaming. They told me that they were my mum and dad, and I had to believe them. "You belong to us, you're our son, we love you." "Oh! How do I know that?"

Being back home was very confusing, because I just didn't understand where this was. It was rather like being in a waiting room, and forgetting what you're there for—you know, when you're left so long waiting that you forget why you're there—or like trying to sign on the dole, that kind of abandoned feeling. I couldn't adjust; it took an awful long time. Why was I here with these strangers? It wasn't making any sense. The only way to deal with me, because I was in a constant state of agitation and panic, was to quietly try to get me to think what it was that was bothering me, and why I wasn't recognizing things, and that I *did* belong there.

Oddly, I never felt out of place with my brothers. I instantly felt right with them; they never acted like there was something wrong with me, which was what all the adults did. It was good—Jimmy would say things like, "Where have you been? You've been away a long time!" And then the answer was, "I don't know." He just thought I must've gone on a long holiday alone.

Once I began to accept my parents, it was like opening the door in my mind. It just clicked in my head, and the memories started popping back. It took an awful long time for the information to come through, but come through it did, in bits and pieces, and always it was a sheer joy. I'd run to my mum—I couldn't wait to tell her that I'd remembered something, and that it made sense what she was telling me.

When I accepted that they were who they said they were—what an emotional breakdown, and an eye-opener too. They talk about Catholic guilt—but having doubted your own parents is a guilt that far surpasses anything religion can plonk on you. An insane guilt. But it was so wonderful to realize they weren't lying. They really were who they said they were. What a fantastic revelation!

I still didn't believe them for years afterwards, though, that I really did have to go to school. I never believed that. I'm making light of it, but I'm deadly serious; this is how an eight-year-old, when he's just come out of hospital and doesn't remember fuck-all about himself, will be. Many a time I'd forget the way home, and just wander aimlessly. I'd walk into shops. Luckily, because of the community spirit, they'd go, "Oh, you're the sick one, we'll show you where you live." But then you build up a resentment to that. "I'm not sick!"

In terms of rehabilitation, the National Health Service didn't supply any usefulness at all—quite literally nothing. My mum and dad told me that all they were advised by the hospital was to never let up on me, never mollycoddle me, or baby me, because if I fell into a lazy-arse way about it, I'd never resolve my issues. And being agitated got me to think. Agitation's a powerful tool sometimes.

You were more or less abandoned by the state, and you were definitely abandoned by the school. So much happens in a year—you're so behind. Regardless of losing your mind, you're behind anyway by a year. Everything becomes an escalated problem. Trying to blend back in was very difficult. That was a friendless first year, very friendless, and kind of lonely, because of the kids' attitude—"Oh, he's sick, keep away from him!"

I hated school breaks and lunch because it meant I had nothing to do. No one would talk to me; the rumor ran around the school that I was a bit "out there," and so that's exactly where I found myself, cast out on the outside. I know what that loneliness is, it's very, very fucking damaging. The only people that talked to me at break time were the dinner ladies. They were very kind Irish women—"We heard you were ill—how are you?" I didn't even really remember being ill, just—"Why am I here?"

Just to give myself something to do, I thought I'd stay late and join the Cub Scouts. Hated it! Hated bloody sitting in a circle and going, "Dob dob dib!" It meant nothing to me. To me it was very antisocial because it was full of rule books and you've got to get this uniform, and when you earn this badge you get so many merit points. I realized within about half an hour that this was an absolutely pointless waste of my life. There was the scout master who was, well, a creep, coming across very much like a priest, dark and shadowy. You know, that smile they all had, you could see the gritting of the teeth. I only attended the one night.

One of the nuns one day called me "Dummy Dum-Dum." That nickname stuck around the school. It's deeply shocking, what them bitches put on you. From the boy who could read and write at four, to Dummy Dum-Dum. It was a real challenge to break through that, but I did. Within a year or two, I was back up in the A grade.

Those fuck-arse hateful nuns made life punishing, so I educated myself. I just got on with it. If there was a book about, I'd pick it up and read. I *loved* reading. Not newspapers, they bore me. It's yesterday's opinion—I've always felt that. No, it was books, books, books—anything and everything. After my illness, I got onto a course at the local library after school, and I'd go there and paint till nine at night, then take home a load of books and read them until I fell asleep, fighting off sleep all the time. I had that constant fear of not waking up, or waking up and not knowing who I was again. I tell you, that's absolutely the worst thing that can happen.

What I learned is, the harder you work, the more you get. That's been my experience, and I absolutely don't mind hard work. In fact, I love hard work almost as much as doing nothing at all. I like my life to switch between those two things. When I was about ten, a friend of the family let me have a go running a minicab service every weekend. Even though I was still trying to remember who the hell I was, I was smart enough for that.

I loved that job, absolutely loved the pressure, the stress. You really had to have a very clear, concise memory. You're running up to sixteen drivers all at once, and you have to remember where all of them are, and ring them up, and talk to them on the radio, and book jobs in advance. I loved it, always in a state of near-collapse. Just on the border of messing up—but, never! The responsibility of it—I really felt proud about myself, and that helped me no end.

I soon discovered that words were my weapons. I learned that I could get out of a tense situation and not be bullied, with comedy. Or the correct formulation of a sentence, that would leave them baffled and amused. And therefore you became accepted, as strangely strange but interesting. Of course, when I turned that artillery against the teachers, who I viewed as complete lazy fuck-ups, that interested the rest of the class very much. I became something of a spokesman of terror, with no viciousness or violence in it at all. I'd always make sure that my arguments were correct—it wouldn't be just disruption for the sake of it. My ambition is to get to where I want, to achieve the correct information level, and then go on to the next problem.

I expected everybody else to tell me what was what, when I had no memory. It was vital to me that what they said was true, as I was desperate for the answer. I'm still like that; I want to believe what people tell me. I'm very open and trusting, but some people can push that too far, as we know in life—people who have their misguided selfish directions that they obscure from you.

The memories all came back, almost photographically, over the years. That's why I'm not prone to exaggeration about actual facts from my life. They're so vital to me. I hinge on them. I don't know what you would call the system, but you siphon out fantasy from reality. There's a significant way of doing it. Even before, when I was having terrible visions and nightmares, before I went to hospital, I'd imagine a dragon at the end of the bed, and my mum and dad would be going, "There's not a dragon there." And I knew they were right, there was no dragon there—I didn't *see* it, but my brain was telling me it was there. You know your brain is tricking you. That's why I would put something like what you would call a soul, as separate from the brain. The two talk to each other, so I see them as separate entities.

Quite frankly, I don't have very much fantasy going on in my head. I don't have room for it. Maybe that's why I'm mistaken sometimes as being a bit blunt. I really don't like time-wasting. It takes an enormous effort for me to get up in the morning, but absolutely tenfold to get to bed. I don't like sleep. It frightens me, in case I don't wake up, or don't remember myself. That will be with me, I suppose, for the rest of my life. That won't go away, so I'm rather prone to the "stay awake and alert" side of life. I may've had some "assistance" doing that, over the years, ha ha.

For a while, after leaving hospital, I'd still have visions—terrifying ones. There was one that reminded me of a priest. To this day, it still comes back every now and again. It's very tall and thin, black hair, black eyes, and very, very evil, staring at me. It's a real challenge: he comes sometimes in dreams—I have to force myself to confront him. If I do that, it goes away. But it's very hard to get myself to do that. In a state of dreaming, you've got no control. But somehow or other I've managed to control my dreams. I've had years of practice.

In short, I survived a major illness that had its effect on the way my brain now operates, and that's part and parcel of the making of me. I don't know what the mechanics of the recovery are, but when I read modern research on how the brain works, or the scientific approach to human life, I know there's a bigger thing in there. There is a personality; it's not just a series of chemical equations—there is a heart and soul, above and beyond the sheer machinery of the soft machine, which is the human being.

I know it was a strange childhood and all of that, but my mum and dad taught me a sense of independence, and an ability to work out what a problem is, and being able to tell a reality from a fantasy. I loved watching this TV program when I was a kid, called *Mystery and Imagination*, and it was pure horror. It used to come on late on Sunday night, and they never wanted me to see it, and that of course made me want to see it all the more. I love a good horror story or a ghost story, but I know the reality of these things to be different—and that's proved extremely useful.

I do laugh at the stuff that comes on TV because they're missing by a mile what's really going on, but I don't laugh at the idea of picking up on psychic things. From time to time I'll see things. I'm aware of atmosphere, and I don't know quite what that is but I'll pick up on a thing and I'll know if the mood or the tempo in a room or a house is a bit off. I will feel presences and I do know the difference between imagining and the reality in that. I can feel the vibe. It's an empathy for the tunings of your surroundings. There's a way of tuning in and out. I can completely ignore it or I can let it happen and then you will see things. Sometimes the visions or situations are forced on you.

Many years later, in this old recording studio, the Manor, I definitely, totally, completely felt what I thought was a cat jump on the bed when I was in it. I knew it and I felt the way it moved. I felt it was telling me it was a cat but I couldn't see it. But I kind of knew it was there. Whereas before I went into the meningitis coma properly, I would imagine a dragon at the end of the bed but my mind would tell me it *wasn't* there. So I do have a good watchdog inside my head and I understand the difference quite clearly. Hard to explain but it's there.

I've seen many things. I knew when my granddad, my mother's father, died. I ran over and woke my parents up and told them. I'd seen a huge flash in the corridor. There was no reason for a big bright light to be there; it seemed to be looking around and searching. I went out and I followed it into my mum and dad's room and I told them what I'd just seen. I'd seen things like that before. "What is *that*?" It's not *Most Haunted*. That's what it's not. For me that's total fraudulence, whatever it is those fools get up to in the cellars of allegedly haunted castles. It's something else: it's clued into a pulse that's currently available to those that know where to dial it in, on the radio that's called your brain. It holds no fear for me; it's one area where I am extremely brave. It either doesn't exist at all or it does and I've found a way of it not presenting any damage to me.

Now, again, back to hospital, there would be images in my head of characters that would stand around the bed or off in the distance in the hospital ward. I still remember them. One of them is the extremely tall priest, that ominous, odd character who turns up every now and again. He seems taller than the space he's occupying; it's not in any dimension I can understand. But I know it's malevolent and I know how to stop it. I'm usually sound asleep when this is happening and I force myself to wake up and stare at that particular area, where I'm imagining this thing to be. By doing that it's gone, it's dissipated. I can do that if I don't like the dream I'm in—I can find the way out, back into consciousness.

It's usual that these incidences occur when you're alone. That's a great skill, to come through that. It gives you a great sense of empowerment that you've conquered the assault on your psyche. It *is* an assault, a challenge. You have to win through it and it makes you feel stronger somehow. Maybe that's just my mind going through daily exercises. I don't do physical exercises but it's clear I run the mental gamut—or gauntlet.

Finsbury Park: it sounds like such a lovely place, doesn't it? Well, it ain't, and there ain't no horse-riding going on around there, except the police on a Saturday afternoon, chasing the youth. I was eleven when we moved up there from Holloway, just before secondary school. It finally came about because of the overcrowding in the old flat, and through my dad pulling some "We're Irish too, you know" to the local MP, who was also of Irish roots. It's about the only time being Irish actually paid off. He was just helping out people of his persuasion, I suppose. It was all very "gangster lean." I imagined there was money under a table, because council flats like our new one were very hard to get.

It was in Honeyfield, a block on Durham Road on Six Acres Estate. There was a horrible, maudlin song out at the time by Roger Whittaker that went, "I'm gonna leave old Durham Town," which kind of contaminated the good vibes, but otherwise I was thrilled. Just the idea of so many rooms! I loved walking around inside, going up and down the stairs, touching the banister. "Oh, I think I'll look out this window now!" I couldn't get enough. Of course my dad would always moan about the rent. That's what all those extra rooms amounted to—a whacking great rent bill every week. That was the end of my minicab job too, when we moved. It was too far to go in the morning. Let's say it was thirty yards further than before.

I was really looking forward to secondary school, because it was a fresh start. I was to attend William of York, another Catholic place off Caledonian Road. I loved the first day—everybody was equally shy and open. All of that Dummy Dum-Dum stuff was, I thought, put behind me. What I didn't know was that the school already had me listed as a bit of a problem. On my first day, which bitterly offended me, they put me in the D stream—D for dunce. Hello! They just assumed I had brain problems, and that was that. But within a week I was out of it. Way ahead of the game.

Soon, of course, the bully system crept in, and then there was the us-and-them nonsenses that young spotty kids can compartmentalize themselves into. Then I hated it. It was all boys, which became monstrously boring as adolescence reared its ugly head. There were no priests, but there was one that came occasionally to give Maths. Again the choir thing was there, and I kept myself well out of that. Really, Catholicism is murderous on potential singers, there ought to be something done about it.

I liked some of the classes a lot, but I hated the physical education nonsense, because they made you feel really poor, because you had to wear certain uniforms for certain things, like a rugby kit or whatever—just unacceptable to me. If you turned up without your kit, it meant you couldn't do physical education—great!—but you'd get, "Bend over!" and get whacked on the backside with a slipper by the PE teacher. So I volunteered to be beaten every single time. It stung like mad.

The resentment I had for them trying to impose a uniform on me made the pain almost enjoyable, in a self-satisfying "Ha! You're not gonna beat me" way. Many other kids did that too, and we ended up the majority, so those classes were very poorly attended, and they just got bored slippering us. We outlasted it. Fine! When it came to that particular class, I just walked straight out the gates of the school and went off to do something more interesting to me.

Around twelve or thirteen, I started to find friends of my own, like John Gray. A fabulously awkward chap was John. He was at William of York, and absolutely didn't fit in, or go along with anyone's agenda but his own, and I loved his individuality. He's a diamond of awkwardness and at the same time has an arrogance based on real knowledge of things. Encyclopedic, very useful. Anything you didn't know, you'd go, "John?" and there's the answer.

He reminds me of that movie *Desk Set* with Katharine Hepburn and Spencer Tracy. It's about replacing the knowledgeable staff in a business with a computer. The computer messes up, and they eventually realize that the human brain is far more reliable and emotionally a better response to things. Well, that would be John Gray.

Dave Crowe was another one—a very odd, dark, ominous fella, a bit Frankensteinian in his body frame too, so a huge bulky hooligan kind of a bloke. Quiet, very quiet, but could turn on the deadly seriousness. He was in my class, but we only started hanging out after a year or two. He's also an absolute mathematics wizard, and Maths was something that puzzled me intensely, after meningitis, you see. I find the mathematical approach to life very confusing. I either understand a rhythm instinctively, or it's not going to happen.

Dave got bored hanging out with the Arsenal yobs at school because he was a Tottenham supporter. Because he was an odd penny in that world, and I was an odd penny in mine, and both of us never wanted to do PE—and neither did John Gray—that's how we all came together. A very odd bunch of characters but all totally resolved, who would rather get slippered than have to strip down into some odd outfit—for badminton.

The presumption of this squalid little Catholic school off Caledonian Road, presuming that they'd be training future badminton players—impossible in a world of brutality. All around us was gang warfare, football rows and thuggery. And then they were trying it on with sissy nonsense like that. How can you tell young chaps from an area like that to hit the shuttlecock *lightly*! Unacceptable! Having to wear white dainty outfits with super-short shorts. Never! No! No! Even the gay kids weren't gonna do that. Just no way.

My brother Jimmy soon followed me to William of York, but the two youngest ones, Bobby and Martin, went to Tollington Park. By that time, my mum and dad had started to fall out with the Catholic Church, so William of York was a no-no. There was no way our younger brothers were going to have to endure that priest shite ever again. My dad was very good on that.

The trouble was that the school he picked for Bobby and Martin was probably the worst hooligan school in London. Tollington Park was ground zero for all the serious Arsenal elements in the area. That's also the same place that my future manager Rambo *didn't* go to, if you know what I mean. Attendance didn't feature very high in that school.

I'm an Arsenal man all my life, so in many ways, me not going there was a sorry gap in my education. William of York was up the Caledonian Road, but that didn't mean that you were mixing with the Callie mob. You were stuck in this isolated Catholic nonsense that was very narrow and insular, and trying to blinker your vision. Trying to suppress you as to the way the world really worked. A hardcore school like Tollington Park was absolutely about "This is it, mate, no one likes ya, and we don't *care*." "Pretty Vacant" to my mind would be the anthem to Tollington Park. It wasn't a school at all.

Just as I was starting to find my feet at William of York, something terrible happened. My paternal grandfather, the Owl Fella, died and I had to identify the body. By now, he had fourteen children and was living with a prostitute. Can you imagine that, how my dad felt about representing that to me?

My aunt, who had fourteen kids of her own, came over from Galway, but my dad had to go to work, so I was left to go along with her to the morgue. They'd had to patch up his skull quite a bit, because he'd fallen backwards and split his head open while shagging a prostitute on a doorstep—that's how he died. When they pulled out the body on the slab, he'd died with a stiffy. And it weren't the leaning tower of "Pissy!"

So I'm there with this auntie—Auntie Lol—and she started screaming and crying, yet that was her father. Her hysterical behavior really freaked me out—how adults sometimes can put so much pain on you when they should be taking responsibility at that particular point. "Argh, urgh, I can't look at it! It's the worst thing I've ever seen!" That's what she said. And they went, "Yes, but we need someone to recognize the body." So, up I had to go. He looked, again, a bit like Frankenstein's monster, with the stitch marks across the front of the skull, but I recognized him all right.

As young as I was, I realized he must've been a dirtier bugger than I ever knew because, for my father's sister to behave that way, when they pulled the body out nude with a big fucking hard-on . . . Christ, I'm not that well-endowed—it really was big. God almighty, that's your own father. What on earth's gone on in this family?

This is County Galway, my father's family. My mother's family had different ways of telling me they died—by flashing through the corridor. For some odd reason, my mum and dad loved each other, they truly did, and had us as offspring, but both sides of their family backgrounds are incredibly crazy. It doesn't make sense. The coldness of my mother's family, the insane fear of whatever, and endless troops of disaster marching in from the other side.

That night at the flat in Six Acres, Auntie Lol was in the bedroom next door—Mum and Dad gave her a bedroom to herself, so that meant me, Bobby, Martin, and Jimmy had to share beds. And we heard her screaming all night long—really terrified screams—and I'd have to go in because that's what we were told to do by my dad, to calm her down. It was too much to listen to her screaming—"He's comin' back to haunt me!"

Something had happened, because you can't be crying about your father in that way. Something evil must've gone on. And that's a terrible truth and reality to know about your own family, just as I was getting over my problems.

My whole world was school and our little slice of London. What else did we know? The furthest I'd traveled was the farm in Carrigrohane, and those periods following Dad's work in Hastings and Eastbourne. That was the extent of my travels, up until the Sex Pistols. There was a school field trip to Guernsey, and a Geography trip to Guildford. Guildford was an awful long way from London in them days—a murderously boring coach journey down very windy little country lanes, and it would take forever. It was a week in these awful huts on Box Hill—which I referenced years later in PiL's "Flowers Of Romance"—and you'd have to deal with the PE teacher threatening to slipper you unless you took a communal shower. "Ah, thank you, I love the slipper!"

It was all about us kids looking for ways to get into pubs. That's what we did. It's a way of growing up, and you feel like you've achieved something—something approaching manhood—once you stretch into those no-go areas.

During William of York, Dad got a job driving cranes on the oil rigs off the coast of Norfolk. It was winter, and we stayed in a holiday camp in Bacton-on-Sea—no one there, just us. That wasn't for too long, but while I was there, I picked up a bit of an "ooh-aaarr" in my voice. When we came back to Finsbury Park, that didn't do me no favors at all. "You *what*?!"

I used to run around in a Norwich bobble hat, without the bobble. My only affinity was I liked the colors—yellow and green. I also had another one-color bobble hat, also without the bobble. What with the way I dressed and looked, which was always a bit different from the norm, it seemed to rub people up the wrong way at the back of the North Bank—the home terrace at Arsenal's old ground, Highbury.

I was wearing that hat one of the first times I ran into John Stevens. Rambo, as we know him, was a mate of Jimmy's from around the flats. He changed the face of football violence forever, with his commitment and organization. You'd never keep up with John! He'd be quicker than a ferret into a "row"—one third the size of whatever was challenging Arsenal, and always coming out of it with a big smile on his face. An eejit like me, I was slightly taller—I'd be the first to be punched in the gob. And always having difficult teeth—oooh, I must've broken so many knuckles just on my buck teeth.

I don't get into the psycho aspect of the violence because I'm not like that. I don't hold grudges too long and my anger is a temporary thing until an issue is resolved. It's just plain and simple. Don't shout up Tottenham or Chelsea or anything at all in the back of the North Bank. Don't. And then once we chase you out, everything's happy! I'm not one for pursuing the issue. But I *am* one for going to their grounds and yelling Arsenal as loud as you like. That's a kind of hypocrisy, but that's the wonderful arena that football creates.

The sense of unity was astounding. Every ground had that depth, and you knew it. You knew that this was from the bottom to the top of the terracing. It was not there for the taking, it was there for the full-on argument. Glorious, really. I loved History in school—the Roman invasion of Britain was my favorite subject when I was younger, and the Saxons, the Vikings—I always wanted to imagine myself in one of those scenarios. Well, a football terrace was exactly that. And it was done in that exact same way. Flanking mattered a lot. I also had this book from the library about the Battle of Agincourt—the tactics were all-important, and Rambo's very tactical, even as such a young kid. Our part of Arsenal's mob was so young, up against these enormous fucks in their thirties and forties, but we wouldn't run.

Anyway, that night, Tottenham had a home match. There was a rumor that their mob may be coming down. Forty of us met in the courtyard of the Sir George Robey in Finsbury Park. I was there in my one-color bobble hat. Rambo had set an ambush for them, and any of their mob returning from the match. He just took one look and went to my brother Jimmy, "Oh no, he's no good, send him home!" Jimmy went, "No, that's my older brother, he's harder than me!" I was working on the building site at the time, so looks could be deceiving. This is, what, fifteen, sixteen. I was utterly fucking fearless. Gone were the days of when I was younger and couldn't really handle a fight at all. But for somebody like John to come up, and he'd back you up—that's like, wow, you don't be turning that one down. Not at all, not ever, as I only fully realized many years later.

At school, I suppose I started to become a bit of a handful in class. Not habitually, but instinctively. If I'm puzzled, I want to know the answer. And if they resent explaining to you what it is they're babbling on about, then fuck 'em, and then of course you will agitate them. You can't expect people like me just to sit there and be nullified. I knew in my own heart and soul that I was there to learn, that's what school was supposed to give me—an education. When that's being denied by rubbish teachers, I'm furious. Not violent, but I always had the right words.

It was punishing and frustrating with subjects like History, which I loved. I'd have to ring up people like John Gray and ask them, "What was the lesson today about?" They got bored explaining to me, so then I'd go down to the library and research it myself. But slowly, left to your own devices, you lose interest. The perks are gone, the novelty wears off, and it becomes just cumbersome to do that.

I was finally chucked out of William of York mid-year, mid-season. I turned up late—tardiness was their excuse, and not wearing a correct uniform, and my hair was too long. They thought I was a Hell's Angel, because I used to wear my dad's leather coat. I couldn't afford the bus pass, and so I cycled to school, and they just put all the wrong things together.

It was Prentiss, the English teacher, who got me expelled Piss-Stains Prentiss. These days I go along with the notion of "Let the dead rest in peace," but in them days I hated that fucker. I despised him, yet ironically he was a brilliant teacher; it was absolutely thrilling the way he explained Shakespeare, I was fascinated. So complicated and in-depth, right down to single-word analysis, and the poetic beat of a single sentence, and the structures—absolutely thrilling! The technicalities of the English language. A really masterfully wonderful teacher, but a complete hateful git.

Because I was still underage for leaving school, and I wanted to finish my O-levels, I had to go to a College of Further Education in Hackney. It was like an approved day school for misfits—when school decided they couldn't cope with you, that was more or less the local state-run detention center. We were all supposedly vagabonds and reprehensibles. I'd take the bus, and then it was a ten-minute walk.

Let's face it, Hackney was never a great place. Let's just say it was a different class of Arsenal fan.

And that's where I met Sid.

ROOTS AND CULTURE

Music was always played in the Lydon home; it was a constant thing. Dad was particularly into it—he used to play accordion when he was very young. At twelve and thirteen, back in Ireland, he was in Irish show bands, all the "diddly-doodly-doo" stuff, but he would never teach me any of it—which I thought was really odd of him. Maybe, as with everything else in life, he wanted me to find my own way with it. He still had an accordion, but he buried it at the bottom of a cupboard, and he didn't want to talk about it or have anything to do with it. It was so strange, this mysterious atmosphere he created around it. He didn't want to pass on any knowledge about music at all.

But Mum and Dad had an enormous record collection. There was music playing all the time, especially at the weekend. They had very varied tastes, and varied friends. Everyone would bring stuff around to listen to, so endless records would come in the house, which I loved. "A Boy Named Sue" by Johnny Cash was the kind of record my mum and dad would like to hear, to challenge their friends and see what their reaction would be. Mum also liked traditional ballads and folk, but she also loved the Kinks, the Beatles, big singers like Petula Clark and Shirley Bassey, and lots of dance music.

I vividly remember my mum and dad dancing to "Welcome To My World" by Jim Reeves on the Dansette in the front room—her with her bouffant and pink Crimplene outfit, and my dad in his suit and tie. It was a very romantic song, but also kind of political, that the world could be a better place—just hopeful, positive. A *wonderful* song.

That was where I learned my DJ skills, because I'd see that as my job when I was young, to put the records on. And serve the drinks—in them situations the DJ had to run the bar, and the younger the better, because you'd serve up big measures to make your elders happy. I loved putting on records. Now, that was the kind of machinery I understood, because I was getting results—"a-ha, pleasant sound at loud volume!" Great, what a payoff! And so I got into buying records myself, and went from there on.

Oddly, though, when I was in hospital, which was for nearly a year, I never missed it at all, probably because I didn't remember it, but there was no music, no radio playing in the hospital ward or anything like that. In fact, I don't even think there was a television.

I was soon guzzling up all the popular culture that came my way. I remember us always having a telly, a small Rediffusion, maybe. It was something that looked English and was, so it didn't work too well—grainy, black-and-white, and small. My dad was never too interested in it, nor my mum—for them it was just something to stare at when you're exhausted at the end of the day.

After World War Two, class was redefined completely in the UK. The landed gentry really were an all-but-dead dinosaur, so things had to be readjusted. So you had the BBC—Tory, upper/middle class; and ITV—Labour, working class. The lines were drawn that clearly. We'd never watch anything on the BBC at all apart from the football, because it was considered posh people doing rubbish. I loved the plays, I grew up with them, but the posh accents drove me nuts.

I hated Sundays bitterly because the TV was always so bad. The religious programs in the early morning. We loved *The Big Match* in the afternoon, but after that you knew it was more hymns and *Stars on Sunday* and all them horrible *Sunday Night at the Palladium* things, which were just grim to watch. I hated them, hated everybody on them—even the comedians; you got the idea that they were watered down. Even at a very early age you knew the humor was just babyish. You'd end up watching *Upstairs Downstairs*—I used to love the mother in the strangled neck clothing—simply because there was fuck-all else on.

It's worth reminding people: how many channels did we have? Three, at that point. And what would there be on for kids? Rubbish like *Thunderbirds* and *Supercar*—urgh!—and *Fireball XL5*. That's my youth. I hated all of them. It was just daft puppetry. You could see the strings! You could maybe laugh at that, but really I had no attention span for it at all. There was *Doctor Who*, which involved humans, but only because of the Daleks would I have any interest. Most of it was those stupid ant people, and you could see the big fat legs and you knew it was a man in a suit. It was a bloke in high waders with an ant job on top. Daft!

What I loved was the comedies—particularly *Steptoe and Son*, because there they were, they were dealing with garbage and trash, but the writing was a jewel, and the characters were so real to working-class people. The portrayal of the characters was not cartoonish, and the dialogue was overwhelmingly educational. The understanding and the comprehensive balance and delicacies of being British were all in there.

Poor old 'Arold trying to be sophisticated was a scream. I could immediately empathize with the pain he was going through, getting it wrong, but I couldn't empathize with the fact that he never seemed to learn—his presumptuousness in wanting to go "poash" and every single time completely misunderstanding posh people. Whatever environment he was trying to sleaze his way into with his sycophantic "oh yaah!," those alleged posh people came over as decent folk who couldn't tolerate him and thought *he* was the snob. He was always being reprimanded for his social climbing because he was the most judgmental one of the lot. This is all what I gleaned from it as a very young kid.

I loved Norman Wisdom films too—a heart of gold, and always misunderstood—but music very quickly became my thing. For me, exploring it alone was the best way. I'd obviously take hints from Mum and Dad, but their taste wasn't always mine. At all. I could never understand the Beatles for some reason, and they loved them. It was all that "she loves you, yeah, yeah, yeah" stuff. Urgh! I hated their hairdo, hated everything about them.

So, when I'd got a bit of money together from different odd jobs, I'd go off to different record stores and deal with the challenge of finding out what's what. My first purchases were all bad choices based on the colors of the covers, but one thing led to another. I loved the shape of records, I loved the feel of them, I loved the power of what came out of the speakers from them. It was astoundingly rewarding to me, to do that. All kinds of noises thrilled me.

When I was ten or eleven, it was 1966 or '67, and albums were becoming more important, but they were outside my price zone. So I'd always have a single off it. Some of them stores would actually let you hear the album, and that was always fascinating. You had to find very select stores with people who really did love music and would share it with you, and could see that you were an up-and-coming musaholic.

For all the "Swinging '60s" nonsense, Britain was still stuck in the Max Bygraves era. We were still being force-fed all that showbiz syrup. Like always, what they play on popular radio isn't necessarily popular—it's what you're told is popular. It's what you're denied access to that's really the most intriguing stuff.

Until punk started, there really wasn't anywhere to listen to new and different types of music. It's been belabored that John Peel or the pirate radio stations were doing that, but they weren't really. It was still music that was above and beyond the average working-class listener. No one around my area gave tuppence about *Sergeant Pepper's.* That was when the Beatles were rich kids having fun. It was still highly orchestrated and highly organized, and a lot of money went into promoting it.

The Beatles—yeah, a couple of good records there, but my mum and dad had driven me crazy with their early stuff, so by the time they'd turned into Gungadin and his Bongos, there wasn't much there for me. The people surrounding them were pretentious, with flowers painted on their faces and rose-tinted oversized sunglasses. The whole thing was too silly for words. I remember watching them on *Top of the Pops* doing "All You Need Is Love," all that "la la la la-laaaa"—oh, fuck off! No, I need a hell of a lot of other things as well. Don't make me feel selfish for acknowledging a truth at a very early age.

My impression of them always was: cold as ice, not made

for sharing. I preferred listening to Slade, which was a bonkers stoopid-looking band. Noddy Holder with that perm—come on, that's ridiculous, and what a great guitarist!

From about thirteen, circa 1969, I started getting really heavily into record buying, and that was my albums period. I was listening to everything and anything, not just pop and rock. I loved Rachmaninov, anything that had a Rimsky-Korsakov banging-of-the-piano was way up my street. That heavy, heavy stuff. I think it was called *Romeo & Juliet*, but there was a part in it that just sounded like tanks coming over a hill to me—loved it!

As an aside, the school orchestra always thrilled me, because it was just dreadful, but glorious. I'd always listen at the side, banging a triangle, while forty of us made this insane row. I'd listen inside of the ringing, and pick out tunes, all absolutely discordant. There'd be the haters in there bashing away, and the arseholes who were trying to do it properly—the Matlocks. And then there'd be the merry-makers, such as I.

It drove our music teacher mad. I can't remember his name. He was so effeminate and ridiculous, and he loved the Bee Gees. He had these silly plastic xylophone things, with just one metal clip and a tiny hammer, and asked us to be twottering away to Bee Gees records. It was great to hear the Bee Gees in school, but they were hardly to my mind the voice of rebellion. At the same time, that music teacher, he had to endure the hate and the wrath of the Catholic hierarchy at William of York, because they viewed the Bee Gees as a negative influence on the youth. There were no youth running around trying to look like the Bee Gees, I can tell you that.

At the same time, we had everything around us in Finsbury Park, which is what "Lollipop Opera" on the *This Is*

PiL album is all about. Reggae was always around—you couldn't miss it because of the Caribbean community living there. Oooooh, the *dirtiness* of some of them early ska records. One I remember in particular was "Dr Kitch"— "I cannot stand the sight of your injection/I put it in!/She pull it out! I push it in . . ."

From a very early age I'd go to my favorite store, the one under the bridge in Finsbury Park, run by a little old lady. People from outside the area used to come just for that store. I don't know how, but she had the best reggae in the world, all imported directly from Jamaica. The shop was full of Jamaicans, and heavy metal heads. There was a lot of Jimi Hendrix in the racks, a lot of hardcore heavy metal, which it wasn't called that at the time. It was called progressive. So there would be the brilliant combination of those two elements that felt right at home with each other to me.

All of these records used to intermingle; I never made any cultural decisions about them, they all just seemed to fit together well, and blend well. So I'd like bits and pieces of anything, and I'd quite happily mix reggae or classical up with Alice Cooper and Hawkwind. I realized that it all exists inside the head, it's another universe entirely. It's as real as anything else; it's the gift that we humans have for each other, that extra special form of communication that goes beyond words and sounds. It's a dreamscape, I suppose, and out of dreams great things do come.

I loved Status Quo, for instance. I loved the way that they found something inside a simple format, to say so much. Their methodology is simplicity, and perfection inside that simplicity. I'm so empathic with what they do, it just sounds like jolly good push-and-shove. Very, very skillful to me—superb, and beat perfect. Fantastic rock. Wonderful, brilliant, beautiful stuff.

I also latched on to Captain Beefheart in a big way. I had no idea what he was about, but I knew I liked it. Captain Beefheart was a comedy act, slightly. He never took pause when he was going into deep comedy or parody. He was a bit like a Tommy Cooper of music at that time. It was wonderful what he did—taking deep Delta blues and all those Southern things, and turning it upside down, and making really, really good tunes, out of tuneless cacophony.

He wasn't liked by many serious blues musicians at all, precisely because of his chaotic handle on it. They would take themselves rather too serious, and were too wrapped up in themselves as historians, shall we say. Which is missing the point and purpose of music, which is to entertain, enthral, and educate. But not dictate. Authenticity? Oh, stop it! That's the devil in music. The people who were preaching authenticity in blues were the likes of Eric Clapton—now, hang on! Apart from coming from the wrong country—there's a few other things wrong there! He's imitating something, then preaching the rights and wrongs of it. He misunderstands that music is written by people, for people. I understand that purity is a very fine thing, but some of us sometimes—we like impure also. Y'know, I like to mix my drinks!

"Progressive rock" was an unfortunate title for all the music that came at the turn of the '70s, because most of the bands under that banner really weren't very progressive at all; they all seemed to be following each other, and there was too much Beatles influence in so many things. I was never one for Yes. I loved the covers and the artwork, but that ridiculous dribble that they released—there's not much in there for me. But I took the Roger Dean trail so seriously. I bought albums by bands like Paladin—anything that he had artwork on. In many ways, that opened my mind to records that I wouldn't normally have listened to. There are many ways to get to the music, and artwork is one of them for me.

I used to love the Vertigo label, when it had the spinning spiral, that was just great, and it was always on the B-side, so I'd always play the B-side just to watch that revolve on the turntable. It was great, it was quite trippy, on their 45s, on the singles. I've been prone to epileptic fits, after meningitis, so any kind of movement like that gets a bit trippy in my head. Watching that symbol circulate, that ever-ever-ever-ongoing tunnel, and trying to get the fucking stylus on the groove—wowzers!

By the time I was fifteen, sixteen, glam rock had taken over. T. Rex's *Electric Warrior* was a stunning album. Again, I loved the cover—the gold, and the power amp—*phwoar*, it was the dog's bollocks! And there he was wisping away over those beautiful underplayed guitar parts—more than a nod and a wink to Bo Diddley, but God, look what he'd done with it!

The productions at that time really, really thrilled me. Pop music in general sounded just great, so slick and groovy, even down to Alvin fucking Stardust, who I adored, and David Essex's "Rock On," and Gary Glitter's "Rock And Roll (Part 1 & 2)." Not much music in them, in a way, but there was something else going on, the atmosphere it would create. It was modernizing rock 'n' roll, taking it to a new level, and it wasn't always gonna be about Yes and bands like that, who were torturing you with their fine-note productions. This lot were "Oh, bollocks to that."

In making his transition from hippie-dippie folk, Bolan was rather disliked by the cross-legged brigade, but he was instantly adored and loved by girls and young boys at the local disco. They were records that formulated a great deal of sexual activity, which cannot be undermined. Tamla Motown did the same. So we had it from all sources. You must let the youth bond with each other.

Then there was David Bowie singing about "man love" in "Moonage Daydream." That'd be Sid's song—he loved that, but he wouldn't explain it. For me, it was all about Mick Ronson's bloody guitar, which to this day is still lurking around inside my head as the most wonderful sound I ever heard. It was smooth, delicious, tonal . . . Ooooh, such a wonderful fucking *thing* to get to grips with. It would empower you.

Before him, if you were looking for guitar heroes, of course there was Jimi Hendrix, but nobody could quite work out what it was Jimi Hendrix was doing because—wonderfully so—it was beyond music. But because he came from an American culture, there was still a mystery as to what that Americanism was, so it was very hard to relate to on a street level. Mick Ronson just seemed like a lad, with a bit of glitter and satin pants, but he was playing tones that felt very, very soulful to my culture and my background.

When you went out, you'd spend all weekend out drinking, drugging, whatever, whoring—except we wouldn't call it that, more like "having mutual-benefit relationships." Growing up, in other words. Just finding out what your body parts really can do. None of this was a bad thing, and in the back of that there was Mick Ronson's guitars. And various other sounds too, but music does that—it kicks everything off in your psyche.

And this wasn't coming from intellectual bands like Emerson Lake & Palmer or Yes; this was from root-core, bog-standard pop. The absurdities of Marc Bolan, the absolute beauty in simplistic stuff, the alleged three-chord wonder—"Oh, that's not music." Well, it bloody well is. There's something in them three chords that hits everybody; that's why, to this day, the bottom-line function I see in what I do is—I write pop songs. I can go into elaborate versions of pop songs but the basic root of me is *pop music.* I love "Storm In A Teacup" by the Fortunes as much as I do—well, a lot more than I do—"Smoke On The Water" by Deep Purple.

Bowie was propagating this man-love imagery, but he was doing it in such a brave way that Arsenal's mob really liked it. Football thugs liked the audacity of it, and the toughness, and suddenly outrageous gay people became warriors, respected by hooligans. It's a good lesson to learn about the way things really work—what you'd think would be exact opposites could sometimes meet at the same place. If you stand up for whatever it is you really believe in, if you really stand up, and be accounted for, people will rate you highly.

A lot of glam rock sounded great, but none of them had it in the complicated class that Bowie did. Bolan had great records, but, y'know, he was still a little whimsical elf. Bowie was rather loud about it, and in a completely antisocial way according to the powers-that-be at that time. And therefore you made great room for him. And a Bowie gig was a great place to meet girls, that's for sure—absolutely full of them, and all rampant!

The sexual curiosity that glam rock kicked up—Bowie standing up for something, saying, "Who are you to tell me what to do?"—it was a great breeding ground for punk to begin. Punk didn't just begin overnight; it came from all of these things. It was a gradual gravitation towards the bloody bleedin' obvious.

2

FIRST INDOOR TOILET

Hackney & Stoke Newington College was full of girls! Problem girls—yummy! There were girls at Eden Grove, but in primary school girls were always bullies. They seemed to be more adult than boys. But William of York was all boys, so Hackney & Stoke Newington seemed great to me.

I was an arsehole, and I'd fall in love with anything that walked past me. Very romantic! An absolute penchant for romance, imagining all kinds of situations, and of course all ruined the second I'd open my mouth.

After meningitis, here came another nightmare and a half around the corner—adolescence! A lot of kids go into that with some form of artillery. All my defenses were down and beaten to a pulp, so everything became doubly antagonizing to me. An awful lot to consider, and consider it I did, because I'll tell you, I had such a fixation on girls' summer dresses. I would turn into such an oogly voyeur. In them days, I don't think they called it "voyeur," there were far harsher terms. But I wasn't aware that I was staring so intently. I'd become completely enveloped in the beauty of that visual—schoolgirls in summer dresses. Fantastic.

At the time, though, I didn't have the words to deal with those scenarios. I was very backward in social groupings of girls, very, very shy about it. I didn't know what to say or do, and there was no one you could ask really, because your relationships with girls aren't a subject open for discussion with other fellas. They just aren't, unless you join those cliques that go, "Yeah, I shagged her, then I shagged her," and you know they're *fucking* liars.

So I became Chinese there for a while. Johnny Wan-King. I didn't get up to much else. There were girls around the flats, and you'd do things behind the bicycle sheds, that you look back on now, and you go, "Oh, I don't want to remember it." I hope *they* don't!

In terms of specific crushes, absolutely any girl would do! I was like a complete parasitic leech—I'd hook on and follow them around, and drive them crazy. There were several—their names I can't remember now. There was a girl who lived above us, she went to a convent school in Highgate, and she just thrilled me. She'd come over in that uniform—just, wow! You look back at it, and she was just a bespectacled, spotty tomboy of a girl with knobbly knees—but good enough for me! But apparently I'm not good enough for her, so there. Rejection is such a terrible thing, isn't it? But it's the making of the man. You need to be told to sod off every now and again. It's useful.

At Hackney, it became more like actual dating—meet, go to the cinema and things like that. Or sit in a cafe, which was kind of good. It was different and I liked it. But I've never been what you'd call Fanny Hunter Number One. I'm just not much good at it. I tend to form deep relationships, me. Flippancy doesn't really work with me, and it takes a lot for me to open up to anybody, anyway. I have to really trust them.

Everyone at Hackney had a social problem in one way or another—that's the reason it was there. It wasn't a violent place. You would think that all manner of bad would come out of that. No, everybody there wanted to achieve, but couldn't achieve under the duress of the system, or the "shitstem." It was basically just school by any stretch, so I wore my William of York uniform still, because I didn't want to wear out anything that I liked. But it was a bit of a fashion parade. Sidney certainly used it as a catwalk.

The fella I rechristened Sid Vicious was an amazingly funny character. It would be midwinter, absolutely bitterly freezing—a typical November winter's day, and you know how to-the-bone ice-cold those winds can be in London—and he'd turn up in a short-sleeved shirt made of cheesecloth, which was the fashion at the time, and no coat, and thin pants—feeling very fashionable but freezing to death, but it didn't matter because he thought he looked good.

I met him around the college, and just thought he was hilarious. He was always brushing his hair, trying to look like Bowie, and it wasn't working. What an oddball. Very funny bloke, great company, but dumb as a fucking brush, and absolutely convinced he was gorgeous, and he'd say so. I loved that outwardness. "Gurls luv me!" he always said. When that ended up in the Pistols documentary, *The Filth And The Fury*, there was a double stroke of joy in it for me because it was something he said right from the very first minute I met him. I know he knew I'd get it. It cracks me up to this day. That's so typical him, he was so *not* gorgeous—brilliant, hahaha!

His real name was Simon, but he never liked it, so he was using his other one, John. The story he told me was that his father was a Grenadier Guard. He'd proudly say, "Yeah, just like Bob Marley!" His mother was an Ibiza hippie, and it was an unwanted pregnancy. The father didn't want to know, so she brought him up. She was a well-educated person, was Sid's mother, but she didn't seem to have an occupation. She'd be one for the long flowing hippie dresses, and the black fingernails. But sometimes I'd see her in what I'd call a nurse's outfit, but in khaki. Very odd. I don't know what she ever did. She probably bagged nails. Somebody had to put all those nails in boxes.

Ritchie was his father's surname, Beverley was his mother's, so how he was registered on his birth certificate I don't know. He couldn't get to grips with it, so he was more than pleased when I started calling him Sid, because that was a new name to add to the repertoire. It was after my pet hamster, a stupid thing, but very friendly, hence it was appropriate. At the time Sid was such a downer name, because, with the direct correlation to Sid James, it meant everything awful, a very bad working-class name, so he loved it all the more, he reveled in it. That was Sidney.

He used to live with his mother in Fellows Court, a grim high-rise in Hackney. At first I thought, what a great place to live. NO!! Its elevator never worked, and it was always up eleven flights of stairs when you went to see him, so I wasn't too eager about visiting initially.

Sid was very witty, and again that was his survival technique—humor. To pronounce *Vogue* magazine "Vogg-you-ee" was very funny. I would've been none the wiser but for the fact that we had French taught at William of York. In fact I, along with Sid, preferred "Vogg-you-ee." It seemed to sum it up much better. But he used to treat it like it was the Bible. Of course, he never bought a copy. He'd just go to the newsstand and read it. Or view the pictures, actually, no reading involved. He liked his fashions to hilarious degrees, and for Sidney, David Bowie was his fashion icon of all time. If Sidney ever wanted to be anyone, it was Dave.

The Sid speciality was getting his hair to stick up like Bowie's. He would get two chairs from the living room and put them in front of the oven, open it, and lie upside down with his head inside with the gas on, and the heat would make his hair stiff. He once caught fire that way too. Sometimes it would frizzle at the end, but it was a good look. You know, "How does Dave Bowie get that happening?" "Well, just like you, Sid!"

It was hilarious to bring Sid into Finsbury Park. There were top Gunners left right and center, going, "*What the fuck is that?*" I went, "That's a brave fella, you've got to admit. It's midwinter and he's wearing a sleeveless shirt because fashion comes first!" "Yeah, fair point!"

One time I took him to the back of the North Bank at Arsenal. As it turned out, he had good mates there—serious mates; I was surprised. There was one chap that years later became a really serious problem—a real battler. He weren't no weak heart, Sid, and there he was with his Dave Bowie quiff that he'd spent two days with his head backwards in the oven perfecting—because the idea of hairspray or a hairdryer never occurred to him!

He turned up at my family's house one day, and he's in a thin T-shirt, but he's wearing this Afghan coat that he said his mate had nicked off a Manchester City supporter, and there was still "M.C." etched on the back. And he went, "Have you got any spray paint?" You know, "Come on, Sid, have Man City really come to town wearing that? Hmmm, I don't know, I think that stands for Maria Cachuba—a girl's name or something." "No, no," Sid goes, "I won it in a battle!" He didn't. It turned out it was stolen off a hippie.

But Sid wasn't a threat to anybody. His thing was, I look better than Bowie, and I'm a virgin. That was his selling point. At that age, that was incredibly brave. Everyone our age, between fourteen and fifteen, was like, "Oh no, I'm not a virgin." You know, when you've got your three weeks' summer holiday, then you come back, and everyone tells you how many women they shagged. I doubt it's any different to this day, except maybe the age has dropped to thirteen or fourteen. But that was the basic principle, and Sid ran it the other way: No, I'm a complete virgin. I loved that very much about him.

I may've taken the piss out of him for the Bowie thing, but then trying to be like anybody else leaves you open for ribbing. At that time, I had really long hair, and I had "Hawkwind" emblazoned on the back of the jean jacket that I wore over my school uniform—with no sleeves—very biker-y, I suppose. The very thing I was accused of at William of York, I'd adopted as an image.

Sid did a hilarious drawing of me: it was this tiny little head with one string of long hair, and huge wide shoulders, looking very much like a brick with a pea on top, and one thread dangling. That was his image of me, so how on earth we ever got to hang out with each other is anyone's guess. Other than, I think, humor, and his preference at the time for being called John when it was really Simon. That was, "Oh, another John—after me, John Gray, John Stevens, etc. How many of them do I need!"

There was another John at that school; he had extremely long hair, but he had a tendency to be psycho-violent. He was a brilliant artist and a great footballer, but very antisocial and he ended up in some criminal alcove somewhere. He was adopted and not liked by his adoptive parents, so he was having real problems, mentally and socially. I learned a lot from him, and nothing at all from the art teacher. So, another John—after the war everyone ran out of ideas. "Call him John, he probably won't live long." And if they did, they could pick their own. "It's up to you now. Call yourself what you want, just get out of the house!"

Friday nights at Hackney & Stoke Newington would be the college dance nights. I ended up running them, and that would be a brilliant juxtaposition of events—lots of Kool & the Gang–type stuff, and then hardcore reggae, and the occasional Hawkwind thrown in, and it absolutely went down a treat. A great mixture of different belief systems in music coming together because it was a chance to sneak in drinks and be naughty, and watch the girls and see how they were when they were "off duty," when the guard is let down. That's what social events are all about: it's being able to drop your guard and be rewarded for it, rewarded with friendliness and openness from others. Music's a great leveler in that.

We started going out clubbing in Hackney, because there were loads of places to go. I'd go down to Sid's first, and there'd be trouble—ter-wubble!—no matter where we went, just because of the way we were wearing our clobber. Many a time we'd have to run back to Sid's place because we'd missed the last bus, and I weren't going to walk through that particular area at night. I'd always stay at his, because there were no buses running, and it was way too long a walk back to Finsbury Park, and very dangerous at night too. You'd go through Hackney, then Stroud Green, and all manner of things could go wrong.

Sid's mother, Anne Beverley, never really spoke to me. She never really understood or liked me. I suppose I might have come across as a very silent character. They didn't know what my potential was, and neither did I at the time. She'd always have a dinner ready for Sid—just Sid, whom she would oddly call Michael, even though we knew him as John, and Sid by nickname. Not even Simon. It was so strange, so dissipated from reality in a weird way. So there I was, the man who'd just saved her son from a kicking, and I wasn't allowed to eat. I'd have to just sit there and watch Sid scoff it all.

The teachers at Hackney & Stoke Newington were really good, some of them, really inspiring; they'd get my mind to open up to all manner of things. For instance, there was one who made us write an essay on the word "encounters," and what that meant. There was no answer to it, and that was the joy of it. It really annoyed me at the time: "I want to know what you mean. What is an encounter? Tell me!" "Nope, find it out for yourself, and put it in an essay." Of course, I was nowhere near it. It was an eye-opener, but also infuriating, and I wanted more of that challenge.

From what I remember, I came away with about seven O-levels. I wanted these things. I'd started those courses when I was young, and I wanted to finish them, as a sense of personal achievement, but also of course out of the foolish belief that by getting all these exams I'd become amazingly clever and everyone would want to employ me. Funnily enough, it didn't work out that way.

I felt like I'd committed to school, though, and I wanted to better myself. I decided to go on to do A-levels elsewhere, but I had to pay for my education at that point, so my dad got me jobs on building sites to earn the money to be able to go to Kingsway College. There was no grant for me. I just didn't qualify. Bad school reports from the previous places didn't help. No student loan. Nothing. I paid for it with the money I earned working on the building sites, and the money was so good that I could do that, and also live off it rather comfortably while also contributing to Mum and Dad's rent at Honeyfield. I thought Kingsway was a very good investment for my future. And it paid off, because no matter what I did or didn't learn there, I learned social skills, how to get on with other people, and how to listen to teachers. When they're saying interesting things, I'm all ears.

Kingsway was about a ten-minute walk from King's Cross up Gray's Inn Road, and when you followed the road right up to the top, you could get into Soho in the heart of town. But the college itself was bang opposite a council estate—poor people's housing all round.

The main thing was, I wanted to continue with English Literature, because I loved my reading, and Piss-Stains Prentiss, however much of a bastard, had got me into Shakespeare—so yippee, thanks to him, not all bad. I also wanted to do Technical Drawing, because I love draftsmanship, but it came together with Maths and Physics, so that was a no-no.

Apparently I'd been quite good at Maths before meningitis, but afterwards it was like that capability had been extinguished in my brain. Stuff like Physics is a literal rocket science to me. I find those subjects mind-numbing because I can't place them in any kind of reality. They all seem to be like complicated suppositions to me. It's like imagining three-tiered chess, without the chessboards. Where's the inspiration in logarithms and binary? It was never explained why we'd sit there like dummies, going "Zero, zero, one, one," over and over again. "X plus Y equals what?!" "Who cares, if I don't know what X is!"

So I did three A-levels: English, Art and History. Initially, I found it extremely difficult to get into the way subjects were debated rather than lectured; previously you were told, "This is this, and that's that, and don't ask a question." But now it would be a lot of preponderance on what your thoughts were, but that was good because it dragged that out of me, and slowly but surely I came out of my shell.

I found that I could actually do what I could do socially now, also in an educational scenario. What a thrill. And to not be shy, to be able to stand up and read out aloud a piece of poetry or a section out of a novel. I learned public speaking, I suppose. That's not what I went there for, but that's what I got from it—the emphasis on words, and sentence structures and all of those delicious things. I suppose I was writing things of my own. I'd tease myself with a subject I knew nothing about, then I'd go out and find as much information on it as I could and put together a thing on it, a piece, to educate myself, and I liked doing that. At Kingsway I could actually share those ideas with other people because they were doing the same kind of thing, and I'd be able to stand up and proudly present my thesis. It was creative writing, really. I was ready for something, I just didn't know what.

The English Literature teacher there in particular was great—loved her. Really proper analysis of poetry and the written word. Even Samuel Pepys' diaries, we'd have a poke at that occasionally. Just loved it. Behind the scenes, I'd be reading everything and anything. Chaotically. Probably the same way I approached music—"I like the color of that electric-blue book!"

I loved Ted Hughes. That was fun. Years later I had a conversation about him with Pete Townshend from the Who, because he wrote the intro—in German!—to a Ted Hughes anthology. Wow! Ted Hughes' poetry was just great. The first one that pops to mind was a poem called "Thrush," as in the bird. No, not *that* bird, *a* bird. It was great stuff at that level, great stuff for kids of sixteen, seventeen, to be reading. It seems quite complicated and confusing, but as you grow older you realize that that's quite a childish level. Small steps get you there in the end—don't rush into Polish philosophy straight away!

Dostoyevsky: there's another hard one at an early age—you

can't quite get to grips with the sheer audacity of the size of it. *Crime and Punishment*, yes, but Tolstoy's *Anna Karenina* I kind of disliked. And I had no tolerance at all for them bloody *Jane Eyre*–type novels. That's Barbara Cartland territory to me, I can't relate to it, can't empathize with the self-pitying woman having to deal with a man's world. It's Presbyterian, that's the word. Everybody's so overly nice in it, and the cruelties are so exaggerated as to be cartoonish, so I have no time for it.

Oscar Wilde I found outrageously funny. Way ahead of the game, that fella, and wouldn't be ground down, and led what was a very dangerous lifestyle at that time. Not delving too much into exactly what it was he was doing, because there are no hardcore details, but it was the fact that he mocked the class he came from so well; he got at all the faults that were there. He was really criticizing himself at the same time, and I liked that, I learned from that. We're not perfect. And if I'm approaching things in my working-class way, I'm damn well sure I'm going to be mentioning all the negatives along with that. And there are many.

Sid went to Kingsway too, and within a week or two, I'd met another John—John Wardle, whom Sid named Jah Wobble one night when he was so pissed he couldn't talk properly. The three of us were all problem children, for very different reasons. One way or the other, we didn't fit into the system, and I don't suppose many people can or do. The system I think should be adjustable to us, and our tastes and needs. If you're not meeting our expectations, then you're going to get these oppositional scenarios.

Wobble, again, was hilarious. He looked so weird, a little warped. He was trying to affect a tough-boy look but it didn't quite work on him. He looked more like someone's dad out of World War Two days, with the hanky on the head and the braces. And a big Tottenham scarf, and a big grin on his face. Hilarious, but a chap full of malcontent.

He lived near that Krays pub, the Blind Beggar in Whitechapel. The first time I went down there, I said, "But don't you know that's an Arsenal pub!" We never rowed about football—there wasn't the need for that. We had much bigger rows with everybody else. I had other mates, like Dave Crowe from William of York, who were Tottenham. Of course there's always been a big rivalry been Arsenal and Tottenham, but it wasn't like something you'd resent each other for, because we had other things going on. I'm not going to want to kill someone over a game of football. And I emphasize the word "game." And seeing as I'm not specifically playing or in the team, I have to remove myself from the brawling. Although I have enjoyed a good football brawl from time to time. Sid of course was indifferent, anyway.

I never really discovered what Wobble's curriculum was. I don't think we ever sat down and discussed our lessons. He was someone else to hang out with at lunch breaks. You had a group of people to communicate with, and of course we were all on the outside of things. Wobble couldn't understand Sid, and I'm the unification, how we all knew each other. Sid, me, Wobble—and there were a few others, like John Gray—we didn't really look like we belonged in this environment at all.

We were all viewed as potential-for-violence people. We understood that, because a friend of ours broke into the files at the college, and they had on file that Wobble, Sid, and I had a propensity towards fighting. Now, we didn't. What we had was a questioning propensity, and then if you talked shit to us, you'd get beaten up. We were in this ridiculous fiasco. It was utter nonsense, the accusations and misunderstandings of what we were and came from—that led us into being violent. Wobble genuinely wasn't, originally. He wanted to achieve, and he was pushed and presumed by the system. He was Stepney, Sid was Hackney, I'm Finsbury Park. It's basically the same manor, it really is, with variations on a theme. The problem being that the school system adjudged us as unteachable, uneducation-able.

Wobble was gone in six months—bored. He'd had enough. But he was my best mate, and stayed that way for a while. Do you know why? Because he stood up for Tottenham. He believed it, as ridiculous as it is. And I have no doubt he believed me, as ridiculous as my Arsenal is. We were forming terra-firma gangs, outside of the regular discipline. That's good roots to punk, mate.

I was a diligent student, but about what mattered to me. And again the authoritarian encumbrance of times and lessons was not very helpful to me in the long run. Or the short run. After about a year I got so mindlessly bored with it. It just wasn't moving quick enough, and there wasn't enough to occupy the head.

I was still working on the side, so I was bringing money in. I had all manner of jobs—I'd take anything that was going. Mainly, I worked on the building sites—Dad got me jobs there for a while. I loved that, the money was fantastic.

On the sites, it was hard having to deal with the threatening behavior of the Paddies. They were definitely always trying to enforce a pecking order, which I would have none of. "Ye need ti knoaw yer place!" "No, I fucking don't!" I resented being given a shovel and told to dig a hole. That was not fun for me. What fascinated me on a building site was working with the site engineers, and designers, because it meant I could look at the technical drawings, and I loved all of that. I didn't mind going out measuring the landscape.

Dad loved his cranes, loved them. He could talk up a storm on cranes of all kinds. He loved any heavy goods vehicle with a jib on it. That was his fantasy. He just loved being in control of machinery, and he was very good at it. Manipulating cranes and moving things about, very excellent, pinpoint accuracy. The workers on the site really loved him for that, because if he was delivering the bales of concrete, you knew it would go exactly where it was wanted. There could be some awful mishaps. I've seen people seriously injured with that stuff. If the crane driver wasn't up to it, there'd be bodies knocked off.

Dad taught me a lot about how to control the cranes. He'd just shut me in the cabin, and—"Get on with it!" The noise alone would terrify me. There was no such thing as earmuffs in them days, and those machines could kick up a noise—solid cast iron, everything. Everything cold and freezing and hateful. I couldn't make out why he loved this. It wasn't at all my thing.

If I was misfiring on the pedal, he'd slam his foot down on top of my foot, and that would hurt like hell. I suppose it had to be done, but Jesus Christ, the technicalities of trying to operate two legs and two arms, and making them all do different things at the same time, was just beyond my reach.

One time, he broke my ankle with a shovel. Yep. I was actually in bed watching *Mystery and Imagination*, and he told me not to watch horror shows, because they give you bad dreams, so he slammed the shovel on the bed, and that's where my foot was, although he didn't realize that till it was too late. I don't remember much, like how my mum reacted, just the pain. My ankle's been a problem ever since. If it gets in any way cold or damp, oh boy, does that ache. I've had a form of arthritis from it ever since. It's just one of those annoying things that don't go away.

It's like when I dislocated my shoulder; it was not for any good reason at all, but I was too lazy stretching out in bed for a glass of milk—I love milk, you see, I drink it all night long, so I always have a glass next to the bed, but I could not be bothered to actually move my body, and so I kind of twisted my whole arm, and dislocated my shoulder. So: hunchback, dislocated shoulder, shattered ankle . . . Now I can't move like a suave Mediterranean, and my life's fucked.

Long hair had worn itself out for me. It was just a nuisance. It was a good thing to have on the building sites, because the old-aged Paddies hated it. Long hair made you a magnet for coppers. But then, because that was the case, many thugs *wanted* long hair. Long hair meant many things. For some it meant, "Peace, man, I wanna look like Jesus, and here's my couch slippers." For others, it was a full-on aggressive act, like, "Fuck you, I'm not cutting it!"

The crop, the full-on skinhead crop, was an absolute act of aggression. I think most things begin with a form of aggression, even for the most passive of hippies. Passive-aggressive was the stance. It was declaring that you don't fit in, just let it grow out, and whatcha gonna do about it? That's going to be the order of humanity, I think, forever and a day; we will strive to be different. By the time everybody catches on, and we find out we're the norm, then it's time to move on.

So I decided to have my hair cropped short and dyed green. Krazy color was genius. It's a shame it's not of the same thickness and durability today as it was then. They've somehow watered it down, and the colors aren't as vibrant. It's pretty damn near next to useless unless you want to look like a faded newspaper. You know the cartoon segment that used to be in color in rancid old newspapers? Them kind of colors—that's all you get out of it now. Or maybe people don't know how to bleach properly. Back then, the colors were really zingy and thrilling.

My dad seriously didn't approve, though, and it was the final straw that got me thrown out of home. Dad's famous quote was, "Get out the house, you look like a brussels sprout!" I never forgot him saying that. I just laughed. Even in the painful separation of child and parent, there was humor. I loved him for it, because it was witty. Up to that point, I hadn't realized it, but it was true—I did look like a brussels sprout.

The only way I'd ever get in the house after that was if I crept in at four in the morning. Except of course if my Auntie Pauline was over from Canada, at which point I wasn't allowed to come near the house at all, because I was an embarrassment.

After I got kicked out, I went straight up to Hampstead, where Sid was squatting. Sid had set up the squat, so well done him. I think he had a lot of help from his mother, so he was the leader in all of that.

It turned out that his mother was a registered heroin addict. I was round at the flat in Hackney one time: we were playing

Can's *Tago Mago* album, and it was Sid's birthday, which I didn't realize beforehand, and she gave him a little bag of heroin to shoot up. I have to say, I was really shocked. Sid said to me, "D'ya want some?" "No bloody way, I don't go *down*." "Okay, it's time for you to leave."

So, there I was in mid-Hackney at 3 a.m., and I had to run the gang gauntlet to get back to Finsbury Park. That was a death walk, a *serious* death walk, particularly how I was dressing and how I was. I gave a toss for *no one*. I knew what was coming. Even local Arsenal boys, they'd still have a reason to row with me, just because of my attitude, and I—don't—back—down. In them days, you get stabbed on the street, ain't no one opening the door to help, because you're not local. Seriously dangerous stuff, but I made it home somehow.

Anne Beverley had the most peculiar relationship with poor old Sidney. It didn't feel like a family. She never offered me anything, not once, not even a cup of water. It was religiously true, over and over again. There were other nights when there were others from the gang-of-Johns there who'd come over to stay, and they all remarked on it too. "What, don't we exist?" "No, I'm afraid not." Strange, strange woman. She wouldn't accept Sid having friends at all; she didn't accept any of us. At the time, Sid's other best mate was a guy called Vince. He said the same thing—"Bloody hell, that's a house of ice."

I told Sid, "You can't live with a mum like that. Look at her, Sid, she's giving you deviled kidneys with heroin sprinkled on top, do you really need this?" Sid had actually been anti-drugs when I first met him, but you know what drug mothers do? It was in the food. Insane, right? She'd go, "Here's your food, Sid, make sure your friend doesn't eat any of it." Sid'd go, "Okay, Mummy," and then we'd go into his bedroom, and he'd go, "Try some of this, John, what on earth's she giving to me?" And I'd go, "I'm not hungry, mate." So that was Sid's inheritance.

His new abode, the squat, was round the back of Hampstead

Station, so I went over there and asked, "Can I move in?" and he went, "Great!" He was there all on his own, with nothing to do, so now there were two of us with nothing to do. Somehow two are much better at doing nothing than one.

There was no electricity and no hot water, but the toilets flushed, so to me it wasn't that grim—that's what was available, and quite frankly it was of a similar standard to when I was younger and we lived in Benwell Road. But this place had an indoor toilet, so I was one up on the Richter scale.

The whole block was squatted by old hippies and Teddy boys, and we were the flotsam and jetsam that fell into the generation gap. Squatters united many different ways of life, because squatting was essential at that time. The government were doing fuck-all in terms of housing. You couldn't get a flat anywhere, and what was available was overpriced and just not worth it. There were enormous amounts of unoccupied old housing; flats with nobody in them, nothing happening. So you weren't depriving anyone of a place to live, you weren't pushing out a family. They were just unoccupied and semi-derelict. There'd be the signs up front: "Derelict Building—Do Not Enter." Great, I'll live there. In we go. That was the promotional flag!

I was on the dole for a very short time, for about two weeks. I didn't want to be lining up in the dole office. I hated that place, didn't want nothing to do with it. I didn't feel I belonged there. The two times I turned up, I bitterly resented it and I swore I'd never go back there. I really didn't like the whole format of it, or the institutionalization it entailed, or the way they make you feel somehow guilty about it all. That's your right—you've worked, or your parents have worked. If the state can't provide jobs, then what the fuck are you supposed to do? In many ways I completely understand people taking to illegal activities, because frankly there's no other way to make any kind of money at all, or get yourself out of the dumps. For me, personally, I could never get involved with theft, I can't do it. What's not mine, I don't want—that's what Mum and Dad taught me.

Job-wise, I'd do anything, whatever I could get. One was in a shoe factory—I loved that, boxing shoes. Another was at Heal's, the furniture store–cum–department store on Tottenham Court Road. At the top was a state-of-the-art vegetarian restaurant, and me and Sid were the cleaners.

The thing to do was to experiment with what vegetarian food was, because there'd be leftovers, you see. That was my first taste of a nut cutlet. In them days vegetarianism was a very new thing, a trendy fad for very wealthy people. And utterly tasteless. It was about colors and shapes, really, rather than any flavorful content. Very amusing. Not much to clean up, actually. A few chopped peanuts on the floor and that would be about it, but we'd make that last two hours every night, because that's what you were paid for.

Then over the summer holidays, John Gray got me a job in a daycare center in North London. I'd be looking after kids of seven, eight, nine, ten. I could play with the younger ones, that wasn't a problem at all. The problem was the institution that ran these places: they didn't like the idea of someone like me near little kids. In a world of Jimmy Savile! That's the bitter irony of it, because I'd be the last person to bugger about with children, yet you're so readily and easily labeled, and so wrongly too. People can't see through to a man's heart and soul, their character.

We'd make balsa-wood airplanes—biplanes or triple-wings. Everybody wanted to be the Red Baron, so that was the favorite one to make. I had woodworking skills, from Technical Drawing and Woodwork at school, and also from working on the building sites. I'd double up there with the site's carpenter, and he taught me lots.

So, instead of saws and hammers, the tools were tiny little Stanley blades and balsa wood. But the principles are the same, and kids love being involved. It's what I loved, so that's all you do. You want to quieten them down, you want violence to stop? Get them interested. All kids love to create, and feel like they've come up with something on their own, and achieved it through their own means. For instance, when any child asks you a question, do not shy off from giving them that answer, because they'll resent you forever for not telling them. At least this one did, and will.

I realized my teachers had forced me into a caricature of myself that actually wasn't me at all. They made me uncomfortable. I was uncomfortable anyway, just trying to deal with life, but they made me feel unwanted and resented for being there, and I of course responded to that. I don't think any child is born like Mr. Nasty or Mrs. Badmouth. That comes from what you're taught by example. And I think I've turned all those aspects into positives. I'm not a self-pitying, nasty piece of shit, I'm not criminally minded. Thank God, they gave me the tools: how *not* to be.

To get by, me and Sid had to sell amphetamine sulfate. We had to do a bit of small-time dealing, one way or another. Money was hard, times were tight.

I'm not the druggy sort, but I like to stay awake—the alertness that certain chemicals can give. In them early years, your late teens, you're well up for it. You don't want to miss anything that's happening anywhere at any time. Indeed, you want to be involved with everything all the time. Then when there's nothing going on, and just indolence all round—grrrr! In those early days, whenever I'd get these attacks of colds and flus and any kind of allergies, and they were coming on me big, the amphetamines that were around at that time would blow them right out of my system. It seems to be very useful that way.

It's not any self-aggrandizement going on in your skull, not with proper amphetamine. It just makes you more alert. I don't think there's a high comes off it at all, it's just you're able to activate yourself. I treat it as a key to the box. Or I used to.

My trouble was, you know how they say a salesman should never sample his own wares? Here's one that did, so I wasn't very good at selling anything. I'd sit by a big bag and feel very glad of myself until it was gone and not think of moving. Speed doesn't make me get up and run out around the world, it makes me sit down and think and enjoy whatever it is I'm doing. Even if it's cutting my fingernails. I'm enjoying it. It gets me into the state of not being constantly tired, which again is all back to meningitis.

I suppose they'd call it self-medicating nowadays. My normal existence at the time would be one of inactivity—just zero, run-down, lack of iron in my blood. Genuinely, my brain couldn't handle too much going on, and physically everything was just too much to endure. Except, oddly enough, on the building site. That was a good ten-hour day of solid manual labor, and I never found that a problem. But I always found a problem of repairing a windowsill, or fixing a toilet. I've learned later that you can turn those things into great adventures. But not then; I used drugs a bit differently. It wasn't so much recreational as a necessary thing to do, to give myself any get-up-and-go. I was very prone to depressions, after meningitis, right up through my twenties.

Amphetamines had been around on the streets for years. It was a throwback to the mods. It definitely kept you up all night and you could go to many clubs and all of those things. In them early days what I loved about it—I didn't like the soporific downside of alcohol and the speed would definitely take that away, so you could drink as much as you liked, and somehow not be drunk. I really love the flavor of beer. I'm not a cocktails person—unless you count beer and speed as a cocktail. We bad rock 'n' rollers are all at it! But then, joy of joys, so are all the football hooligans, or they were then, so we had common ground.

So that was the backdrop for the Hampstead squat. For a while there, Mad Jane moved in with us. That crazy cow! She was like one of those voluptuous women from a 1940s film noir. Her hair was wavy and long on one side, and she'd wear those dresses from that period. Very strange girl. Movie-starlet-from-the-'40s kind of imagery, lots of that "come up and see me sometime," kind of Julie London-y—a very hard image to pull off in the shittiness of London Town at the time, so I admired her bravery. I don't suppose we got on very well, but well enough. And from time to time, future PiL guitarist Keith Levene would come round. There may have been a chemical thing going on with him, but no worse than anyone else.

Drugs were everywhere, probably because of the mods. The mods were very into uppers, and that passed on. The skinhead thing was a bit purist, but not by Arsenal way.

I'm not talking heroin here; that was a great unknown to most of us—it was just something the Grateful Dead did, and by God, didn't they sound it! The dullest band I've ever known. What a waste of four and a half hours! I saw them once at Alexandra Palace when I was young. No! No! I couldn't relate to the crowd that would be digging that kind of stuff. To me, it was life-threatening. Comatose.

These were very difficult times, in 1973 and '74. *Everything* was flared. Please: how to avoid flares! We had no relationship to hippies, they just seemed to be spoilt rich kids. That's probably what drew me into the second-hand demob suits and the Paddy look. We were coming out of the '60s, and that for me was more in keeping with the skinhead approach to clothing than the hippie lot, so I headed straight into that one.

From an early age, I'd been hanging out on Sundays at the Roundhouse. John Gray lived in Kentish Town, and the Roundhouse is not far away in Chalk Farm, so what I'd do is, get the bus up to Kentish Town, pick him up, we'd walk to Chalk Farm and then we'd spend all day, way into the late hours, watching about twelve or fifteen bands. As I got a bit older, I'd probably been out since Friday night, so what a perfect way to end the Sunday.

It was astounding, the diversity of music. I'd see Roxy Music, Judas Priest, Queen (when they were very young), T. Rex, the Seeds, Mott the Hoople—the variety was fantastic, and there was no snobbery about who was top of the bill or whatever. It was just whoever turned up at that specific time—they'd put their equipment on the stage, and off they'd go. The audience was mostly hippie, lots of floral prints and girls dancing barefoot, and bongo players, the scent of joss sticks—all of that. I kind of paid no attention to that, I only liked what was happening on the stage, and I just soaked it all up.

Punk history later dictated that music was shit all through the mid-'70s. Not true, if you knew where to find it. It was the making of me. I could quite happily spend a whole weekend—*alert!*—going around to all these late-nighters around town. The Roundhouse scene was full of insane bands. People like the Pink Fairies were full-on, hard, heavy, loud, aggressive—absolutely the opposite of the hippie vibe. There they were with their long hair, but throwing it back at you in such a noisy destructive way. Fantastic!

Likewise, the Edgar Broughton Band had the longest, filthiest beards and hair, and dressed like bikers and sang songs called "Gone Blue," whose classic line was, "I'm all undone by the things she said, but I love that little hole in the back of her head." Hah! Wow! That topic for that time and that age was like—"Oh, they're going somewhere here." That's not your hippie message at all, is it? And their album cover was sensationally hilarious—racks of dead cows hanging on hooks. I don't suppose the music would bear up too much today, but that isn't the be-all and end-all.

Black Sabbath were the same—a very different approach to music, and different drugs, more on the up-all-night variety. Ooooh, yeah, you were completely aware what this lot were prepared to do with themselves. When you listened to bands like that and the Deviants, you knew the chains were off. Rules are for fools—that's what you were gathering from them. At least I was. You know—"Oooh, don't do this, it's bad for you!" "Bollocks! Go forth, create chaos, and begin in your own head!" What's wrong with being off yer nut every now and then, you know? It's a healthy thing. But these bands, it was a very youthful contingent—it was all about us young bloods who were made to feel unwanted by the sit-down mob. I went to concerts to dance. I shoved as much down me neck and other areas as I could possibly get my hands on, and got up on the good foot.

One of the people I really liked a lot was the Crazy World of

Arthur Brown. He'd walk around in the crowd, and I went up and said hello, and it was great. His band was billed to support Alice Cooper at the Finsbury Astoria, before it became the Rainbow, and for some reason Alice Cooper canceled that concert. I'd bought tickets for John Gray, Dave Crowe, and a few others, but I was such a fanatic of both bands, I kept the tickets, I never went back and cashed them in. In fact, I've still got them.

I even joined the Alice Cooper fan club and had a box of chicken feathers sent to me, and this silly letter of information. The whole thing just struck me as really funny. There were people that really would take this a bit too seriously.

So I told Arthur, "I've still got the tickets from that gig that was canceled," and he went, "It weren't me!" So, that's how our conversation started. I was just some awkward kid that was giving teachers a bad time, and he was good enough to talk to me as an equal. I won't hear a bad word said about him, because there's all too few of those kinds of people on this earth. Anyone who talks to me openly is fine by me. It's the ones that leer down or flare their nostrils that drive me crazy. But that man was bonkers, seriously out there. Lunatics make good records, oddly enough, and they make good paintings too, and write good novels. They just can't seem to fit into the shitstem.

Another great band I saw at the Roundhouse was Can. They used this equipment that reached bass tones so low you wouldn't hear them—you'd *feel* them. Well, so did the stage, which vibrated and collapsed. All the scaffolding crumbled. Afterwards everyone waited hour after hour for it to be rebuilt, and finally at the end of it all—the most amazing drumming I've ever seen! Thank you, Jaki Liebezeit! Just the sound and the audacity of it, and where it was coming from. It was way beyond the trippy-hippie bongo crowd in the audience. This was coming with a far harder message, and it wasn't the dull stupidity of love and peace.

Also from Germany, Faust earned my love by selling their album, *The Faust Tapes*, for 50p—a bargain, even in 1973. I actually saw them at the Rainbow in Finsbury Park, and they just basically made their noise, which was made up of very interesting, hypnotic, trancey electronic-box-produced noises, while they were wrapped around a pile of old TVs in the middle of a huge, empty stage. I must admit, at the time I was really angry because I didn't have a TV. "What are they doing with all those TVs—I could definitely use one of them!" Then they kicked them to pieces, and rewired them. It was an appropriate backdrop for what they were doing musically, but at the same time—forever the practicalist, me!—I tried so hard to get backstage to nick one.

Everything and anything, I was into it. I went to free festivals too. I even went to one of the first Glastonburys. I think Audience played, maybe Atomic Rooster, and possibly even Melanie. I really don't know. It was nonstop alcoholic faze, perpetrated by wonderful amphetamines. It was a texture of gloriousness.

I don't think bands were even introduced. It just seemed to be that one lot would mélange into another. There wasn't a great turnover of equipment, road crew, or DJ activity going on. It just seemed to be who turned up, turned up, and then things swifted over, and before you knew it, it was a completely different band. It was quite wonderful for that.

And in the middle of that kind of affair, I'd also be off to sit cross-legged listening to Nico waffle on about the "janitor of lunacy." Fantastic, completely Queen Vampire! It was John Gray who said, "Oh, we must go see her!" Everybody knew she was a smackhead, like that'd be an enjoyable concert experience, but it was. It was the creepiest thing, her and her harmonium for an hour and a half, groaning away slightly out of tune, which made it even better, because you could feel the angst in her. The tragedy in the voice was just overwhelmingly powerful for me. I've learned a lot from them very early years of going to concerts, that it really isn't about perfect pitch, it's about the emotion.

I'm not one to sit cross-legged for more than three minutes, and I'm quite happy to dance to "Janitor Of Lunacy," I don't care who's looking. I loved dancing. Loved it, loved it, loved it. There I was—me long hair, Hawkwind embroidered on the back of my jacket, and Teddy boy shoes, because I found them the most comfortable to dance in. I wouldn't wear flares. Any gig, anywhere, anytime—get up and dance! But by God, the best guy for that in them days was Jesus.

Jesus was a guy who hung out with the two girls who used to dance for Hawkwind. Sasha and Stacia were their names, I think. He'd strip naked, he had the smallest willy in the world, and he didn't give a toss who looked. I loved him for that. I thought, "He doesn't care, and look, he's completely happy. He's got bongos which he doesn't know how to play, no sense of rhythm—none!—but a total sense of joy!" He certainly wasn't what my mum and dad had in mind as Jesus.

But his message was good and, years later, when punk started and the Pistols were gigging—I think it was at the Marquee, when we were supporting Eddie & the Hot Rods—he was there! He looked completely different, he had a suit on, but he still had the same ludicrous hairdo, which was a very deep fringe and a long mullet at the back, and deathly blond, a natural blond.

It was very hard to bring Wobble into these kinds of situations. He was hateful of the Roundhouse, straight away. "I hate these people!" He'd think everybody at the Roundhouse was a love-'n'-peace fool, but that's his lack of insight. What he didn't realize was the building was full of really oddball characters and that he was one of those oddball characters by the fact that he was there—although it was just the once. That's all it took, I knew not to bring him out anymore. Wobble's angle would've been soul clubs—that would be his thing.

Around that time, though, I'd been going out to Ilford in Essex to soul nights at a club called the Lacy Lady. I wouldn't be alone; there'd be a mob of us—the Johns. Sid would be there, John Gray, a couple of others. Proper mob-handed, in fact.

The other clientele out there was very interesting. There was the semi-gangster-ish local toughies. They'd look at you threateningly, there'd be no two ways about that. But we were a pretty mad bunch ourselves. That's where pogoing really came from. That's how we used to dance, jumping up and down. We didn't know the moves, so we invented our own, and good times were had by all. As a result, you weren't then perceived as a threat, because you were up to your own universe and enjoying yourself in your own way and not there to nick the birds—although, girls love "different." Did they want to mother us? No, but that's good too! I was young for the age I was, there was never a chance of "Come back to my place." I tried hard, though. Anyway, you couldn't do anything; you couldn't go off on your own, because you had the responsibility of the collective.

What they'd play there was a good root course in where soul music in America was going. It was beginning to split into different angles, after Tamla Motown. There were more interesting and exciting varieties coming out, it wasn't all so orchestrated out of Detroit. They played kind of West Coast funk which was really interesting, a lot of Philly and Chicago stuff that later turned into all kinds of different things.

It was early disco, really, and I loved all that—"Hi-jack your love, hi-jack your love!" etc. The DJs out there were great. Some of them were BBC DJs, but they played the stuff they liked, outside of their regular broadcasting playlist, like the hardcore stuff. Loved it. And in them days you could go up and ask, "What's that record?" and they'd tell you. That's a lesson modern DJs could well learn from. I'd be ferreting out future purchases, that's what Ilford was all about—ooh, must get that, and then I would.

Disco sucks? You never heard that from me. Whoever wrote the punk manifesto wasn't listening to the actual punks—them what started all this. No one was paying *any* attention, it was all negative two-steps-backwards Dumbsville. A great pity. I still have a deep love of The Fatback Band. They had a great way of catchy little dance-y singles about them. Love 'em. Kool & the Gang, love 'em. What more can I say?

The only drawback about going there was, there was no way home after, and the only person we knew out that way was called Tony Colletti and he wouldn't let us stay at his house. Three or four times of that, freezing to death until four in the morning when the next train came, and the fun was gone from it. We were all under car-driving age, and probably under the influence anyway, and we definitely didn't have money for a minicab back. So we had to go elsewhere for our fix of that kind of music.

If you ever went into Soho in the middle of London Town, the gay clubs were the only ones that would welcome your different imagery, and, again, you wouldn't be pestered by the boot boys and the yobs. You didn't have to deal with that whole angle of life, of, "Who are you? Arsenal or West Ham?" Again, there were lots of girls dressed really well, with different clued-up ideas of fashion. So, it was thrilling to watch and be in amongst, and frankly, there was a better class of drugs.

The music was usually dance-orientated. There was always exciting things coming from odd little bands from up north, and I don't mean Wigan Casino, because that didn't entirely sum up the northern scene at all. There'd be many different angles on things, remixes of Bowie tracks, whatever. Just great fun. Not too much orientated towards sensationally eye-opening music, it was more like a social gathering where you'd have a bloody good laugh, and if you got out of your nut you wouldn't be beaten to a pulp for it. People were genuinely helpful. Very open and friendly, no judgment going on.

The macho stance that progressive rock had adopted for itself was repulsive to me. I loved Status Quo, I always will, but the audience were just the same bunch of long-haired, waving-it-in-the-breeze dullards. They were just identikit, from the front row all the way up to the back. I had no time for that. I didn't want to join an army, and I felt that none of these fools were really listening to what was going on at all. Whatever they were masquerading as, was fuck-all to do with the band.

It was just hairy students in RAF coats, that was the look—kind of a Led Zeppelin cast-off thing. Those Great Coats—very big thick blanket-y things with silver buttons—were everywhere, thanks to Army Surplus stores. Now, I don't mind dressing up, I go for a bit of this one day, and a bit of that the next. But as a lifestyle thing? No, never.

It was just really exciting to find like minds who dressed differently. For instance, John Gray and I used to go out to Canvey Island to see Dr. Feelgood. Again, Wilko Johnson—what a guitarist! Man alive, that bloke thrilled me to death. Like, how on earth are you doing this? So fantastic. And the lead singer, Lee Brilleaux—oh my God, the seediest, tackiest, harmonica-playing sleazeball, stains all over the white dinner jacket. He looked like a vagrant trying to look classy, a great image. The whole thing about them, they were outside of the agenda, and they were really kind of grubby.

It may've taken a bit of courage to dress differently, because I suppose it's the way society is. It's always trying to regiment a thing. Give it a uniform and a label and thereby contain it. Containment doesn't interest me. I want it all.

At that time I was wearing demob suits. I liked the look on the building sites, the Paddies would come to work, and last month's Sunday best would now be what they'd be shoveling shit in a ditch in. I liked the look of it. In a world of flappy flares, which I bitterly resented, that was what was instantly available. I liked gasman suits too, at the time. They came in this electric blue, so I'd wear that a lot. It was a short little jacket, a bit like a Harrington, with matching pants, and it looked great with a pair of red steel-toe-capped boots. Then hair violently cut short, which I decided needed to be taken to the ultimate—from "brussels sprout" into "mad hedgehog."

Anything that comes from the streets is about "short of cash."

There were times when I could afford expensive items and I would, but it would be just the one thing, like an astoundingly amazing pair of shoes, which fitted nothing I had, but I liked them shoes, so Johnny Rotten's happy-go-lucky mismatch style was developing nicely. I even bought platform boots, but it was the solid wedge ones that had no heels. It was just a huge wooden block. You were seven inches off the ground—very dangerous to be going around London in. But I loved them, because they were in sky blue and electric blue, a mad brogue-y pattern, like the old skinhead shoes, but taken to outer space and back again. Dangerous too, because hard to walk on, definitely a nightmare on the escalator on the tube. Also, if a mob of lads spotted you, these were not runaway material. You'd have to stand and take whatever came, and hope that your wit won the day, which it normally could, but not always.

I don't know, maybe I was a style pig before my time, but I set to work on those demob suits. I thought, "The idea of it's good, the style is shit. Let's try and change that." To start, you cut off the lapels. "Nah, it was better with them on. Maybe I should take the sleeves off an' all, but—nah, they look better back on there . . . Cue safety pins!"

3

JOHNNY WEARS WHAT HE WANTS

Ever the fashion victim, it was Sid who'd heard about this outrageous clothes store called SEX, and suggested we go and check it out. There were a couple of expeditions walking up and down the King's Road before we actually found it. Someone could have said, it's at the farthest end, away from anything useful. Us being young and silly, we didn't put the dots together. But once we got there . . .

This would've been mid-1975, and in those days it was still selling Teddy boy gear. That was the main financial gain—really special Teddy boy outfits, and of course the brothel creepers. Other elements were creeping in there, though, like bits of rubber clothing for the pervs and Cambridge Rapist masks, and quite quickly they phased out the Teddy boy side of it, which I thought they should have kept up.

More than just a shop, it was a social center for all kinds of strange, fascinating people. I mean, a few months later, I worked there briefly, and selling Reginald Bosanquet a skin-tight rubber top was a genuine thrill. He was a very popular newscaster of the time. You'd get comments like, "Do you think it fits?" and you'd reply, "Yes! Perfectly!" The man was large, like a roly-poly squashed into it, with bubbles coming out at the bottom. It was hilarious, but I liked his bravado, that he had no shame about it. That's what he wanted, and that's what he bought.

It was Vivienne Westwood who did all the designs, and Malcolm McLaren was the mouth—he'd come in with the intellectual codswallop to justify it all. Vivienne was basically a born-and-bred shopkeeper, of the Margaret Thatcher kind. The absolute dictator. "You can't buy that, unless that goes with that—I'm not selling it to you unless you buy the whole outfit!" "Er, what?!" I was well mix-and-match, so I was a big annoyance to her. She never liked me, and I never got on well with her. I kept myself quiet around her, because I knew I could flare into a row with her in three seconds flat—just a ridiculous person but thrillingly creative. All of her obsessions paid off for her. She means what she does; it's just sometimes it's too meaningful and she does too much of it. I once said to her: "Methinks thou doth project too much!" Ha—sacked in the morning!

The jukebox had stuff like the Flamin' Groovies on it—abstract garage-y bands from America, a couple of English mod bands, a lot of rock 'n' roll, because of the Teddy boy gear they were peddling. They were very into "Yeah, we firmly believe in street movements—which would be made ever so much better if we've dressed them!"

The shop was so brazen and antiestablishment, I loved it. I loved working there for the short time I did. It was only a couple of weeks, but it made Sid very jealous that I managed to get in there and he didn't. He worked there after me. For me it was like, I can really go to town on myself here. I'd turn up in tight lavender T-shirts, but I'd have the tit-holes cut out. Very repulsive, and I'd be wearing winklepicker shoes, drainpipe jeans, and a big gold belt that I bought in the shop, and gold wrist bracelets, and a dog collar. I liked that look, I thought I looked well 'ard.

It was a look that was challenged every weekend by the various soccer hooligan gangs that would be roaming around Chelsea every Saturday. Nottingham Forest come to mind—they tried to raid the shop, but I stood my ground outside and the door closed behind me. When this fiasco of a brawl ended—it was just a slapfest, really—I went inside and Vivienne went, "I told you not to attract that kind of people." She basically hung me out to dry, left me to deal with them, then blamed me for that kind of people being there in the first place. "Okay, Viv!" But that's her.

Malcolm was incredibly witty and well read. He understood the dilemmas of the time, but he was an English teacher who didn't quite know English. He came from that attitude of presumed knowledge, and that position will never work on me, ever. It just automatically makes me think that guy's suspect, what he's saying is doubtful—by the by, a bit like Arsenal's present manager, Arsène Wenger.

Oddly, his favorite books were those kinds of *Jane Eyre* novels. He would always go on about them. It turned out that his mother used to read them with him when he was younger, so, hello, that was the bigger reason for it. It's not that he fancied himself as a young emaciated lady, no, I don't think it was that. I think he was missing some kind of love from childhood. It was preciously rare for him.

All his friends used to tell the same story about going with him to the Grosvenor Square student riots in 1968: they said, as soon as the trouble started, he vanished. He was all for yelling and screaming the big-mouth slogans beforehand, but as soon as they all led the charge, he was mysteriously absent. That's what I wrote about in the PiL song "Albatross" a few years later: that lack of commitment, him always running away.

Malcolm had been managing the New York Dolls in their dying days. When they came over to London, it was great fun chatting with them because they'd always give us their little insights into what their idea of Malcolm was, and of course none of it was favorable. If ever there was a man that actually really never did much at all, it would be Malcolm. They'd be saying,

"He just puts up a few silly ideas but nothing that actually ever really stood a chance of being helpful."

He'd covered them up in red vinyl, but Russian stylee with hammer and sickles all over the place. Just stupid. "Ah yes, this'll terrify the world!" The only people that got terrified by it was the New York Dolls themselves, because they had to pay for all that nonsense. When I saw the photos in the shop, I loudly proclaimed, "That reminds me of the Beatles doing 'Back In The USSR,' Malcolm!" A clear, straight dig. But they'd allowed that to happen and got charged for the privilege. They never stood up and said, "No, I ain't wearing that." For all their tomfoolery, I suppose they just had an ambition to be popular finally.

Malcolm and Vivienne's relationship was very weird. I still find it hard to believe and rather peculiar that they manufactured a kid between them. But it did have ginger hair! There was another child from a previous marriage of Vivienne's; when the one they'd had together was young, it felt to me like he was being ever so slightly ignored. I felt really sorry for him. There he was, this normal little bunty of a kid, quite plump, running around their house, and there was all this bondage gear and rubber-wear on model stands, and Vivienne sewing jockstraps and things. A strange upbringing, by any stretch, but no stranger than mine.

Their house was an absolute mess; no one had ever thought of washing anything, and that kid always seemed to be hungry. It used to really annoy me. Yet he was so large! To be fair, honestly, I don't think they ever brought the kids up wrong, there was none of that nastiness in there. It was all in Malcolm and Vivienne's innocent and naive way, of how they were trying to project sexuality through other people—which wasn't anything that either of them were particularly doing themselves. Hence the need for both of them to manipulate a pop band around to their way of thinking, and in there lies all the problems with the next band they tried it out on. Because Johnny here wasn't liable to lay down—not for *no one*—EVER.

I believe it was Malcolm's mate Bernie Rhodes who spotted me amongst their clientele and said, "That's the one!" Not Sid, the fashion victim, because that'd be too much of the same old thing. "You want the one with a bit of gusto." We were an odd bunch of fellas that had started hanging out there. John Gray was very effete, shall we say; Sid was like an oafish model, and me—I don't know how I came across—probably a bitter, twisted fuck. Quiet but fuming. An "angry yooong man," as Morrissey would say.

At that point, there was no concept in me at all of being in a band. It took Bernie to say, "What about 'im in the 'I Hate Pink Floyd' T-shirt and green hair? He looks like a singer." This must've been about August 1975. Bernie was one of the intellectual ideology people that hovered around Malcolm—they knew each other from college.

Bernie's a troublemaker, there's no two ways about it, but he was also one of these fellas that could get bitter and twisted. He took it all a stage too far, did Bernie. Politically, he was extreme left-wing, almost communist in his approach. Of course, the way he later steered the Clash was about raking in huge amounts of money, in spite of his communist leanings. So, a curious juxtaposition of events, is Bernie Rhodes.

The element of talent-spotting always gave us the reputation of being a boy band, but Steve Jones and Paul Cook already had a group, going back as far as 1971/'72. They hung around the shop, and Glen Matlock worked there, so he fitted in because he was learning bass. And they needed a singer, and through Bernie to Malcolm came I!

Bernie was very aware of how to manipulate a public, which you could see in the early T-shirts he designed for them. There was one—and to this day, I still have it—which had a triangle on each side, and a line in between. The upside triangle contained all the good things, and the downward triangle all the negative things. That already tells you a lot about Bernie's black-and-white kind of thinking. On the good side—somewhere near the bottom, haha!—is "QT Jones and his Sex Pistols." That was their modus operandi before I came along, until they discovered that Steve couldn't really sing, and didn't have that kind of personality to project, in that way.

Our first meeting that day was so bad it really shouldn't have worked at all, should it? No one ever believes me, but I was playing the diplomat there. I wanted to be revolting and disgusting, but I also wanted the job. I was working at cross purposes. Now that they'd asked and told me to come back after the shop had shut, I really wanted to be in a band with them.

For me nowadays, I know exactly what I'm doing it for, I have a need to explain how I view the world, and I enjoy doing that. Whereas meeting the band for the first time, I had to come to grips with all that inside myself, pretty much on the spot. There was no more reserve—it wasn't acceptable. If I was going to get on with these people, I had to be completely open.

When we went back to the shop and I sang to Alice Cooper's "Eighteen" and various other records off the jukebox, I really wanted it. I was up for it. I instantly had the mannerisms, the characterizations of the words. That I could do, I just couldn't sing. A minor thing . . . Fair play to Malcolm, he said, "We can fix that." Which was a big wow, because that was the hard one to get over—but he was right: if you've got all the other things going, that'll come. That's a matter of disciplining yourself. For the record, I did do singing lessons, but they didn't help, because the approach was different to how I wanted to be. "Do-re-mi-*fa*" was *far* beyond me—I couldn't relate it to how I wanted to write. But it was useful to see that's how singers normally go about it, so I found my own way even in that. That's how I ended up with that very sweet dulcet tone that the world has fallen in love with.

Afterwards we went to a pub up the road, the Roebuck, and we really didn't get on. I turned up at the interview—shall we call it that?—with John Gray, a gay guy who hadn't yet come out at the time. I thought that would be very audacity-minded to them, and it was, it really perturbed them. They weren't expecting that from what they perceived as, I don't know, a weak-link art student, possibly. He really off-put them. This effeminate fella just drove them nuts. But John would talk pure sense, and has an absolute wizard knowledge of all things music—very much a librarian, in that respect, and so between the pair of us, we had the answers to every situation possible. That kind of astounded them. Glen couldn't pull out any rare Kinks records that would surprise me, because I'd be fully endowed. I love the Kinks—love 'em!

I'd had a couple of pints beforehand to steady my nerves, so the whole day was a huge alcoholic infusion, and it only ended because we had to catch the night bus home. I really didn't think I'd got the gig. I just thought it was a great night out, and we laughed all the way home in a complete drunken state on the night bus, which, again, could be very dangerous in them days. The gangs would tend to hop on, and muggings and beatings were the order of the day. But we managed to get home completely unaccosted, so it all looked rosy to me.

I didn't think they'd ever ring, but then the phone call came in a couple of days later. It was Boogie, aka John Tiberi, one of Malcolm's people, telling me there'd be a rehearsal in Rotherhithe. So down I went, got there, and no one was there. It'd been canceled, I still don't know the reason why, but nobody even bothered to tell me. And that's a tough thing for a Johnny Rotten at that age, to go to Bermondsey Wharf and wander around them docks alone. No one there to say, "Sorry, John." I was fucking furious.

So I rang and told them where to go, and then there was a whole series of phone calls, and I thought, "Oh, bloody hell, this is worse now, because they're really telling me they want me—not the band, this is all the underlings of Malcolm—really, really pushing for me." But I thought, "I ain't got the tools to handle what I've talked myself into."

Another rehearsal was arranged at a pub in Chiswick. I didn't learn, did I? I went to the next one an' all. Why did I go? I'd bitten the bullet, hadn't I? I wasn't gonna let go at this point. I wanted to push it a couple of stages further and, fair play to the lads, not one note of any kind was in tune. They were just running through what would now be termed mod, the classics of the 1960s, by the Who, the Small Faces, and what have you—the usual verse-chorus pop hit kind of stuff.

I was only used to choir practice, where I'd perfected the art of un-singing just to get chucked out. So I had to break out of that very quickly and find my own voice, and not sound like I was imitating, or trying to be this, that, or the other. That was my approach. It was pure hell and torture for me.

That night ended in a row—a good row, a healthy row, because it was about the messing about, and really trying to get into doing this properly. I think I got my point across: that no matter what you think of me, give me a chance, and don't be rude like that. You don't just not tell someone and think that's funny, because it bloody well ain't. In them days if you made an enemy of me, you could live to regret it.

They showed an incredible cowardice, to my mind, from that point onwards. I was very resolute that I was gonna make this work regardless. But I was so impressed . . . There was a piano in the corner, completely out of tune, a typical pub upright piano, and Steve and Paul played "Good Golly Miss Molly" on it. Steve had the rhythm, and Paul played the ding-ding-ding afterwards, and listening to the two together, no one else but me would've worked that out to be—I wanna be in this band, I like the way they're destroying yet creating.

They were trying to do early rock 'n' roll, but it *weren't right.* The notes were wrong, but the patterns were right! The emphasis, the energy on it, was excellent. I loved listening to it, nothing to do with discordancy, or accuracy of notes, and obviously the wrong placement of fingers—it was, the energy was right, and Paul Cook always had brilliant timing. And timing is EV-ER-Y-THING. If your drummer is out of time, nothing makes sense; it's the root core to music. From that point on, I listened attentively to Paul Cook, and I had my anchor.

That night I went up to the Sir George Robey in Finsbury Park and tried to repeat what they'd done. I'd watched where they put their fingers, and I remembered the hand positions—really simplistic stuff now, but for me that was thrilling. It was like, "Wow! I'm beginning to understand the chemical formula!"

The first time I met Malcolm in the shop, I didn't know if he liked me or disliked me. He was never a friend, he never in any way related to me, and possibly only introduced me to Steve and Paul as an act of spite. I never got to grips with any of that, but somehow or other he must've assumed that I was a character above and beyond the band's very dull activity up to that point. He almost certainly saw something in me that he didn't have in himself. But how far that thing could go, to Malcolm's vision, wasn't too clear. But he did, he backed me in that agenda: I was an ideas fella, and I always will be.

I came in with the concept of lyrics and that punchability, to break the boundaries of dullness. Without me, I suppose they might've been a Small Faces imitation band, at best, possibly a pub-rock band, and they might've liked it that way. And Malcolm might've liked that, because of his love of the New York scene, which at the time was all happening in very small clubs—places like CBGB's were tiny. It was all so precious—"Oh, we're the creators, we're special, nobody else counts!"

Well, I'm sorry, I'm a fairground attraction. I like the funfair, I like the chaos, and I like the aspect of being able to break into the larger majority of people, to challenge.

I'm not sorry, actually. I resent that I actually said "sorry" there. I'm not sorry about that at all. What I'm sorry about is that most people actually don't understand how to change, and readily accept the formats that are given to them. And Malcolm—and the band to a minor extent, because he was their mentor—were fully loaded with that: "All we gotta do is wear nice clothes and look ridiculous, and we'll make some money, right?" And that's what it would've become. Steve, love him. Paul, love him. Glen, love him. But their attitudes were not one of the bigger picture. Not at all, ever.

QT Jones and his Sex Pistols became, upon my entry, the Sex Pistols. Rehearsals, for me, were: no monitors. There's them bashing and playing away on their amps, making an enormous noise, and they cannot hear me. They've got no idea what I'm like, except that Steve would decide, "You can't sing!" I'd go, "What do you base that on?" "I can't hear ya!" "Well, I'd like a microphone then." We get a microphone from the pub downstairs, then I also realize I can't sing, but Paul Cook stood up for me. "Oh, you know, you've got to give the guy a chance . . ." Paul was very friendly, and helped a great deal in that respect.

You could say, however, and you'd be dead right: "What a fucking weak heart! Why don't you get a proper singer, who walks in with a monitor system all of his own?" But they didn't, they stuck with me, and Paul stuck with me—he secretly backed me. Against all of it. That made me more crazy, wilder. I'd start to turn up in the clothes I really wanted to be in. I'm no drag queen, right? I'm full-on, hardcore, a lunatic male, and it's like nothing that had been happening in pop music at all, since I don't know when—since maybe the Teddy boys invented themselves? And, by the way, I have a huge affinity with the early Teddy boy movement. Or any street gang movement, or street culture. I understand completely what that is.

At the time of those first rehearsals in Chiswick, I was wearing a ladies rowing jacket. I only found out later that's what it was—it was white, so I always thought it was a cricket jacket. No. It was a woman's jacket, and I dyed it pink accidentally by putting it in a cheap washing machine with a pair of pink trousers that I bought at Vivienne's shop. So, I wrote "GOD SAVE OUR GRACIOUS QUEEN" all over it. That eventually put me in mind of, "Hmm, that'd be a good song title . . ."

Any of the gear we wore from the SEX shop we absolutely had to pay for. If not the full whack, then as close to it as possible, and no yelling at Malcolm, "That's absurd, we're promoting your stuff, you're using our name." The reply was always, "Oh, but I'll try to get you a slight discount." Later on, some of the northern bands putting themselves up as punk would bitch about how it was easy for us because we were dressed by Vivienne. No, babies! And while I was paying through the nose for it, I'd tell her what I wanted and didn't want, regardless of her assumption of good taste. She's a designer who needs to be told a thing or two from time to time, otherwise, like all of us, you could end up crawling up your own big bottom.

Twenty or thirty quid for a sweater was a lot of money back then—huge money—but everything in there was a one-off. There might be a line on a similar theme, but every print was slightly different, each one was special in its own way—but just don't ever attempt to have it washed, because then the inks would all run, and the seams would come undone.

The stuff Viv made back then wasn't exactly built to last. The buttons just popped off—they flew across the room like they were allergic to you. The necks of her T-shirts ended up down between your breasts after one wash, because of the way they were cut, so a nice tight T-shirt ended up like a man-bra.

I already had the safety pins going before the Pistols, but now they really came into play. Actually, you can see in the old pictures, on everything I wore, I'd always have a set of safety pins hanging off the collar. It was about fallout, having an instant repair kit for when Viv's goods fell apart.

There was never any sit-down discussion of direction with the band, or Malcolm, or anything like that. We were just shoved into a room, and bang, crash, wallop. No matter what Malcolm may have claimed after the fact, it was just the four of us bashing it out in a room. All the hindsight in the world did him no good at all, because he wasn't dictating our pace, tone, or content in any way—and he was miffed about that.

I was into Captain Beefheart and Can, but that didn't mean that's what I wanted the band to sound like. Not at all. At the same time, I'd be the chap telling you that 10cc's first album was one of the greatest things I'd ever heard—and that was so overstructured! I thought, by containment comes perfection. No, I had it all going on, all of it. I had no expectations other than that's what Steve's good at, that sound, that angle, that's his universe, and that gives me a lot to work with. It gave me a huge kaleidoscope of possibilities. Things I'd not considered; things I'd never heard from my record collection or indeed anywhere else. That's how I perfected myself, really, through Steve and his apparent faults, which weren't faults at all.

They would all give Steve a very hard time about his lack of musicality and I'd be telling him, "Sod them, there's no such thing as a bum note. You've got the balls to stand there and play the thing, that's good enough—with time, all the rest will fill in!" Malcolm was really pushing him in a wrong direction, I thought, it really screwed with his mind. Steve needed encouragement, not smug dissatisfaction. In many ways he's a bit like me, he can get very distracted very quickly and then lose the center of a thing. I recognize those things in him; those are traits we share.

He seemed a bit of a handbag snatcher to me—a low-rent thief, crooked. He had a really saucy sense of play. A completely untrustworthy character, a proper Dickensian street urchin, like that Jack character in *Oliver!*—you know, "You've got to pick a pocket or two!"

But at least that's what he was. It was real. Malcolm was on his case all the time about "Urgh, look at your hair, it looks like a perm!" And indeed I was too, because it did. It really did! He looked like an old woman with that curly hairdo. A curly mullet was a crime. Against nature! You know, "We don't want Robert Plant in the band, thank you!"

Steve had had a hideous upbringing, but we were all damaged goods; we were all soiled kittens. By that time, all of us had been locked up for one thing or another.

On the surface, Steve was a bit of a fly boy—but not too fly, and not too bright either. He wanted to give the impression that he was on to something. But he's forever trying to get away from being asked a question—very noncommittal, slyly judgmental, and difficult to get close to. We've had our moments where we've been very close and had a great laugh, me and Steve, but he would swing straight back into that alienation thing in a heartbeat.

He can be hilarious, but he doesn't like another comedian in the room. And, well, if I'm there, it's gonna happen, innit? And then with people like Sid around—well, that's just too much for him. If he'd just bothered to open his heart, we could've helped each other along there quite brilliantly, but it wasn't in him and we were young. Oddly enough, we all viewed Steve as the older member. He was a year older than Paul; they weren't in the same year at school. So he was like the elder influence that, well, really wasn't an influence you wanted at all.

Glen was, the—quote—musician of the band, and so his approach was, "You can't do that, that's not music!" "*Pardon?*" Right from the start, there was an argument with him, because he wanted us to be these dandies, these Soho ponces, a throwback to the mods. That was never going to work, pretending to be something which we plainly aren't, so I laughed that one right out of court. "Look, we're not dandies, why would we want to fake being that?" For me it had to come from a real hardcore, felt emotion. You can't just pluck a fantasy out of thin air and think you can cover yourself in that and that'll be good enough for the rest of the world. That's contemptuous behavior, for me.

Unwittingly, Glen was very helpful. You do need negativity thrown at you, it's a great driving force, and it makes you work harder. When this flummox of a situation started to weld its way into tunes, it was fantastic. Our attempts at other people's songs, particularly the Who, were great. I really started to feel like I was in a band, and I loved that feeling.

We could have, and should have, hung out more together socially, but that never happened. If Malcolm came to rehearsals it would only be to pick up Steve and Paul, and possibly Glen, who usually had something better to do. He'd take them off to these very nice eating clubs that he was so prone to going to himself, and I always understood and knew that I wasn't even considered as part of that. So I never really bothered to ask after the first one or two rejections. Listening to the fumbly-arsed lies about, "Oh, uh, no, there's not room for one more," or whatever, you get the message and you move on.

I don't think Malcolm ever liked music at all. To him, it was just the noise that accompanied his exotic-clothes future for the human race. He didn't quite comprehend its importance or its social significance at all. And indeed, why should he, because really there wasn't much of that at all up until we arrived. What would songs of social significance up to the Sex Pistols mean? It'd be some dreary-arsed folk singer prattling away on an acoustic guitar. Oh God—*urgh!*

There was one meeting in a pub at the top of Tottenham Court Road. We were having an argument, and Malcolm introduced us to Bloody Marys. "Wow, they called a cocktail *that*? And it tastes great, you don't have to suffer the vodka tinge, but you get all the effects." Brilliant evening! But in the background was John Lennon's "Working Class Hero," and I knew the song well. I said, "There! That's social significance, proper, my style." And then he started to grasp it, what I was coming from. I don't know if that left a good impression on him. From there on in, we never really spoke—very odd. He really didn't want to move mountains at all, he wanted to rearrange piles of glitter.

I, meanwhile, homed in on the lyric-writing almost immediately. That's what Steve gave to me, that burst of energy. He was like a diamond to me. Stunning, gorgeous, brilliant, beautiful opportunities. I was quite the dynamite once the doors had been opened to give me the chance to sing. I really went at it. I went straight into the writing and just wrote all the time. The words just flowed out of me, all this pent-up stuff that I'd had no place to aim at before, that I had no ambition with, suddenly found its mark. Fantastic! Magnificent!

There were many, many things I was toying with. I don't remember this, but apparently one of the first things I tried out was about the Archangel Gabriel, slagging off the Catholic Church. I might have done something like that, but what was released is where the energy and effort was. Everything else I view as the scrag end, best left on the cutting room floor. Anything that hasn't been finished is that way for a very good reason.

The first one we actually rehearsed, I think, was "Mandy." I'd been at this girl Mandy's house for a party, and she made a punchbowl using Southern Comfort, Martini, and some kind of fruit juice which also had alcohol in it, a liqueur, and ice. I drank so much of it that it took me two days to wake up—John Gray, Dave Crowe, and Sid dragged me back to my family's house. I can from time to time be a creature of excessive stupidity. I'm well aware of the warning signs and yet I'll dive in and just go with it, but overdo it. I tend to lack subtlety. Maybe in later years I'll catch on to that one, the idea of being subtle. But anyway, that was the subject of my first song—best left unrecorded, eh?

In rehearsals, someone would play something, and I'd go, "Oh, I've got some words for that"—an absolutely open-minded, spontaneous approach, which, I'm well aware, can also hideously not work. But for us lot, it did, and I've kept that methodology ever since. It used to be like that in the Pistols, you know, for a good two weeks. There'd always be somebody in a corner playing something from their heart and soul, and I've got the ears for that moment. And I'll zoom straight into that and go, "Empathy with that!" I write with a melody in my head, and I'll just try and stand there and sing it, right off the bat. Sometimes it's impossible to sing what I write, and I'll scratch for any idea that comes into my head.

Of the early songs, "No Feelings" was a character analysis, satirizing the way rock 'n' rollers were trying to present themselves, as being hard and above it all, and yet being a bunch of wimps in reality. There were these nitty-gritty bands flirting around, all trying to make themselves out to be harder than they were. I wasn't presenting myself as a hard man, just a full-on honest one, and dealing with people telling you things that they're not. Liars. Liars will always be a good subject matter.

"Lazy Sod," which later got more tastefully retitled "Seventeen," came from a Steve Jones idea. The lyrics killed me! Sorry, Steve, but to this day, I didn't mean to humiliate you, but you should have felt humiliated, those were bad words. Fabulously dumb lines, like, "I'm all alone, give a dog a bone!" I suppose he was writing about a self-pity kind of thing and not being able to find a proper attachment to another human being. A good idea, but I wasn't ready for that kind of wimp-out, so by the time I got to it, well, look what I did.

"I don't work, I just speed"—that was all I needed in life! It was very much an anti-hippie thing, drawing up the battle lines with them. "Let's all live in the forest together!" "Fuck off!" Listen, I'm well into passive resistance, I understand that's a vital way of bringing down empires, but the hippie thing of peace and love was vacuous. It didn't actually mean anything, and it was never conjured into anything solid.

When I went to the festivals, I'd watch them squabbling over where to park their Volkswagens and pitching their designer tents and arguing about peg-holes. Come on, where's your peace and love now? But in all of that there were generous souls, and they tended to be the people that weren't wearing the uniform. The "velvet loon pants" lot would never help you.

On "New York," I used the New York Dolls as a reference point and played around a bit. Personally speaking, I've got no problems with the New York Dolls at all, I thought they were great, but we'd already had men dressed up as tarts quite a lot in British rock. And then, my God, when they came over—what a mess! There was a band that fell apart. The other New York bands—

Television, the Ramones—we couldn't believe how old they all were, and how much more loaded. They could afford the things that we desperately wanted, but had no taste, so they'd come over from New York and they'd look terrible. They'd be trying to dress down—black and leather—just depressing, trying to look dirty. And yet at the same time I never knew an American that came over that didn't have a walletful. I thought we were better than that.

Chrissie Hynde tried to help me on the music side. She used to hang around the shop a year before I did, maybe even a couple of years before. She and Vivienne used to be close but they fell apart. One of the most delicious lines she said to Chrissie one day was, "The thing I don't like about you, Chrissie, is you go with the flow—well, the flow goes that-a-way," pointing to the door. Chrissie would be in fits of laughter. The delivery was so funny, that she had to go, "Fine." Vivienne can definitely deliver a good one-liner—no doubts about that mouth.

I can't remember what the brawl/row/scene/whatever was. But—*ooff!*—when women decide to not like each other, wow, us guys have got lots to be proud of, because we could never take it to that level. Well, actually I know a few guys who could. We'll talk about them later.

But Chrissie—what a lovely nutter! We never had a physical relationship but mentally we were very attuned to each other. I respect many, many things about Chrissie—a very smart girl, who went through all manner of trauma as a child, and was wayward in her youth, shall we say. She'll always be an awkward person to get on with, she's very difficult, but I think she deserves your space and your time, because in that difficulty she's looking for answers. And she does find answers from time to time, and that makes her to my mind a very important person in the world.

She was great fun to hang out with. She'd take me to Clapham Common, and we used to walk for miles talking about music, and our understanding of things. She tried to teach me how to play guitar. I'd moan about being left-handed, but of course her line was, "Jimi Hendrix was left-handed, that's no excuse." "How about this excuse, Chrissie—look, I had a bottle jammed into my wrist!" Which was true, I had two fingers that had to be sewn up, and I'd lost a lot of control in my left hand.

I was hanging out with Paul Cook one night, and he took me to see some of his friends out in Chiswick, and a couple of them took offense to Paul because he was in a band. And a fight ensued, and it was all chaos and very hard to remember, considering the amount of booze we had plied into ourselves. We had to walk from Chiswick and eventually ended up at the Hammersmith Hospital, where I had to have my hand stitched up, because the fingertips of my two pointy, fuck-you fingers were ripped to the bone and bleeding profusely.

That's what you get when you're left-handed. The first thing you stick into an affray is your left. It's the hand no one expects, because everybody's expecting a right-hander to come in.

So, yes, I'm left-handed, and playing is difficult on that level, but beyond that I'm not the kind of person who could ever sit down and learn how to—and I use the phrase correctly, in a negative way—"play an instrument." I don't have the time or patience. Actually I don't have the mathematical brain that can absorb that kind of thinking. Making it in the wonderful world of music in any way is something of a really out-there achievement. I've managed to break through that barrier, I'm still capable of putting a song together without any of those—what shall we say—laws of logic?

Chrissie really had her work cut out. It's just a shame I didn't feel I could ask for guidance amongst my so-called colleagues in the band. Right from the off, things got really harsh between Glen and me. It shouldn't have got that way, but when Glen dug his heels in, it was very difficult to deal with him.

Malcolm had this notion that Glen was "the musician." He also must've identified something positive in what I was doing, because one day he took us two to the pub to patch it up and write songs together. Legend has it he gave us twenty quid to spend, but we were smarter than that—we wanted twenty quid each. Then he went off and left us to our own devices.

We giggled. We were on the same plane at that point—there was a moment of truce. Although we were really bad enemies, the commonality was, "What's *that* one about?"

And we had a great evening. We knew that we'd have to work something out here. If there was going to be any progress in the band, it would have to be coming from us. It wasn't going to come through Malcolm and his alliance, because that was a dead end intellectually. We really got on with each other, and "Pretty Vacant" was one of the ideas we put together.

On his way out, Malcolm had said, "I'll give you some ideas—submissive, as in the bondage theme, if we could have that kind of topic?" Me being me, I took it literally for a laugh and then put a twist in it. I called it "Submission," but the line went, "I'm on a submarine mission for you, baby." Anybody who suggests things to me, I'll sneer, but I'll see a possibility in it. And off we went.

We knew full well what he was doing, trying to use us to flog his new S&M line in the shop. They'd ousted the Teddy boy stuff and gone into the full perv—from two people, Malcolm and Vivienne, who were eyeing the world of perversion like the odd couple from Tring. It was just a means to an end—they weren't actually part of a pervert scene. They were observers, then praising themselves that they were somehow manipulating the wonderful world of fetish, when really they were just floggers—clothes floggers. They'd always be on the wrong end of a whip, and we knew full well they wouldn't like what we came up with.

In fact, we never really got a comment on what we wrote. There was no conversation at all with Malcolm and me. From the initial outburst and a sense of backing, to suddenly nothing. Just cut dead. And I suppose he eyed me somewhat as being a problem to his art-movement theories. His interpretation of the artistic leanings of the band—mine, or indeed anyone's—were very different. I didn't think we needed to try and skillfully craft an image. For me the words were creating that, and my own persona. I just expected the chaps to stand up, and they could be whatever they wanted to be themselves, just so long as it was genuine and not crafted.

That night in the pub, Glen and I both understood that we had to amalgamate our two different perspectives without concessions into something even better than either one of us had conceived independently. I think we did that, and I know I wanted more of that, and I know Glen wanted more of that, but again these other issues kept creeping in.

People don't believe me, but we'd hardly ever see Malcolm in those early days. He wasn't at rehearsals, and fair play, that would have been a hard thing for me. Too many people in the room that are not actually contributing is a no-no. It would've been, If they need backup then I'm going to bring backup too, and then it would be my friends versus theirs. And my lot would win, but it wouldn't have got us anywhere.

On the rare occasions we did catch hold of him, he'd be like, "Oh, right, yes . . . Look, I'll book some gigs . . ."

The early gigs were nerve-wracking and terrifying. As a band, we all felt very inadequate and fearful. It's that scenario—you're facing judgment. Indeed, *negative* judgment was all we ever seemed to achieve. But we grew to really like that! Or at least I did, and I began to expect resentment just because we were so refreshingly different. Although "refreshing" wasn't the word some of the audience members would be using.

The very first gig was at Central St. Martin's art college, in November '75. It wasn't even something Malcolm organized—Glen was nominally supposed to be doing a course there, so he sorted it out. Kudos to him that he had the audacity to show us off in front of his friends. It was just across the street from where we rehearsed in Denmark Street—I've gotta say, renting that place was a stroke of genius from Malcolm, right by Soho, in the heart of town—but that meant it didn't feel like a proper gig. We just walked across the road with our gear, bit by bit, and bingo! That's not exactly how you think it's all going to begin.

Of course we were all nervous as hell, absolutely terrified of what was about to happen. Could this be sink or swim? It's so fantastic when you come out the other side, and it's swim—although it probably didn't feel like that on the night itself.

Bazooka Joe was the main band. They were terrible, exactly as you would imagine a band called Bazooka Joe would be. To name yourself after an American bubble gum was just—*eeuuurgh*! Nowhere to go with that one, boys. And they were matchy-matchy, in that they all wore Converse white sneakers. Hi-tops, at that. Even though Adam Ant was their bassist, it was hardly surprising they took such umbrage at what we were doing.

I'd never sung all the way through for fifteen minutes. Well, it was probably more like twelve and a half that night, with all the nervous energy. And when you account for all the Strepsils I was chewing for my throat, it was probably more like ten minutes. And then repeating the set, just to try and fill up half an hour!

I quickly realized you just have to rely at that moment on your ego resources. Stamina would be a word to pick up on, but not, "Oh yeah, I can walk ten miles, me!" It's not that, it's a mental stamina—that you can endure, no matter what the problems that come up, you've got enough going on inside yourself and enough self-belief to win through. And all that without the normal approach to training and technique.

When you look back on it, our rehearsals were obviously some kind of training. Something was garnered from those moments that we took on the stage, and finally produced what we were doing in rehearsals in front of a live audience. And how to deal with the first boos in a positive way, rather than go into woe-is-me mode. In that respect, I think I pulled the band through on an enormous number of occasions. The more negative the response, the more positive my reaction. I never minded a bit of banter with an audience. I could trade one-liners with the best of them, all of which, I suppose, simply amounted to "Fuck you!"

It went by in a weird haze, just trying to become accustomed to complete strangers staring and judging us. I felt very protective of my chaps up there. There was literally no clapping—silence is golden—and there was a huge scrap at the end. There always was, but Christ knows what about. Most of the scrappy situations were always about the other bands. Always, there was some sad-sack two-bob fucking jealous cunt going, "You can't play, you're shit, that's not music!" All those clichés! These days, they seem almost quaint, but in them days, those were apparently insults.

After the gig, I just had to go home, there was nothing to do. We had no money, there was no great drinking celebration, no "yee-haw, us together" about it at all. It was like a very dull muff-dive into a solitary subway ride home. I might have had some friends there, but the thing is I was inside my own head. There were friends with me, but I wasn't connecting to them. I was really worried about all the things that weren't right. I became for that moment very self-absorbed. Well, not self-absorbed, more like a commitment to making these things work better.

After that, for the next few months, we'd play every college and university we could, in and around London. I became very used to the fact that the student body was not the volatile hotbed of rebellion we were led to believe. They were a very conservative bunch, but they had loads of money at the time, and they'd throw it at anyone and everyone.

We were just doing it to earn enough cash to be able to buy something that would improve what we were doing. I suppose the concept was very much like—you know the free apps you can now get on cellphones, the video games and things? You get drawn into it but then you need to buy things in order to progress to an extra level. That's what touring does: you need to earn more, so you need to work more, in order to get the equipment you need in order to play better. And at the same time, hopefully be able to take at least twenty quid out of it for yourself by the end of the month, and have to be able to survive on that.

We loved the college gigs because there was always free sandwiches, and union prices on the beer. It made perfect sense. We're playing to people who don't like us, will never like us, don't understand anything we do, don't clap, don't even have the bravado of booing, but they *do* feed us well.

In High Wycombe, we supported Screaming Lord Sutch—what a thrill for me that was. I'd loved Screaming Lord Sutch for ever such a long time. That man had it—he understood reggae at a time when that was a no-no. And there he was up there, absolutely bang on the money. It was so great playing with him, and to say hello to him. But zero response. He just turns round and goes, "I don't get what you lot are doing."

For me in those early days, the bum notes from Steve were great, because he wouldn't stop. He wouldn't panic—like, "Duuuuh, where am I in the song?" He just went, "Fuck it!," and carried on. If you get back in time with the rhythm, it doesn't really matter about the notes. It's actually more thrilling and exciting, because you can shapeshift with that. It helps you develop your craft way more than going to singing lessons with Tona de Brett.

After those beginnings, I got into taking gigs really seriously, and doing quite a large amount of them, and then very quickly came the bannings, and having to go abroad to play.

But we never wanted it to be just hocus-pocus and press trivia. Malcolm couldn't stand the good reviews: "That's not what we want! We need these old farts to hate us!" "I don't need them to do anything, Malcolm. I'm not doing this for old farts!" On that, there was a unity between the four of us. We liked our songs and we wanted them to work right. If only . . .

THE BEAUTIFUL SHAME

As an armchair critic, I would put it this way, but I am even too lazy for that: as far as anything physical goes I am inexperienced. As many of my friends have pointed out, I don't even know how to walk, and running is a no no no.

My tactical understanding of football is practically zero. I have no concept at all when people talk about all the different formations and tactical masterstrokes. Overanalysis of football is a modern-day problem. I'm sorry, but the middle class have introduced that, and it's a load of nonsense. The players should be able to play anywhere on the pitch, all over the pitch, otherwise what are they doing in football? A player who can't pass, tackle, or shoot—and my team, Arsenal, have had a few of them over the years—is completely bloody useless.

Football—or "sacker," according to my American friends—is a game, after all. It should be chaos out there, and it really is, actually, no matter how much they try to plan it. No matter who's kicking the ball it's fifty-fifty where it will land. I don't care how overpaid or underpaid they are, it all comes down to the same thing. You can over-strategize, or you can purchase all the best players, but still that might not work. It's something about the personality blend, and the confidence the manager can instill in a team, that makes a team successful and thus exciting to watch. The art of football is that you can lose and enjoy it, because you know your team did good. But not really . . . I'd rather have a dodgy goal!

I've been supporting Arsenal, which was my local team around Finsbury Park, since I was a very young boy. I used to hang around with a particular bunch of lads at the back of the North Bank, the home terrace at the old ground, Highbury. It was our territory, our manor, but Arsenal above all else has always been antiracist. There was always a mixture of colors and creeds. It's really sad to see the way football's now gone, what with the middle-class kids and everything else, how it's all got misinterpreted. Everything that was good about going to football has disappeared. From the chants and the atmosphere onwards, the whole thing has become sterile.

On the pitch, all this rubbish about zonal marking, etc., is confusing the format they're playing. I mean, come on! If you leave huge wide areas unmarked, thanks to your ridiculous obsession with positioning, you'll get caught out. If you have weak, slow defenders, you'll get humiliated.

Football's the kind of game where, if your team's doing really badly, it gets you into the mode of having a laugh at losing. You can actually enjoy looking forward to the next tragic defeat. And there's nothing else that gives me that ability. It serves an absolutely brilliant, beautiful purpose. It's the theater of emotions, not dreams.

The biggest joy of being a football fan is that there is

ultimately no joy in it at all. It can always get worse. Years and years ago, when West Ham got kicked down to the second division, I remember their fans singing this glorious chant: "Que sera sera, whatever will be will be, we're going to Bu-uuurnley, que sera sera." The humor was fantastic.

That's the joy of football—it's total fucking pain, and when you do actually win anything, it doesn't last long enough. The pubs close too early, and it's all over. Everybody goes home, and you're left standing there—whaaaaa-uuurgh! It's like trying to get through them apps on your iPad. They're so unsatisfying, they should just be called "soccer." Guaranteed to disappoint, and they all require you to put money in to get anywhere.

Playing properly is when you see the team enjoy themselves, go for it, 100 percent commitment. In that respect: win, lose, or draw, it doesn't really matter. It's just like a gig—you win some, you lose some. But when you see heads going down, and slowness, and inability to make a tackle or create an interesting new idea—that's unimpressive. That's the passion killer.

Living far away from Finsbury Park now, I can't keep track of when the next game is. Every Saturday morning at the crack of dawn I'm ringing Rambo, who also lives in America, twenty minutes before a game, panicking, trying to find out if it's on satellite TV. If I denied myself the torture of watching Arsenal being run ragged, I'd still be very, very angry.

When I'm back in London, I'm definitely not one for dealing with the celebrity boxes there. I've never been invited and I don't want it. I'm much happier watching it down the pub with real people, proper football fans, listening to the banter. Football's fantastic for that. Listening to the one-liners from people, the humor, it's sensationally, genius-ly British.

But £75 a ticket, these days, to go to Arsenal? For that money, you should get to have sex with all the footballers' wives! I'd rather go to a smaller place like Torquay United. How much is it there? Twenty quid? That's the price of a massage—and, on a good day, maybe a "culmination."

4

INTO THE INFERNO

The Sex Pistols is a subject that's been very well hammered home, although inaccurately by most. Everyone's had their say on what we were, or what we weren't, and it's reached a point now where all four of us remaining have no interest in counteracting that. If people are going to be foolish enough to believe other people's versions of our history, then good riddance. There's no point in us weighing in. The work itself counts for everything.

We came on, and we came on very strong and very quick. We became, I think, the world's most powerful band. That's a very hard thing to recover from, and regroup, because it was so dynamic, and it crossed every border imaginable. It opened up people's minds, all doors were open. And unfortunately for most, *open and revolving*—aaah, a PiL song reference.

The Pistols was an amazing coming together of a group of individuals who instantly didn't like each other, who were very suspicious of each other, but somehow managed to make that work for the best. It became a runaway train of thought.

All these ideas had been rattling around in my head for years, but I'd never had a format to put them together and present them to the world. So the opportunity in the Pistols was just fantastic; it all made sense. The volcano did erupt, and out it all came.

Apparently, the lyrics to "Anarchy In The UK" were astonishing words for a lad of twenty to be coming out with. I don't mean that bigheadedly. I mean it in terms of, I never got a chance to stand back and observe what it was I was doing at the time, because it was all so hectic and quick. Everything was happening all the time, my brain was imploding with all manner of pressures. If I listen to that song, or any of them, I'm astounded that I came up with those lines. They're from somewhere deep down inside me, and heartfelt.

Living in Britain at that time, it was like being stuck in the 1940s, with all the energy shortages, power cuts, and garbage sacks out on the streets uncollected. The country was still in massive debt from the war. Unlike Germany, which was built up afterwards, Britain, for winning, got fuck-all. A lesson to be learned. War brings economic disaster, but it does bring wealth for the arms manufacturers and the corporations. The oil industry profits greatly from wars, and that's really who benefits. It's us who are expected to be loyal cannon fodder.

When my aunt visited from Canada, she'd be reading press reports about the shortages and offering to send food parcels to us. I've never forgotten that. My mum and dad were furious. "Never! Never! We won't accept charity!"

It definitely inspired revolutionary thoughts. I was noting how bad it had got, and that was the fuel for what we were doing—the sense of energy waste. All that capability of an entire young generation, just absolutely ignored. Depressing as that was, and hard as it was to endure, at the same time everybody I knew was in the same boat. We had nothing to do. That was the fuel that fired the engine of what became punk.

With those opening lines, "I am an antichrist/I am an anarchist," I wasn't trying to set myself up as some kind of bogeyman. I never thought of that at all. No, no, no, somewhere deep inside me, I was thinking I'd be seen as the victim of all of this, and great sympathy and outpourings of love and joy would be bestowed upon me! Honest! I had no concept of being the naughty bugger.

It wasn't about that, and to my mind it certainly wasn't just about *me*. It was about *us*. We're being given an opportunity here—let's tell it like it really is, shall we?

Of course, everyone around the band at the time was saying, "Why don't you just write a love song? Why don't you just write a hit single? It'll be great then, everyone'll love you!" "What, don't they already? Oh." To this day that's all I keep hearing from the business end, and it's utter nonsense they're talking.

That line, "I wanna destroy the passer-by"—I'm full of pleasantries, I know—I was talking about all those kinds of people, the complacent ones that don't contribute, that just sit by and moan and don't actually do anything to better themselves or the situation for others. The nonparticipating moral majority. I just thought "passer-by" was a better phrase, gets to the point quicker. Rather than use twenty-two words, just one nailed it rather well.

"Your future dream is a shopping scheme"? That's turned out pretty accurate, didn't it? Seems to be the way of the world now. I seen it coming.

The "IRA-UDA" bit wasn't so much about the terrorism and political shenanigans going on in Ireland, it was more "Hang on, I thought this was the UK." The alleged United Kingdom, not so united after all, huh? With all these political intrigues, divided we fall. I always viewed the United Kingdom as good people getting on with each other, not the Empire. I think, clearly by my stance, I wasn't promoting British colonialism—the opposite.

But no, I wasn't an anarchist. I found that the written word could achieve far greater disturbance than planting a bomb in a supermarket. The written word's a powerful thing and I don't think that was too well considered, at least not in pop music, until I started to write that way. There's no personal spite or viciousness in what I'm writing. It's absolutely about demanding a clarity from politicians. As long as I know what's what, and what it is you're expecting of me, and what it is I'm expecting of you, everything's fine. I will not be anyone's cannon fodder. If it's not a worthy cause, I'm with the opposition.

• • •

As we progressed and songs like that started to come out of me, I was really into it, 100 percent committed, regardless of Steve being in a huff. He'd be in a huff quite a lot, downing his guitar and walking off. Then you'd get Paul going, "Don't worry, he's always like that, he's a moody bastard."

I'd catch Steve from the corner of my eye, going, "Oh my gawd, he can't sing, he's no good"—without realizing it's like, "Well, Steve, *you can't play*! We're all doing the best we can, mate . . ." We were together, what, less than six months, and we'd already decided that we were all crap.

In truth, I was inspired by what the others were doing, by the three of them. I was really, really impressed. As for musicality, who gives a damn? It was the bravery of what they were attempting to do, in a sea of only spoilt kids in bands.

I've always loved Steve Jones' approach to guitar. It's borderline falling apart, which I find thoroughly fascinating. How he just manages to pull it back. The closest I can connect to that with any other musician would be Neil Young on *Zuma*, where the song is just teetering on the edge of total collapse and that's a most dynamic point. For me to be able to fit into that, it was great. I could instantly think in stanzas. It gave me that opportunity, and I don't know if he ever realized it to this day. I don't think he quite got what a good deal he handed me, and I'm very, very grateful, regardless of his contempt—in fact, that's just icing on the cake.

Paul Cook's timing—astounding, always has been. He doesn't rattle your cage, Paul, it's solidly there—he would give me the backup. When I made cock-ups onstage, Paul would always deliver the most painful line and make me bite that reality, whereas a dig from Steve or Glen was like water off a duck's back. Paul would confidently produce the sentence that was telling me—"Curtail that, get that right . . ."

I quickly realized Paul was always malleable to Steve's will. Steve would ask Paul to ask me, "Erm, uhhh, do you not think this would be a better idea?" All of those things, no matter what avenue of approach was taken, were met with derision, and in many ways created a problem between me and Paul, because he'd think that was a personal fuck-off to him. It wasn't. I meant it back to the people who were moving him about. Because Paul was all right. I always got on well with Paul. But I never got on well with Paul when Steve or Malcolm were in his ear.

Very strange, the band itself. Paul was a quiet fella. Paul's not one for the fuss, he's very much like his father, quiet and compromising. I don't know what Paul thought. Paul's someone who doesn't want to be noticed. But it's the combination of these characters that creates it and turns it into what it is. I mean, if we were all nutters like me, it would have gone down the toilet quickly. I've got to be honest in that. Left to my own devices, 100 percent completely—*catastrophe*! I don't know how far the push button can go. But: "Yes, let's find out."

I thought I found my voice very early on, but I only really got to hear myself back the first time we ever had monitors, which was the gig we did with Eddie & the Hot Rods at the Marquee. That was a big shock. I accidentally "broke" the monitors because I couldn't bear the sound of my own voice. That was the beef on that occasion—they were foolish enough to let me hear myself.

The others would all talk in terms of counting beats to me, and I had no clue what they were saying. Or, "Come in on the G!" What?! Utterly clueless in that regard, yet I had a huge musical knowledge, just not about the actual construction of it.

It was fascinating—it was like coming to grips with a Meccano kit for the first time. Ever get one of them for Christmas? You open the box, you've seen all the pictures on the outside of what you could make, but once that box was open, you were left clueless. That's what joining the Pistols was. I had to quickly put the pieces together and catch up with them.

Steve and Paul and Glen did amazing things for me, I will love them till the day I die. They're more than welcome to slag me off—that's their right, and that's the way I feel about everybody I work with. Good things come from it, or else you find very quickly if things aren't working, you're not working with them anymore.

The space in Denmark Street should've been our HQ. Steve, however, wasn't welcome at his mum's and had nowhere to live, so he stayed there. He kind of turned it into his apartment, and created the vibe that we were butting in when we wanted to go there to work. Everything has a dark side. Because the walls were bare, I turned to drawing vile cartoons on them. We spent an awful lot of time playing animosity, different ways of winding each other up.

We all obviously hoped that having the band to focus on would distract Steve from his thieving. As it happened, it was six of one, half a dozen of the other. It was very hard to break through Steve's mood moments. Only Paul Cook really ever understood him. Steve was particularly good at playing dirty tricks on people, little moments of spite, and Paul had to endure them as much as everybody else. All of us were at the receiving end of a prank or two, some of them borderline dangerous. This is young men at large.

As for Malcolm, he would have preferred it if I hadn't had a lip on me, because he still wanted to be accepted by trendy art society. One of the early gigs he arranged was at this guy Andrew Logan's loft. It was called the Valentine's Ball. It was very bizarre, a fantastic place to play, but they didn't like me. Maybe it was something I said.

That sort of nonsense was what he groveled and hankered for, hanging around with all these fake socialites, who all self-assured each other of their own importance and indulgence. Rubbish artists, poser bloody statue-makers. It was very arty and therefore very fake.

In a nutshell, Malcolm and his clique were hippie art-wankers, they really were. They changed garb but their mentality was still that same vacuous dead end, all kissing each other's bottoms, and achieving nothing, offering nothing, doing nothing, self-absorbed, self-fascinated, and closing their doors to the outside world of reality, which is what I was firmly entrenched in.

That cliqueyness of trendy London doesn't make room for people like me, and I'm so grateful, because you can get caught up in that swirl, or you can think that personal adornment and self-gratification makes you the savior of the universe. Indeed it doesn't. They're all supercilious and superficial and full of praise for each other, and limited and narrow-minded because of that. They have no real understanding of how the real world works; it's all self-serving. That's what I found out. Outwardly they're very impressive: "Look at these people, their mad clothes and their apparently mad lifestyles, and they all look like they're having freaky-deaky sex." Really, it's just a bunch of suburbanites.

Malcolm was an eternal art student. I never saw a single piece of art he ever got up to. He surrounded himself with people like Jamie Reid, who was an artist of sorts, but more in the commercial aspects of art, in presentation, packaging and selling things. Of Malcolm's crowd, I liked Jamie a lot, and I liked Sophie Richmond, his assistant. She was always morose—"Oh, everything's dismal and I'm bored and I'm depressed." I found it entertaining to be around someone so despondent. Nothing that would happen in the world could ever make Sophie smile. I hope she reads this, because that'll put a big smile on her face.

Malcolm started renting an office off Oxford Street. There were two rooms, and Malcolm would always be locked in the back one. "What's going on, Sophie?" "I don't know, he's locked himself in again!" The biggest expense in that office apart from installing the telephone, the table, and the chair, was the double locks Malcolm put on the room to make sure we didn't break in. He'd be cowering in there because he couldn't confront you on issues. He didn't want us to get in the way of his secretive agenda. He'd keep his ideas very close to himself because he knew they'd be shouted down, and I suppose, in his irrational, egotistical way, he couldn't cope with having to defend the indefensible. Maybe he somehow thought that kind of behavior, that flightiness, made him artistic. It didn't; the end result was *autistic*, it just caused fractures between all of us.

So there were loads of rows. There was one at that pub, the Nashville, in Kensington. We actually played there twice, and the second time was a horrible vibe—Vivienne slapped someone in the crowd. It blew up into a huge row of "It's either me or him" from Glen, and well, guess what? In the end, it was him!

Over the years, there's been plenty to be bitter and twisted about between me and Glen, but I don't want it to be that way, because actually I really like Glen. No, I do! Sometimes. Hahaha! I understand where he's coming from, but for me that's two steps back, and that's uninteresting. I don't see him as a dark, evil creature. He just wants a happy world, where everybody gets on. Unfortunately for us, that happy world would be according to his rules, and that's unacceptable.

There were other problems too, like the "wanting to be like the Bay City Rollers" rumor running around, and no one actually declaring it to me. The proposition was put to me slyly through Malcolm, and that wasn't what I wanted at all. The lyrics got tougher and tougher. Steve—I didn't know at that time that he couldn't read or write; I put down his lack of interest in what I had on a piece of paper to plain negligence. It wasn't; he actually hadn't got a clue. The only one who really read it was Paul Cook—he'd be fascinated by what I was coming up with. Glen's approach was, "It doesn't fit the pattern of the music. You're one beat off." In your mind, you'd be thinking, "Yeah, and you're one off a beating too!" But we never had any violence between us. There was pushing and shoving, but never anything brutal or nasty.

It could have fallen flat on its face, but we stuck at it, and it all worked out. We really did work hard, and that's a thing that's not noted. We worked very, very hard, with no money. One of the people who did note it was Chris Spedding, who was Bryan Ferry's guitarist at the time. He took a lot of time out, to teach us a few things. He took us to a proper recording studio actually and helped us record a demo which was fantastic. It opened our mind to the possibilities.

We did other demos with the road crew, which was a couple

of hippies that had a few stacks of PA, but that wasn't quite the same as somebody who actually has made records showing you how it was done. It was fantastic; that only strengthened our commitment.

At Malcolm's end, there was no strategy involved. It was all happen-chance, "fly by the seat of your pants." Picking up the phone and making a call—which he didn't even do himself, that's what Sophie would be doing—I hardly think that that was strategy. Malcolm would have a way of annoying anybody he ever approached. There were many agents who chased him out of offices, some actually even physically abusing him. They didn't like his arrogance and his tone and his superficiality. You can be prim, but be proper. These are blokes that don't want to waste their money on what they see as a bloke goofing off on his own ego.

He got us a load of gigs up north, but booking the Sex Pistols into a Teddy boy reunion in Barnsley wasn't clever, was it, Malcolm! That could have been a very deadly evening. If you're in the wrong place playing to the wrong type of people it can be an absolute warzone. We still had our fifteen-minute set at that point, which we had to play twice. These were some very angry middle-aged Teddy boys. They were trying to remember their youth, and we—the new youth—weren't having it.

Scarborough, the next night, was unbelievable. The hatred! It was very much out of season, and there was no one around but locals. Outside it was blisteringly rainswept, freezing-cold, stormy-sea conditions. You could see the ocean from onstage, because there was a wall of windows at the back of the club.

In front of the stage stood this solid firm of 300 full-on beer-monster yobs, who were obviously thinking, "You softies from down south, you rate yourselves?" I'm trying to sing while battering people with the wedge end of the microphone stand—the iron bit at the end that holds it solid to the ground. And you get into it, you don't skip a beat, you continue the song, and that becomes overwhelmingly interesting to the people that have set themselves up as your enemy, and they start to pay attention then, because you're not there for them and their silly little fiasco brawl, you're there for a bigger issue. You end up making mates with these kinds of people, just because you've got the bottle to stand there and deliver. It is respected.

Every gig was like a battle, but we came through. The music press started to give coverage to what they called "violence" at our gigs, but you have to say they'd have to have been really going out of their way to find those incidences. Go to a Justin Bieber gig and there could be fights in the audience. It's par for the course. You put a lot of human beings into one building all at the same time, then there's going to be an affray.

Never, never, ever, ever have I preached violence. I've got a very intense working-class background where violence was the order of the day and I, possibly through childhood illness, had to find another way out of that.

I may have said things up there, but—oh, that's banter! Don't ever mistake that in any way as opposition, unless of course they're trying to kill me, then I'll have something more serious to say. I like the give and take. I like when people shout things out. I respond. It's a connection. These are human beings, they're trying to say hello. They're not trying to be offensive, although some of the remarks could be swung that way.

Shit happens sometimes, and every now and again in an audience there's one particular person that absolutely hates the ground you're standing on, and there's nowhere to go with that other than deal with it directly. The very last thing you should do is leave the stage, because there are 99.98 percent of the crowd who are absolutely there for the right reasons. But you have to meet the challenge.

Because of all this nonsense, it had got to the point in London where no one would consider letting us play their club apart from Ron Watts at the 100 Club. I liked Ron a lot. His background was jazz. It was a jazz club, but he's a very fair, open-minded fella and would give anybody a go. As he said at the time, "It's either George Melly or the Sex Pistols, what's the difference?"

I don't think he liked Malcolm very much, but then again, not many people did. He'd come in with his "I'm the Royal Highness" approach, and manage to offend the staff and the owners with sheer pomposity. Always trying to overlabor his position, which of course opened him up to no end of ridicule, not only from outsiders, but Steve Jones—he'd be on that, *phwooarf*, like butter on toast. Those two had a very odd relationship. Steve didn't want to do anything without Malcolm's say-so, but then would spend all day and night ribbing him. Deeply wicked ways, but Malcolm seemed to like that. Go figure.

I couldn't be wasting my energy being sarcastic to Malcolm. The further I kept out of that, the less likelihood of any audacity along the lines of "I think what you should do, John, is . . ." Only once in his life did Malcolm approach me that way, and he never came back!

During the heatwave that hit Britain through the summer of 1976, my God, I did everything not to get a suntan, but it was irresistible. I liked my death-white complexion, because I was much more a creature of the night, but the appalling dead heat of all day long trying to sleep in it was a no-no. I kind of reversed my process around that time and became more daylight-driven. You had to, you couldn't sleep in that. It was unreal hot, regularly in the nineties. So strange, so un-British. I didn't see any Labour council member trying to complain about it.

The weather will influence all, that's an act of nature, and you learn to go with the flows of nature. You can't stand up against them, you will be flattened. It was a gift, and things changed in London from then on. I noticed that the restaurants started to put chairs outside because it was too hot to sit inside, and London became very much like Europe. Once doors like that are opened, society-wise, they don't close, because they're enjoyable.

I really enjoyed playing the 100 Club through that summer,

as our reputation grew. Yes, it was a sweaty basement, but there was something approaching an air conditioner, even though it sounded like aircraft engines revving. I've only got one vague memory of being onstage, and that's prompted by a photo of me kneeling on the floor in my torn sweater, screaming into the microphone. I remember doing that, and feeling, "Oh, look what I can do. I'm really enjoying this, I like this, I like being in a band, I like the songs, I like the power and the energy of it, and by God, look, happy faces. All things are possible!" We were often portrayed as speed-crazed maniacs onstage, but that was far from the reality. The other three weren't "up there" at all. Regardless of Steve joining AA years later, he's never been big on anything, barely a pinch of salt by equivalent. And me, I stopped. I wasn't going to become too infected with drug pleasures, because I really wanted to conquer the opportunity. Also, you can't be singing on any kind of upper, forget it, ain't gonna work, with your heartbeat racing . . . You'll finish the song before the band have hit the stage. It makes you far too agitated to concentrate on the job in hand—not exactly what you want onstage. Heart attack, panic—you get all of those things naturally. I don't like being in a condition of angst, so for me drugs were always strictly extracurricular.

My preshow tipple of choice in them days was something Nora, my future wife, introduced me to—Liebfraumilch, a German wine, although she said it was awful. Mother's milk. It was terrible, but gosh, it got me through . . . how many gigs?

Another punk legend is that we were all on the dole. Well, we weren't. Paul kept his job at the Fullers brewery in Chiswick for quite a long time. He was a trainee electrician there. I tried to make whatever money and bits and pieces I could. I was still doing Kingsway, but as I got more into the Pistols, it was impossible, I couldn't focus my mind on the two angles. It was like, take a risk, take a leap, dive in, get this together. And I committed fully to the band. Which was infuriating because Paul wasn't quite doing that, so rehearsals would have to be scheduled around his job.

We were clearly showing anyone who saw us that it was

possible—that it was possible to do it without enormous financial backing. In fact, we managed to get this far with no backing at all. We did it pre–record company, pre-everything—we broke so much ground there. We had no one to teach us conceit. Conceit is something you learn, you just don't naturally waffle into that area.

Once there were people there to see us, it wasn't just the gig, it was afterwards I liked—talking to people, finding out what their interests were, their assumptions and opinions, and to find that they're all coming at it from many different angles. The Bromley lot would be very different from, say, what you would run into up north, but all equally interesting, and equally equal.

The so-called Bromley Contingent were really made for the Sex Pistols. They came out of liking Bowie and the Roxy Music scene, which was very much about dressing up. A Roxy Music gig, particularly the big ones, was all about being seen in the foyer. Not so much about the gig anymore. And different cliques would have different fashion sense; there'd be a camaraderie but also a sense of competition going on.

When it looked like Roxy were falling apart there was nothing left to fill the vacuum. Then we popped up, and it was, "Ooh, ha, yes please! Now we can all really *be ourselves*!" Everyone could admit to liking all manner of music, and create their own imagery; not necessarily having to tie it down to a deliberate copy of some fella onstage. It gave them the new door, that you can break away from the competitive element and the restrictions, because it was all becoming very much like ballroom dancing—the rigidity of overdressing. We allowed a break from that, and bingo, there it was. Many, many doors opened, and sexual orientation didn't matter one iota. Nobody was judging anyone.

In fact, in the beginning there was lots of little glam-rock-type bands opening for us. The more different they were, the more interested we were in having them. But that kind of scene was done with. The world had moved away from high-drama anxiety music, and quite frankly Roxy Music did it best, and why try and be Part Two of that.

So it was great when people started catching on for themselves and forming their own bands. If I can do it, you can do it. It was kids our age, but not only from my own background—from many different backgrounds. Fantastic, that amalgamation of differences. Girls standing up for themselves in some of those wonderful girl bands, like X-Ray Spex, the Slits. There was so much of it, and so much difference between the bands musically, it was an incredibly interesting and exciting period, to watch and listen to these groups. The Roxy was a great club for that. Even though the Pistols never actually played there, we had a place to show off our different approaches to each other. I always thought there was no sense of competition in that environment at all, before punk became available to the masses through the media, and the clichés started to arise, of every song at a hundred miles an hour, screamed.

All that came from the Clash and the Damned, more than us. And that was very much like Joe Strummer's R&B bands before he joined the Clash. They would be all this kind of breakneck rockabilly hundred-miles-an-hour stuff, which I never liked then. Don't get me wrong, I loved Joe. At the beginning, he was friendly, friendly, friendly, but that soon changed once he started taking the Clash too seriously.

For me the shining light in that band will always be Mick Jones' personality—lovely person, really warm. The same with the bassist Paul Simonon—he's a posh kid, from a good background, shall we say, but he talks like 'e don't know tuppence from thruppence. I put that down to shyness. Individually I loved them. They'd only been playing a couple of months when they supported us at the Screen on the Green in Islington, and they turned up with so much equipment—oh my God, they had a huge PA, which of course they wouldn't lend to us. So when the support band left with these enormous bins of sound, like, there was us back onstage with our little boxes.

A band I loved was the Buzzcocks. Great fun lyrics, a totally different approach to music, and unfortunately, with everything being lumped under the banner "punk" at that time, people didn't really notice they were a little bit off the beaten track.

They played their first gigs with us, up in Manchester. All I really remember is arguing, about anything you care to mention. It was the singer who left not long afterwards, Howard Devoto, and the one who became the singer, Pete Shelley—they took us to a pub called Tommy Ducks, whose big gimmick was underwear stuck on the ceiling, and I misbehaved accordingly. I just found the whole thing rather silly. Because you're nervous before you do your gig, and you're not up for, like, these kinds of absurd environments, so I was a little unfair to them, but I think they understand, that's how we are. You've got to be yourself. Many years later, I got to say sorry.

Malcolm loved to paint himself as the orchestrator of this burgeoning movement, but it was all happening above and beyond his control. His every move was clueless. That September he took us to France for the first time, and we ended up playing to a packed crowd in a Parisian discothèque, for a diehard disco-dancing crowd. You know—"Zees eez not ze Bee Gees, eez eet?" "NO!"

What a fantastic time we had, though. I was wearing my Basque beret, the whole thing. Malcolm took us to this great show-off outdoor bar, which was apparently where all the cool *célèbres* in Paris used to hang out. He introduced us to some multimillionaire friend of his who took us out to this five-star French-nosh restaurant. I loved it—the biggest, fattest, juiciest, rawest, bloodiest steak possible, and I hammered that home with a quail—a whole quail, which looked just like my pet budgie. Tiny little thing, absurd. Where's the substance to this? I never grasped it, but I grasped the deliciousness.

So that was opening your mind to things, and not letting that change you into an effete culture whore. Quite the opposite. I think that that level of quality should be available to all of us. That's my way of thinking. Open the doors. And if you get a chance to get a crack in the door, kick it open a little bit further, so someone can follow in behind. Make the world a better place, not a worse place.

Glen thought along those lines too, but his idea of better comes with rules. You shouldn't swear because, you know, the children, it's a bad influence on them. It was a major row then, and it's one I maintain, that there's no such thing as a swear word; it's a matter of interpretation. They're just sounds that humans make that have effect. *All* words do, and once you start banning words you're banning what it is to be a human being. How on earth can we restrain ourselves in that way? These are not acts of violence. They're opinions, and it's all the more fun when your opinions are wrong. Wow, don't you go down in flames! You're fully welcome to talk shit, but if you talk shit, you're gonna get shit in return.

Another legendary gig came not from Malcolm's planning, but by invitation. How odd that Chelmsford Maximum Security Prison invited the Sex Pistols to play there—to killers and psychos. Fantastic gig, and fantastic prisoners. The amount of gear them fellas had! They were all long-termers; there ain't no one in there for nicking a handbag, at least not in the crowd that came to see us. These were real people contaminated by a shitstem not of their own making and caught up in the problems there accordingly. It's very easy to become a criminal without understanding the guidelines. I see everybody in jail as a victim one way or another, that's my sense of empathy. I loved it.

These fellas got the songs, I'll tell you. Halfway through "Anarchy," I go, "Cough, cough, the smell of marijuana's slowing me down!" But when they're locked up like that, all those poor fellas had was drugs. The pain I felt for them locked up with no hope of getting out.

Talking to them after, there was no kind of control, we weren't separated from them. They meant no harm to us, and quite a few of the fellas were expecting me to become a fellow prisoner. "You're on the road to ruin, you are!" I've since proved differently, but they understand society and how society can turn against you.

When we put a bunch of bands together over two days at the 100 Club Punk Festival, the word "festival" was used totally inappropriately, with a huge sense of fun, and by no means meant to be taken as seriously as it has been ever since. Calling it a festival was a hoax on ourselves, definitely tongue-in-cheek. The reporting at the time—and ever since—saw it as a deadpan, serious, dour affair, rigidly adhering to the cause. No, nonsense! It was a bunch of bands having fun, being entertaining and somehow informative.

A lot of the reporting got taken over by events surrounding the *NME* journalist, Nick Kent. He'd thought he was a Sex Pistols member at one point—he'd partaken of the odd rehearsal situation long before I joined. He'd subsequently set himself up as the mouthpiece of dissension about the Pistols, and it just felt a bit like "Get John" when he turned up.

That night, he got beaten up by Sid. What can I say? I'm amazed he'd even be mentioning it, because to be beaten up by Sid is pretty near impossible. If you're gonna be a bad-mouth, sooner or later, someone is going to try and tell you to shut your bad mouth. A lot of people had a lot of issues with Nick Kent. You can't go around just being that spiteful and inaccurate in what you write, and think that somebody isn't going to do something about it. Sid wasn't in the Pistols at that time; he was just angered by what he was reading.

But I enjoyed the 100 Club "Festival." For me, the diversity of the bands was great. You could stand there in the crowd listening to all of them, and then go on and do your bit. That was hugely enjoyable. And of course all of us would be drinking together, so everybody was inebriated. There was no hard-weight pressure on you to be superb. You were just being yourself, and then there you were in the audience mixing it up. That was a very close feeling. Liked it. A sense of camaraderie, and there didn't seem to be any supercilious inter-band jealousies going on.

By now, the girls that would come to the gigs had their own

creative genius just in the way they'd be dressing. There was a whole mob of girls that started wearing trash bags, long before the press caught on. Because of the strikes, the garbage on the streets, it was the natural thing to evolve into. The authorities had run out of black trash bags, so they started to make bright green and bright pink. Astounding colors, and perfect if you couldn't afford top-notch alleged punk—you'd wrap one of them on, a few belts on it, and studs, and bingo, ready to go! "Right, where's the boys?"

Vivienne's bondage suit, on the other hand, was the most restricting, disgusting, annoying thing to be in. I felt really hateful in it. I loved it! Now, the zip on the pants ran from the arse to the front. The trouble with that was, it was too tight, and she cut them always with a feminine design, and that made male genitalia inside feel incredibly uncomfortable. It was like a "u" cut, rather than a u-bend and then up. There was nowhere for your meatbox to gather. Oh man, so you'd have to swing it either to the left or the right. You know the old tailor's quote—does sir dress to the left or the right? But even then, there wasn't enough material . . . So uncomfortable!

She never ever understood the human shape, and, I think, bitterly resented it. She certainly never had any concept of where men's goods are supposed to go. That's what happens when you live with Malcolm as a lover: she performed expensive castration on her adoring fashion-worshippers!

The strap between the legs, that's one thing, right? For me, having a football culture background because of where I came from, the idea of "I can't run in these so I have to stand and fight" was a very good one, but that's not what she was doing this for, because when it came to the zip, listen, my testicles were unfeasibly bothered. Her answer would be, well, then, leave the zip open! Now: I tried that at one gig, I think it was Leeds. No, Middlesbrough. I wore a pair of bondage pants, and what happened was, because I left the zip open, the chafing at the zipper each side of my genitalia was like a pair of saws cutting in from either side, and led to a really major infection. Within two days, you're doing an interview with an incredibly important music magazine, and they're asking you, what's it like to be a rock star? Hahaha! Come on, John, all the girls must love ya! You're thinking, There's no way I can show two half-sawn-off nuts to a woman. Me meat and veg were jeopardized.

Ah God, and if you were unlucky enough with that zip, that ran from the arse to the front, that was very painful, and you had to watch the way you sat down. They were completely cut for women. But the way Vivienne would explain it was, "That's all part of the bondage experience." I quote directly. The woman was hilarious!

But soon everybody was tripping over their leg straps. All over the world! It's an aside but it's a great story: a really good friend of mine, Paul Young—not the pop singer—bought a pair, and his mum ironed them! Creases down the front! She meant well, someone was caring, but my God, is that painful on a teenager? Don't worry—an hour in a pair of bondage pants, and the sweat would have ironed that crease right out of them.

I've got to say, however, her T-shirts were amazing. I liked the idea of two squares sewn together, that was a Vivienne idea, and then the dialogue written on the front or whatever, but the concept of breaking up a T-shirt into two squares sewn together from the neck to the arm, and from under the armpit to the bottom, is excellent. What do you need it shaped for? I liked that, except onstage, when you're giving your biggest, hardest running-around bit, and then there's a photo of what looks like a beer belly, when you're only twenty. You know, *too short.*

The consistent problem I had with Vivienne's designs was that the aesthetics counted more to her than the actual physicality of a human being. And also the unraveling, because seams would never be finished. Then again, she didn't really have the money to pay for proper seamstresses, and proper finishing, so it was all happenstance. I mean, she really did survive a universe of adversity there. I've met her since, and she's said very bad things about me over the years, and I've said very bad things about her. I still respect her. Who else is, like, walking that edge? Who else? That's the joy of what we do, that when we talk about each other, we're teasing each other into the next element. But it gets mistranslated in the press as a bitchfest.

Vivienne was always a very difficult character, though. Very unforgiving and judgmental, and very hard to get on with. At the same time she'd be standing there, yelling abuse, looking for all the world like a turkey! One particular outfit—she was yelling abuse at me before some gig, and she was wearing a full-body, one-piece, zip-up-the-back rubber suit, and it was flesh-colored, and where the nipples were there were red rings, and it went all the way up to the neck, and she had her hair stuck up, and she *did* look like a turkey—an emaciated, plucked turkey. Her wrinkly old neck was trying to hover out of the top of this thing. It was before Sid joined the band. I didn't know what she was going on about. Who cared? I was just looking at this absurd thing in front of me, and Sid went, "Oh, shut up, turkey neck!"

I was now living in an apartment with Linda Ashby in St. James', just behind Buckingham Palace, almost opposite Scotland Yard. Linda hung out with some of the Bromley Contingent. She was a working girl, basically, and all her friends were that way too. I really liked their company. I found them to be really open, honest people. Once they'd got the drudgery of sex, the daily grind, out of the way, they were great fun to hang out with, because you didn't have to have any secrets about yourself with them.

We met through some of the girls who used to come to Pistols gigs, through the lesbian connection. There was always a great lesbian attachment to the Pistols, and I liked lesbians very, very much. They really are touchy-feely, warm people. I understand that what they give to each other is something men don't give them. More power! It's wonderful to be sitting on the sofa between two lesbians—you've never known such warmth in your whole life. It's incredible how open that emotion can be. You know, where you don't feel ashamed of yourself, that's the key to finding quality people in life—always hang out with people that don't feel ashamed about themselves, whether they be lesbian, gay, straight, black, white, whatever—fucked up, mentally insane, twisted, or just straight normal. If they mean what they say, and are what they are, it's a very comforting environment.

I really, truly enjoyed Linda's company very much. I loved her to pieces. We had no "relationship" of any kind, other than equal nutters, I suppose. She was a lovely girl who got me into some really great situations.

For instance, Linda once introduced me to Jeremy Thorpe in the bar at the Houses of Parliament. He was the leader of the Liberal Party in Britain at the time, but his career was soon ended in a gay sex scandal. There was a late-night drinking thing in Parliament and she had access to it, so she took me and a couple of other people. A few pennies a pint—outrageous, brilliant, what a great place to drink! There we were, looking out at the River Thames under an umbrella, the Houses of Parliament overshadowing us, surrounded by all these MPs who all day seem to squabble and hate each other, but there *they* were, discussing who's going with what escort.

The idea of dirty MPs was always tenfold compounded into me, watching them when they were off guard. I suppose they thought I was a male prostitute, with a slightly different way of dressing. I've never dressed overtly sexual, so I'm lucky—I don't attract that kind of attention. I'm just off-putting.

That night, I was free of MP gays. I certainly wasn't what Jeremy Thorpe might've been looking for. And look at the scandals that unfolded just a couple of months later—for both of us! He was so famous for his tweeds—you know—British tweeds, that was his thing—and his silly little hat. When I wore tweeds years later in the Pistols in 2007, when I waltzed out in that outfit, I was thinking of Jeremy Thorpe in the back of my mind.

By the end of the summer, the gigs were getting very pressurized. The one at the Screen on the Green was full of music industry people, A&R men and what have you. It was a very strange gig, everything about it. I wasn't happy with Malcolm's idea of screening Kenneth Anger films before we went on. "Oh, bloody hell, it's gay boys dressed up as Hell's Angels sucking each other's willies. Really, Malcolm, is that Art?" But the Screen on the Green was run by a really nice fella called Roger, and that was his job, to promote these insane far-out movies. I suppose I'd rather they got an airing at a Sex Pistols gig than in between Donald Duck and Bugs Bunny.

I sat in the crowd, watching these films, wondering, "What are people thinking?" People weren't thinking anything other than, "Look at these old farts trying to be impressive." That was the general attitude of the youth crowd; it all looked somehow jaded, like a James-Dean-gone-wrong vibe. I remember saying all this to Malcolm. It was so funny. John Gray was sitting next to me, and he went, "Where's the girls in this film, Malcolm?"

I had no concept of how you put record deals together at all. Of course, Malcolm was doing all this behind the scenes with *his* lawyer who I've never seen eye to eye with. From the start I had great doubts about the value of that EMI contract, me not being legally represented by my own counsel. I always kept that in mind as a reference point for later, that if problems arose I'd have some kind of legal backing here. That there were holes in the contract. I got it all wrong as usual, when it comes to the law.

Be very careful what you sign, everybody in the world! Even though you think you know what it's all about, you'll find out you didn't know nothing! The wording in contracts is so riddled in tangles and lawyer-ese—it might as well be Vietnamese. What you think you're clearly understanding is not right at all, it's something completely different. You get caught up and tangled in these things and then for years later you're trying to unravel them. You go through that and here it is, it's your first record deal, and you're absolutely thrilled. No two ways about it, you think you're made, you're set up for life. Yippee! Achievement Number One. But it isn't. That's what life is, a series of setups and kickbacks.

There was talk of other labels like Harvest and Chrysalis, which I'd loved when I was younger, but they were still very entrenched in hippiedom and obviously weren't the place for the Sex Pistols. This is a whole different genre and you'd be asking everyone from the toilets in the basement to the attic storeroom to change everything to accommodate a completely different approach to life. So, a no-go, really.

It was very hard to tell what was going on, other than how jaded and old EMI was, and how lost. They had no concept of how to invest in a future. They probably just saw us as "Oh, look, that looks like it could be a movement. Let's get in on it!" We weren't the first punk band to sign a deal. The Damned did that some time before us, which was bizarre—using our punk moniker and beating us to the alleged punch. I don't know if they were very happy with their situation either. Not much was said.

I suppose EMI thought it would be a gigglefest and they really, really couldn't cope with what it actually was. The hardcore edge just rocked them to their foundations, so it was get out of EMI quick. And in many ways it was great, because the recordings that we did for EMI were shit, really were, they were so badly demoed, we'd have buried ourselves into a hopeless corner. What came over was just talentless noise.

There was an infamous early version of "Anarchy" that was such a balls-up. It never got released at the time, and thank Christ for that. It put our tails between our legs because we all felt ashamed at just how awful it was. So the next outing to record it, with Chris Thomas, was bang on—you know, "Get this tight, get this right." And you could do that with bum notes, it wasn't about that—it's the timing of the thing.

My voice went crazy in places: "Is this the I-R-*Aayyeeaaaye*," but that was the magic of it. That's exactly how I was feeling it at that precise moment, and it wasn't going down well with the studio guys. "*Iiiyiis thii-iis ver* . . ." "Oh no, you can't do it like that, John." "I just did. THE END." One take is all I needed. And many producers, when I worked with professional producers, have insisted—or asked kindly, which is nicer—that I go back in and try it another way. But I haven't got another way. That's the way I wrote it. I'm not Roy Orbison. I found my own voice, my own way, and my own style, and my own set of scales, and I'm gonna stick to that because it's where I feel healthiest.

When it came out, there was no change for us, no sudden influx of cash for yours truly. I paid no attention to chart positions or anything like that. I was just fully involved with the daily grind of trying to find the next sandwich. These sarcastic bastards would have us on TV shows—and this includes doing *So It Goes* with Tony Wilson, later of Factory Records—and there'd be an intellectual wit—Clive James comes to mind—absolutely trying to slaughter us before we did our rendition of our song live. So I'd have a verbal row with that fella, and point out what's what, and I became known as ferocious in that respect. That's how I am: I will stand up and defend what it is I do. So rather than them admit that, you know, the boy's making a good point, it went into foul-mouthed rants.

We didn't even ask for the Bill Grundy *Today* show thing. It came as a surprise. We only got it because the band Queen canceled at the last minute, and they were on EMI. Half an hour later, I'm in a TV studio, and I'm enjoying my days off here, and I'm challenged by this fuckwit drunkard. I'm not having it.

If you look at the Grundy interview nowadays, you've got to understand the context of it then, how disciplined everyone was, so overtly, Britishly polite, and everybody knew their place. That was the thing at school: know your place, you were always taught that. We didn't know our place, and we were the real deal. There's no showmanship in that, that's an accurate portrayal of young men trying to make it in a world that's absolutely dead set against the truth. If I do anything, it's by truth. All's I want is the truth. John Lennon. Thank ya!

We had to be there at four, then wait around—the show aired

live at six. The green room was full of free alcohol, and I've got to say, Bill Grundy led the charge. "Drink up, everyone, drinks for everyone!" He had a few himself, and he wasn't shy of overtly leering at Siouxsie Sioux and the Bromley girls, because we'd rung them up and said, "You wanna come to *this*?" We turned it into a party, thought it would be a bit of a hoot, and it turned out to be exactly the kind of hoot that we needed—a severe dose of "There may be trouble a-*head*!"

It was actually me that swore first. Grundy goes, "What was that?" "Er, a rude word!" I didn't really want to be the first arsehole out the door with it, but there you go—he goaded me into it, so there it is. "You asked for it. It's not my fault at this point onwards, your honor. I am innocent." If you really understand the way the conversation's going, it's deeply fascinating. It should be in a psychology course, because of all the different things going on in all of our minds at the same time. It amounts to this Harold Pinter kind of scene.

Quite frankly, you look at it, and you see the spots on my face? You see how deathly white I was? That's telling you I'd been up for two days speeding. Poster boy for amphetamines!

Grundy, on the other hand, was the representative of the moral majority, and showbiz. Let's just say he was robust in his cynicism, and yet very corrupt and clearly not giving four working-class lads a fair shake of it. And he should've done because he was from that kind of background, and so in the long run his bitter resentment towards us really just renewed the public's faith that we're all right after all. We weren't up there selling you no crock. We weren't trying to pretend we were from outer space, or flogging you an esoteric angle.

Malcolm totally shit himself in the studio. He said, "I overheard they're going to call the police. Quick, everybody run!" Ha! From what? I've always known Malcolm to be a back-down coward; he'd never meet that final hurdle. And maybe that's because his ambitions were different. Maybe I'm being a bit unfair to him there, but he wasn't prepared to go to the lengths I am, and Steve clearly at that time was more than able for it. He was magnificent on that show, his confidence was superb.

Afterwards, all the phones at the studio were ringing off the hook, and we all piled into a car and drove off. They dropped me off at the nearest tube station and probably went to a party or whatever, but I wasn't invited. I had to squeak into the car just for that lift, otherwise I would have had to walk down the high road outside.

But it was all exploding and I soon realized I couldn't be anywhere at all. There was some girl I knew who lived off the King's Road, so I went and stayed at her place. I had to stay out of the press's eye, because the paparazzi were fully on to us and I always say that was the birth of the paparazzi, British-style. There were hordes of them from there on in, wherever you went. It absolutely ruined any kind of social life. You couldn't be normal, couldn't sit in the local pub with friends. There'd be twenty arseholes there rewriting a story, always unfavorably, whatever, looking for a scandal where it didn't exist. A total nightmare.

I don't normally read the press, but around that time it was astounding what was being written, and I would gleefully read through it, thinking, "But I've done nothing, and you get all this! Who needs to pay publicists? You can get all these lies for free."

Overnight, though, things changed into a chaotic mess, because we didn't have a good captain at the helm, and so it spun out of control. Chaos is a very fine tool but it's one that you have to craft well. Just bubbling along, and bouncing from one incident to another, which is what I felt was happening, was not right for me.

Here we were, capable of making really big significant social changes to many things—not just the wonderful world of music, but to society itself, which was suddenly paying attention—and Malcolm was messing that up. He was scared of that next level, and he was always scared of being arrested and locked up. All of those foolishnesses, that when you're young you're kind of looking forward to. You've got your brave on, you've got your youth, and you've got your kind of ignorance too, because you're not fully aware of the consequences. With me I felt I could take whatever was coming 'cos I could justify it. I could stand up and back up what I was doing.

The tour we'd set up to promote "Anarchy" became anarchy! A pointless futile mind game, with no results. Just being banned everywhere, and Malcolm being quite incapable of backing us up on that. We should have gone for it at that point, and really flooded the networks and the media with "Why aren't you backing us?" But we didn't look for allies. We felt like we were in the Moonies, some secretive religion rather than an accurate, well-oiled machine—broad, wide open, transparent.

The original idea had been to pair up with a circus, and tour that way. I loved that idea and had a great contribution in it. I loved funfairs and circuses when I was young, loved watching the Teddy boys who worked on them. To me, that would be the next level. There was an agent who was very interested in pushing that but it fell apart. Again, Mr. Manager let it fall by the wayside.

Instead, it was three weeks of uncomfortable coaches, gigs canceled left, right, and center, and the prospect of no hotels whenever we got anywhere. That was because we'd have to wait for the money from the gig to pay up, but then if the gig was canceled, we'd have to drive on to the next one, try to blag ourselves into a hotel by telling them that we'd pay in the morning—knowing full well there was no money!

Because there were other bands involved on that tour—the Clash and the Damned—it became like we couldn't have a camaraderie vibe about it amongst ourselves. The other bands became competitive in their attitude—all except the Heartbreakers who weren't like that at all. They were just looking for the next fix, and the further up north we got, the less possibility there was of that.

Very soon the Clash were traveling in another coach, and bits of the Damned went that way, and that was that. It started out all together but it went pear.

At one gig in Caerphilly, there were choirs singing outside, and presuming that we were quite literally the Antichrist. "Hello? No! I'm actually viewing myself as a bit of a savior here, I think you lot have got it wrong! The real Antichrist is religion!" It's very hard to make people understand that, when they're manipulated by newspaper headlines and read no further than the first big block-letter words with a well-chosen photograph that puts you in a bad light. The press really would write what they wanted.

There was such a worthwhile lack of activity on the tour, that I actually tried heroin. Indolence creates incredibly negative situations, as we know from any teenager's experience.

I never wanted to see heroin wrapped around any band. If you like music as much as I do, then you'll see all these problems coming way in advance, because you're learning from the escapades of famous rock stars, and the calamities they get themselves into. Try as you can, you can't get the people off it. It's the kiss of death, and it was something I watched happen to Eric Clapton. I wasn't basing my aversion on fear of the unknown. I was basing it on, look what happened to that burnout.

Sitting around with nothing to do, I thought, "I want to know what the big taboo is. I know all the warning signs, but still I can't be preaching against it unless I've sampled the goods." So I tried it with Jerry Nolan, the Heartbreakers' drummer—I thought, "With a name like that, he must be the Irish one, I can't go wrong here."

I hated it. It makes you sick. What's the point of that? "No, man, if you keep taking it, you get over the sickness." Why would I want to get over the sickness? Others have told me that they go straight into it and they love it, they love that false sense of security that heroin creates inside your brain. That's a lovely thing, but what you're really doing there is running away from creativity. It absolutely kills that aspect. You make yourself pointless. You have no love of the world anymore. All your attentions 24/7 are drawn into where's the next fix. And that to me sounds a hell of a lot more drudgery-bound than any possibility of working nine to five. And then ultimately you've got to face the dilemma of, how are you going to pay for this situation you've become so accustomed to? That's where a chap like me will go, "Well, it ain't for me."

The "Anarchy" tour gigs that did go ahead were horrible. In Plymouth, there was a war going on in the city at the time between the local boot boys and the sailors, and so of course our gig was used as a backdrop for that. We'd waltz in and we'd take the gyp as if we instigated a riot and it would be far from the truth. I was there to try and make these warring elements take the night off. Start enjoying yourself instead of squabbling and being divided.

On many occasions at them kind of gigs I did manage to do that. I gave them one particular target to all hate together equally. That would be Johnny Rotten. If needs be, that's what you do. You come out firing and you quite literally antagonize every single person in the hall. It stopped the brawling. To me that was success. The one thing I really feared when I got onstage was silence from an audience. That's the hardest weapon to fight against.

It all started from the gig in Caerphilly, I think, when all the Christian people turned up, with their religious "When The World Turns To Rottenness" banners. Suddenly our name came up in the Houses of Parliament. It was Tory GLC Councilor Bernard Brook-Partridge who led the charge. I watched him pontificate on the *Six O'Clock News* one evening. "Erm-er, this has got to be stopped . . . er-erm, it's the downfall of society . . ." And/or whatever else he said. What a foolishness. The juvenility of it all.

The hilarity of it is that years and years and years later, I had a friend who joined the Freemasons, and Brook-Partridge was apparently leader of that particular chapter, and had nothing but good things to say about me. It's like all politicians, they get up on their high horses, but they're actually talking red herrings. They don't believe in nothing. We were just an easy target, a bunch of saucy boys from the wrong side of town who were making a racket and were easy to shoot down.

Obviously this was after I'd paid the place a visit for late-night drinking sessions with Jeremy Thorpe. Maybe *that's* what spurred this on, but the idea of Brook-Partridge declaring us as public enemies and trying to plot our downfall was preposterous, because, actually, the laws they were discussing us under were so arcane. How can Parliament re-enact a hang-'em-high Traitors and Treason Act in the late twentieth century? They couldn't. So they backed themselves into a corner and looked pretty damn foolish for it. And quite frankly gave more power to my elbow. Strongbow!

I knew from there on in that these institutions we're all so frightened of are pretty much headless chickens and there for the taking, if we ever formulated ourselves together accurately enough. But it's the dissipation and the personal animosities and jealousies that stop movements happening. I'm not talking violent movements here. I'm talking, if you want to change a situation for the better, it is possible, it really is.

In many ways, you become ultimately fearless of it. At times you'd kind of be wary and you'd still be thinking, "Oh God, they're gonna lock me up," or whatever, but you know, learning from those experiences you reach the point where you don't care if they lock you up or not. It doesn't change anything, in fact it just makes them look sillier—you know, don't play the victim. And don't allow yourself to be victimized or dictated to by what, to my mind, are ill-educated, spoilt children.

"What are you going to say to me? Why aren't you supplying us with jobs and a decent lifestyle, you fucks? You're gonna tell me to shut up because I'm finding the economic situation you put the country in a problem? And using that very thing that they just love to espouse in the West, democracy! Ooooh—the right to say what you have to, to stand up and be counted." Wow. Didn't I blow a hole in that bubble. And seriously, a BIG hole in that bubble. I found that to be an absolute non-truth. I wouldn't tolerate it. And still won through. So there you go, boys and girls of the world, Johnny did his bit for ya. Fucking say thanks, cunts.

I never lectured. Just pointed out the flaws in it all. A song like

"Problems" is telling you, "Too many problems/Why am I here?" It's all just problems, so, one by one, resolve them. And you can't do that by sitting silent on the sofa, or pontificating nonsense from a soapbox. Your songs have to be reasonable in the way you communicate the message. They're not lectures. And so my songs don't lecture, they give you freedom of thought, inside of the agenda I'm pushing.

I think it was understood that Johnny Rotten weren't no back-down cunt. This boy don't surrender. And I won through. I did. I took 'em all on—this is my message to these "punk" bands who don't quite understand it, and they're so busy inter-fighting and trying to bash each other's heads in through jealousy: the bigger enemy is out there, go pull that one down, and fight the cause for all of us, not just your selfish little angles. It's ever so much more fun. Listen, my enemies are not human beings, regardless of people liking me or not, my enemies are institutions.

HUGS AND KISSES, BABY! #1

Punk opened the door on the universe of sex in a really nice, innocent, open way. I didn't realize until then that sex was readily available. Literally from the very first Pistols gig, it was, "Oh, hello?"

Going back a bit further, going to clubs like the Lacy Lady in them early days before the Pistols, it was the way you used to dress that was an attraction of sorts. It was also a problem of sorts, because it could attract massive hooligan attention, and I had many of those agendas to deal with.

Girls found me interesting, and always—which is the way I like it—in a motherly way. I'm a sucker for the soft-bunny touch, even though my imagery, the way my persona came over, was one of "cold, hard, spitefully indifferent," so I'm six of one, half a dozen of the other, really. Girls seem to have a natural understanding that all the signals of someone who's trying to be an outsider are really the actions of someone that wants attention and love. This whole process reaffirms something in your psyche in a healthy way—that you're not really as ugly as you envision yourself. There is hope.

Once you were up there onstage with a band, you didn't have to pursue any longer, and you didn't have to feel embarrassed because you didn't have the right chat-up lines. Very, very interesting. Great nights on the early punk scene, lots of fun, and the girls were as tough as the boys in that world.

I once dismissed sex back then as two minutes of squelching noises. I was telling it like it felt. It was an honest statement. Or was it two minutes and fifty seconds? Yes, maybe it went up with inflation. Well, sometimes it did. That was the round-about best-of figure, the average norm. But there wasn't any depth in it, and therefore, ultimately, no interest. It's the same thing as, I can't be a drug addict because the repetition of it would bore me to death. I'd die of boredom before I died of the drugs.

As the so-called King of Punk, I was almost getting too much sexual attention suddenly. Me being me, my reflex was "Is it me you're looking for? Or is it the pop star side of it?" At which point I withdraw, because I don't like the feeling that I've been treated as a commodity. I saw the change as the Pistols started to happen. From "Urgh, who's that ugly thing in the corner?" to "Oh, 'ello, gorgeous!" in, basically, a heartbeat. But listen, the humor of it wasn't lost on me.

I first saw Nora at Malcolm's shop in 1975. She came in with Chris Spedding, who was playing guitar with the likes of John Cale and Bryan Ferry at that time. He was very shy, and Nora wasn't. He was worried about his flamenco shirts not quite fitting. Nora was fussing around, and somehow the screen in the fitting room fell, and there was Chris Spedding with his belly bursting out of a far-too-tight shirt. That was very typical of Vivienne's clothing. She would never make them to fit, so you'd always have to order them a couple of inches bigger.

Nora already had a daughter, Ariane, who'd been born and brought up initially in Germany, where Nora originally came from. Nora used to promote gigs in Germany, people like Wishbone Ash, Jimi Hendrix, and Yes. Then she ran away from the confines of German society, which was far too restricting and nosy. Everybody's in your business.

During punk, Ariane became Ari Up, the singer in the Slits. Her father was Frank Forster, a very popular singer in Deutschland, in a Frank Sinatra way. Germany after the war was very influenced by the American air bases, and that dictated a lot of the music that was popular. Over here, Nora brought up Ari really well, and got her to learn all sorts of musical instruments, which were always lying around. Ari was only about thirteen or fourteen when I first saw her bouncing around.

Nora, I soon discovered, is a guiding light, and a creature of utter chaos. She was a very odd and different soul. Not at all like one of the average old hippie birds, who weren't quite sure what punk was about. There were loads of *them*. That, or working-class girls out of the estate, full of "fack yous." None of them seemed like options to me. But Nora—God, she shone in a room. From way across the other side, she shone, she *glowed*.

Nora loathed me at first sight. At least, that's what I thought. It was because of what everyone was saying to her. "Oh, you don't want to talk to him, he's awful," propagating a myth around me.

She was short, sharp, brutal, and very intelligent with her remarks, and a lot of that was based on what people had told her about me. But Nora being Nora, she was inquisitive. If people are telling her not to talk to anyone, she'll talk to them, and I'm exactly the same way. I was told she was stuck-up, and so I found her deeply fascinating. Once we started talking, all of that nonsense came to light and we realized we had both been lied to. Everybody told lies, then. Shocking.

I always loved the way Nora understands how to dress. She has a completely individual, incredible style, and that style is reflective of her personality. That drew me in. To the point that I never smoked cigarettes until I met Nora. She used to smoke Marlboro, so I started smoking Marlboro too. So the afterglow ruined me for life. But then Nora gave up smoking completely, and here I am, still to this day!

It was a topsy-turvy situation, for sure. We didn't waltz straight off into the stars of romanticism. There were all kinds of heated arguments, but in those heated moments we discovered each other as human beings.

I've got to be honest, before we met both of us played the field, but we found the field to be full of moos. And those moos turned out to be nothing more than muses, and that's nothing to base a solid lifestyle on. It's too vacuous. I don't personally get the rewards of one-night stands at all. Just don't get it, never did. I always left those situations feeling empty inside, and rolling over and going, "Oh my God, do you really look like that?," and knowing that's exactly what they felt too.

I'd gone through the one-nighters period, but there was a point where it became a futile, boring, repetitive procedure. I didn't know it at the time but what I was really looking for was a proper relationship, and that was slowly forming with Nora. There were girls leading up into that, longer than a week, shall we say, but something really good happened and clicked with Nor', very seriously. We learned to really know each other, and that's the best that any human being can ever look for, I think—the right person who truly accepts you for what you are, warts and all, and doesn't make you feel ashamed of yourself for any reason at all. So self-doubt is gone, and that's what the right partner teaches you.

At first, Nora had a flat right near Chelsea football ground in West London. It was a basement, and it was cold and damp and dark and very unenjoyable. I never liked that place much, but then she moved into a little house in South London, off Gowrie Road. That's where things began popping. We were firmly bonding then, and that's where the likes of young Neneh Cherry came over from the States to stay, and hung out with Ari.

This is what people don't realize: Nora's really the "Punk Mummy Warrior" figure. Without Nora, there wouldn't have been the Slits. She's the one that funded it and held it together, regardless of what anyone else has to say. And Nora has been like that with many people and many situations. We're not talking money here, we're talking benevolent guidance.

You've got to bear in mind that none of us were conceiving of the concepts of punk as such a lasting force. Or indeed what a contrivance it has become for the lesser mortals who now dabble in it. We weren't doing this for titleships—it's just the way it was. Common sense prevails, and so, a very well-led house.

Ari was only fourteen or fifteen when the Slits started up. It was very much like watching a St. Trinian's film. You know, hooligan schoolgirls—fascinating! Oh, I love the lyrics, because only two of them spoke English with any skill. Palmolive and Ari, by contrast, were a contradiction, language-wise: Palmolive had more Spanish than

English in her, and added to Ari's cross-juxtaposition of badly-learnt English, badly-learnt German, and even-more-appallingly-learnt Rastafarian patois, it made for very bizarre songs. I know that Nora helped out there too. Deeply, deeply hilarious.

I was always very proud to have Ari as . . . well, she'd call me Granddad! I was always very chuffed by it; it felt we belonged to each other, even though later on we rowed like cats and dogs over sillinesses like religion. Can you imagine religion dividing the punk movement? Anything is possible. So I always got on with Ari; me being with Nora was never really a problem, and Ari rated me because I wasn't a parasite. I came fully loaded.

So, Nora had been around, and I'd been around, and good things came of that—a slow progression into something incredible, and that's the best way. It doesn't all happen at once always, and particularly when you're connecting with other human beings, it's best to spend some time on it, as I've learned with various band members . . .

5

THIS BOY DON'T SURRENDER

You couldn't miss me in a crowd. I was wearing that pink women's rowing club jacket, the one I'd scrawled "GOD SAVE OUR GRACIOUS QUEEN" all over, a leopardskin waistcoat that I ended up giving to Sid, and a pair of gray pants from Vivienne's shop, which ended up on Paul Cook. That's how we were, mix and match. On top, I had a spiky ginger hedgehog going on.

What I *shouldn't* have done is gone out for a walk in all that clobber with a packet of gear. It was just into the New Year—1977—and we were still all over the newspapers as a result of Bill Grundy, "Anarchy," and being bumped off EMI. We were rehearsing in Denmark Street, when I nipped out with Nils Stevenson, who worked for Malcolm. I had a mental lapse. I thought, "Well, if I leave the speed behind somewhere, even if I hide it pretty well, Steve Jones'll find it." He was very good at "finding things," that boy. Steve always wanted to know everything about anyone. He was a continual rooter. We should've nicknamed him Drano, or Domestos, in that respect.

The bust was a very odd one, because the emblem on their police helmets wasn't one I recognized. They had pythons on them—like, what the hell is this lot? They were mercenaries, a little bit more hardcore than the usual bobby. This lot were a notch up. I later found out they were some kind of special patrol group.

They sat me in the back of the van and drove around until they'd arrested enough people. I was sitting there, thinking, "I'm screwed, I've also got a knife, what on earth am I gonna go down for now?" I was looking, in my mind, at a long, long term inside. And then, bearing in mind the record I'd put out recently, I knew that weren't gonna bode well in court with any bloke in a wig and a long dark cloak.

I didn't consider myself particularly criminally minded, but the police were heavy-duty in those days, and the slightest infraction would get you round the neck. I was given a choice: "What are you going to go down for—the drugs, or the knife?" I thought, "Easy, the drugs!" A very stupid move, in the long run.

The main officer put the gear in a plastic bag, stuck it in his back pocket, and then sat on it. They then drove around Soho for a while, going, "Look at that one, *whurr-urr*, let's grab him!" So, what would the amphetamine sulfate do, next to the heat of his sweaty flesh? Given my detailed knowledge of speed, I was fairly sure it would evaporate. Therefore, as the minutes went by, I knew my charges were diminishing.

At the station I was strip-searched in front of a woman officer, which apparently they're not supposed to do. But what are you going to say at that time: "Stop that, it's rude"? That won't work. When you're in custody, you tighten down the hatches and get wise with your words. You try to diminish yourself in stature, and lessen yourself as a target.

When it came to bail, I didn't know anyone that didn't have a conviction of one kind or another, to get me out. It dawns on you just how under the cosh working-class people are. Everybody's guilty of something, according to the institutions. So that was a bugger. Malcolm could've got me out, but he didn't show up.

I don't know how my dad did it in the end, just to get me out on bail. And then there was an error on when I was supposed to turn up in court—they got the dates wrong. The form I had didn't tally with the actual court date, so they raided my dad's house and caught me trying to jump out the upstairs window, so I was done for evading arrest. But because the report card was filed wrong, I got off. You can't evade arrest when you're actually not even supposed to be getting arrested.

For the hearing, Malcolm was given a specific time, and of course he turned up late. He couldn't actually raise forty quid to pay the fine, so they set another pay deadline a few hours later, and he turned up with it with minutes to spare. Otherwise I would have gone off and done time.

I don't think I got specifically targeted as a direct result of all the fuss with the band. I just should've been more careful.

By now, I'd finally turned that "GOD SAVE OUR GRACIOUS QUEEN" jacket into a song. I was waiting to go to rehearsals one day, and it was a long wait. In them days, I'd get up about midday, and I sat at the kitchen table, made myself some baked beans, just took a piece of paper and wrote down the lyrics—a very rough guide to it, but the absolute crux of it was there.

What I liked most about it was there was no verse-chorus in it. I was impressed with myself when I read it back. The hook lines were really not about chorus effects at all. They were to emphasize and set up the next set of lyrics. I think "God Save The Queen" is a powerhouse example of how pop can be turned upside down, on its head, and still be pop. It breaks all the rules of the pop song format.

Unfortunately, Glen misinterpreted "God Save The Queen" as being a fascist song. He just picked up on that word, but he didn't grasp the overall context. Where can you go with that? It got really bad with Glen, he refused to play it onstage, and that's where it all really came to a head. Before we went on, I had some chalk and I wrote on his Ampeg stack amplifier, "The Boo Nazis."

I mean, it's obviously not favorable to Nazism, is it? But Glen took it as some kind of right-wing statement. Oh Lord, lest we be misunderstood! My kind of sense of fun and Glen's are very different. For my mind I don't think what I was doing was nasty and evil. That was me being really childish and silly—it was supposed to be met with a smile.

At that point, me and Glen had both been in and out of the door quite a lot. "I'm off"—"I'm gone"—"You're gone," but I think it was finally mutually agreed between Malcolm and Glen that artistically this wasn't going in the right direction for Glen, and that's how that was buried.

So, bingo, he had to go, and because we had nothing better to do, we shot a bit of film walking around town, and in it I was asked about Glen. I said, "If you look like a duck, and walk like a duck and talk like a duck, you're a duck." But I changed it from "duck"to "arsehole."

The real problem, with hindsight, was that we weren't playing any gigs. We allowed the boredom to get the better of us and so we turned on each other. I should feel ashamed about being like that.

When I first got into the band, it really shocked Sid. He didn't know I had it in me. It shocked Wobble too. He didn't know at all. Sid was fascinated by it and drawn to it and became our biggest fan, but Wobble bitterly resented the band and was very violent in his approach to them, and they were quite frightened of him. He'd come over really, really hardcore. So when it came to replacing Glen, I instinctively said Sid, although Sid was tone-deaf.

I didn't think they'd ever take me seriously on Sid, but at that point in the band I felt like I needed an ally on the inside. I felt like it was "them and me," and that was not a good position. You mustn't be the knocking boy—you do need backup. As friendly as Paul was—and I used to hang out with him on occasions—he's swings and roundabouts, he bobs from one thing to another, and his commitment with Steve was ever so deep. I always got the feeling that Steve would be the one to say, "I just can't work with him anymore," and that would be the end of that. And Malcolm gave Steve that power, because it was Steve's idea to have a band in the first place. But that wasn't the real reason. The real reason was that Malcolm paid Steve's bills, and therefore Steve would let Malcolm do anything he liked.

Of all people, it was Lemmy from Motörhead, amongst others, who tried to teach Sid to play bass. Lemmy was really funny about it; he said, "Sid has no aptitude at all, no sense of rhythm, and he's tone-deaf." Sid always fancied himself as a drummer. I think that was the Can *Tago Mago* influence, because that was Sid's favorite record of all time. He'd always be making *psssh-shut-pfft-pfft-pfft* noises, and pretending he was doing a drum roll. That would be his frequent behavior, which not many people understood. They thought he might just be a bit backwards.

We assumed that he'd just find his way with it, like we had. And there's the danger in that word: when you assume, you make an ass out of you and me. As it turned out, Sid wasn't actually plugged in at most of our live gigs, and he barely played on the album, if at all.

What I didn't find out till much later was that Malcolm was not only *not* trying to book us any gigs, he was actually turning them down. He'd say, "Oh, no, you've got to understand, John, what I'm trying to do is create a scene that you're a man of mystery, and no one knows anything about you." He didn't want me to be seen at too many public events because it would destroy the imagery he was trying to create around me.

That was how he explained a horrible evening when I couldn't get into Andrew Logan's annual party. I turned up with some mates, and they wouldn't let me in. I was like, "I played here last year!" Malcolm and the rest of them were already inside. I saw Vivienne, and I said, "What's up? Why can't I get in?" and she blanked me.

I became well aware that these people would not stand up for me. Hard lessons in life. I could've barged my way in with no effort at all. No! I wanted to be accepted, and I never was, not within the Pistols contingent or any of those socialite scenes that were using the Pistols to thrive off.

With nothing to do, no gigs to play, me and Sid were going mental. We had to do something, anything at all. I came up with the idea of the four of us going to Jersey on holiday, because I'd been to the Channel Islands before on a school expedition from William of York, and I had fond memories of it somehow. I just imagined us getting off the plane and having a lovely time in this wacky different kind of world.

But no, the whole band got off the plane and we were met at the airport, and strip-searched. As soon as they opened Sid's suitcase they found his smelly socks on the top, and they gave up. What they did do, however, was cancel our hotel booking, so we ended up walking around on the beach, with a cart without the donkey, with all our luggage stacked up on it. Luckily a local villain who'd befriended us found us somewhere to stay.

The next morning we buggered off to Berlin. Malcolm didn't trust us on our own, so his associate Boogie came as some sort of mentor. Boogie was a bad bunny himself. A fun time was had by all in Berlin. Wow, what an eye-opener.

We hardly saw the hotel. We didn't want to. It was the Kempinski, and the rooms, you couldn't even think of sleeping in them, they were so rigidly German. You're supposed to sleep in a straight line and the quilt doesn't come any higher up than your chest. All the wood was very dark, and everything was at right angles. No time at all in there, thank you.

You couldn't escape the vibe: the war, and then the Wall, with the Russians staring over the top. West Berlin was all set up to annoy the East. It was glorious, but a bonkers, crazy universe. Readily available was everything and anything that would keep you up all night. Between the British and American soldiers, they had the place well sussed. They had it amped, so to speak. I fell in love with Berlin, and I've always loved it ever since. The word "decadent," how applicable. Well done, the West, that's what you're tormenting the Russkies across the border with. This is freedom! What *you* got?

So that was what inspired the lyrics to "Holidays In The Sun": "I don't want a holiday in the sun/I want to go to the new Belsen"—from Jersey to Berlin.

The first nightclub we walked into, we were astounded by what we were hearing; the music was exceptional. It was kind of early House, by any stretch of the imagination. Very deep bass drums, a stripped-down Teutonic dance code, so rhythmically structured.

Then there was Romy Haag. She was a drag queen, and our only connection was that Bowie had mentioned this person in an interview years earlier, and Sid remembered, "She's got this great club where all the perverts go . . ." To find it, me and Sid wandered for hours around the streets of Berlin with no idea of where it was, and finally it was just this horrid little door down some steps into a basement. But a really wacky place with loads of British soldiers in there.

And they weren't there for you-know-what. They were out for a good laugh, and in them days these drag bars were very sociable places, they were great fun. It wasn't as separatist as you would think, they were very welcoming, and it was a great place to go and get plastered, and you wouldn't be manhandled inappropriately. And in them days, you've got to remember, being gay, particularly a transvestite, was a very harsh life. It was not accepted, and yet I always found them to be very accepting, open-minded people.

There have been rumors about me and Sid being that way inclined. Just, *NO!!!* There was a fantastic line in a song by the Slits called "So Tough": "Sid is only curious, John don't take it serious." That says it all.

Maybe it was true for Sid. I don't know if Sid ever worked out what he was. He was an exceptionally strange, different person. Very open, very happy, nothing challenged him. He couldn't give a monkey's what anybody thought about him; he just thought he looked beautiful like Dave Bowie. But once he got in the band, all of that went and he became a very dour, serious misery, trying to act tough where before he'd never bother with any of that stuff at all.

The "I'm a complete virgin" line ended when he met Nancy Spungen, a heroin-addicted groupie from New York, who I had the misfortune of passing on to him. I thought it would end in disaster, but not in the way it turned out. I thought he'd fuck her and go, "Ouch, what an ugly old bag!" in the morning. But he liked the idea that she looked wasted and ruined.

It goes back to years before—how do you translate music? How do you translate *Berlin*, the Lou Reed album? Do you translate that as the falling apart of a relationship, or do you translate it as an accolade to drug addiction? That's the problem. "Walk On The Wild Side," to Sid, obviously didn't mean, "go gay," it meant "take a lot of drugs." That's how he's seen it, and he was very overwhelmed by a person like Nancy talking: "Oh ye-ah, in Noo Yawk, we can get it all the time, it's gawnna be *great*."

Well, they got it all the time to the point where it killed both of them. I lived in New York later on and I know the difference, but my poor friend Sid didn't. I can't imagine him in heaven being any cleverer, other than he will ignore his previous existence. He was addicted to the addicted lifestyle. His mother was a registered addict, and he thought that was the road to cool runnings—and I'm not talking about the Jamaican bobsleigh team. I'm talking a real serious understanding of how things were, and how human beings perceive. Sid's perception was very minimal, and desperate and immediate. He was not by any standards unintelligent, but the inflection his mother put on him limited his narrative as a human being.

Heroin users will steal anything. They'd steal your toenails—

anything that's got a dollar value on it, or a pound, or a penny. It goes straight into the arm. And you can't trust them, they've lost their soul. It's a very odd thing to be in the company of someone who is a long-term addict; they just feel lifeless, and there's nothing in the eyes that shows any human kindness or empathy, or anything at all. Ultimately, they are the true vision of a zombie. They are the walking dead.

Signing for our next label, A&M, outside Buckingham Palace was a hoot and a holler. Sid was wicked when he found an angle on someone—he'd keep at it and up the ante with really humorous but negative comments. His big thing with Paul was, "You're an albino gorilla," and in the limo on the way to the signing that morning, he finally earned himself a smack in the mouth from Paul.

Suddenly, everyone was punching everyone in that car, God knows why, but that's how it was. And all of us took a whack at Malcolm. In fact, that was where we bonded—once we finished rubbishing each other, there's the perfect target.

We signed on the dotted line, all grinning and goofing around, just seconds after trying to smack each other senseless. There were so many pent-up problems, thanks to Malcolm's alleged "orchestration" putting us in a world of perpetual chaos—it wasn't pleasant. So that became a great moment of relief. Then we did a press conference blind stupid drunk. Sid threw a custard pie, the tough lad, which just about shows the jolly frolics of it.

There was nothing at all for us at A&M's offices, they'd got no drinks in, so we insisted that they send out. That took forty-five minutes and in came a crate of crap lagers, the usual that we'd been used to every time we signed a record contract—a shortage of inebriations. I've never known anything like these record companies. They don't know how to do a welcome wagon. I'm Johnny, you come knocking on my door—mate, there's a beer in your hand. I'm loaded and ready to go. I entertain my guests.

Failing to lay on the hospitality can lead to all kinds of fury, and of course what ensued was a situation of their own making. I was sick in a plant pot—oh, yes—and they accused us of breaking a toilet. "Look, Sid was never potty trained, all right?"

If you're gonna make people feel uncomfortable and unpleasant and unwanted, then they're gonna hang around for a hell of a lot longer. At least that's my way, and Sid's well up for that cup. And Steve and Paul had nothing better to do, you know what I mean? We were very cooperative with each other, all of a sudden. "This is a Viking raiding party and we're all in it together!" I love that sense of camaraderie in a band.

Ah well, we didn't even last a week at this one, did we? I'm surprised it took that long for them to chuck us off. Apparently, it was Herb Alpert—the "A" in A&M—that sent a communiqué from L.A. to the UK label's offices saying we had to go, he didn't want our sort of undesirables on his label. Simply put, we were a threat to the hamster wheel that they'd become so acclimatized to putting their acts on.

These old-fart bands had found their comfort zone, and they were irritated at having to rethink the agenda. That's terrible because in no way was I setting out to replace them, just remove the flotsam and jetsam that was blocking the drainpipes so the rest of us could have a flush. I don't put roadblocks up for new bands, and in them early days we definitely had roadblocks, and seriously negative attitudes from quite a few alleged musicians, demanding that the record label sack us—the likes of Rick Wakeman from Yes, and Steve Harley of Cockney Rebel. Like, actually, who are you to make such demands? I didn't care who my label-mates were, that's irrelevant.

I found the whole thing very humorous indeed, this arsehole Wakeman who was playing Ice Capades Wurlitzer music, telling me I'm not worthy. How am I supposed to take that, but unseriously? The days of Yes were gone and he had nothing new to offer anybody except criticism—a spoilt fading memory. But it did create problems, and we got a bump because of it, that just fueled the engine of negativity.

From the beginning Malcolm had been fending off overtures from Richard Branson to sign to Virgin, because they were a hippie label. My draw to Virgin was their astounding record stores. The first one was on Oxford Street: it was absolutely awe-inspiring, the things they'd pack into that tiny little one-room place. To just look around and go, "Oh, the possibilities! I could have it all, but I can only afford one." They made music seem fantastic, diverse and limitless. You flicked through all these different album covers and just—the potential of it all, the wonderful creativity that music really, truly is.

So, after A&M, the pressure was on Malcolm to get us a deal that would actually work. "Can we have a label? It'd be kind of interesting, don't you think? Here we are, the ultimate primo numero uno punk band, and we ain't got a record out?"

In the meantime we started recording our album with our previous advances and severance pay. I got my words in succinctly and correctly pronounced, so I was happy. I did one or two takes, and that would be it. There'd be no overdub work at all, so I'd have to be bang on when it came to my turn. I couldn't bear endless guitar overdubs, but the sessions quickly turned into a jolly little joyride for Steve and Chris Thomas, the producer, to "experiment with guitar possibilities." It was infuriating and indeed I left the studio for large amounts of time because of that.

Chris Thomas drove me nuts. I thought what he was leading us towards was too elaborate for us at that point. To be pushing the singer aside in any band so you can have more guitar overdubs is nonsense. The only thing that made him interesting to me was that he went out with Mika of The Sadistic Mika Band, a group I loved. Any conversations, it was always about "What's she really like?" I don't suppose it endeared me to him at all, but it was a very impressive band with a Japanese woman upfront squealing away in a Japanese way. It turned out he was deaf in one ear. Nobody told me until the middle of recording when he'd be leading in with one ear. "What you doing?" "Oh, I can't actually hear with the other ear."

Around the making of the album, of course, Sid went and got hepatitis. Fantastic, huh? I almost think he did it deliberately just so he wouldn't have to 'fess up about his musical inadequacy, or step up to the plate. He was just confused; he never got it, on many levels. He never got the rally, he never got the neighborhood connection, he never got the understanding of the bigger issues.

Sid was introducing an angle into the Pistols which, I immediately realized, fucked us royally up the wazoo. He introduced the drug angle, and I never thought he would do that. I thought he was smarter. I never realized how insecure he really was, and he used drugs to cover up his sense of inadequacy, and he introduced this warfare of heroin into us in such a calamitous, arsehole way. It was difficult, difficult times, dealing with him.

He was really lost—and I should have realized this far earlier—because of his mother, the woman that gave him heroin as a birthday present. He'd always said, "Hurgh hurgh, I'm not getting like my mum." He was always proud of the fact that he could do that—dabble, and then be all right and not require more. But when Nancy Spungen came into his life, it became different; he bought totally into his Lou Reed schtick.

Poor old Sid, he couldn't have sex with anything. He was rubbish. But I loved him *because* he was rubbish! He wasn't a big-stiffy kind of fella, he was just confused and funny and hilarious and brilliantly comedic. He could parody anything instantly. But the shame was, because of that quality, he was now trying to parody a New York lifestyle.

I didn't even know the Queen of England was having her Silver Jubilee, until they rang me up about holding the boat party on the River Thames. Seriously, genuinely—I'd stopped reading the press or paying attention. Sid even more so. He never paid attention anyway. To us, it was, "Oh shit, what's *this*?" "It's the Jubilee," says Malcolm on the phone. I'm like, "What does that mean? Can I have some money?"

That was the drawing power to get involved in our legendary boat party, both to me and to Sid. It meant we could pick up £25, and then it exploded into the beautiful fiasco it was. There is something beautiful about creating a fiasco and, believe me, it really was a disaster.

Virgin had finally put out "God Save The Queen" and, lo and behold, it was a hit. Oh no, hold on, it wasn't, there was no Number One that week—well, not for us anyway! What that did was show—ever so much more than anything we were doing and saying—that the institutions are corrupt. The fact that our band cannot have a Number One on a chart system in Britain showed . . . shenanigans, right? Illicit behind-the-scenes operations, manipulations, telling the public what they could and couldn't like. And that really gave energy and fueled the head of steam behind the song.

Come Jubilee Day itself and our saucy boat party, I was freezing, I was bored, and I had hardly eaten for a week. It was all too much of a circus, with the handpicked guest list, and all the press herding around. It was funny to see Richard Branson on there with his beard and long hair, looking like Guy Fawkes, and fair play to him he was up for the fun of it. But Richard got a bit lame when we ran out of booze. I said, "I'm not doing it unless a crate of lager magically appears." And by some miracle it did, and all twenty-four cans vanished in twelve seconds.

When we played, we couldn't hear what we were doing, which was very much like all the early gigs, but we didn't really care at that point. I suppose it was the only way to get warm. It was a chilly evening—and beyond chilly for me, because I was definitely undernourished.

We did our best to be disruptive. We'd been up and down outside the Houses of Parliament three times and nothing happened. So we went back for one more—"Oh God, really? Must we? All right, here *goeszzz*!"—and it was just at that precise lucky-for-us moment that the police decided to pull up alongside, and go, "Stop that! It's not British!"

As soon as we docked, and I've seen all the police, I thought, Right, it's either going to happen or it's not, but I'm not going to get caught up in this crowd of fuss and confusion, so I stormed to the front, walked down, and was the very first off the boat. The first copper I come across goes, "Which one's Johnny Rotten?" And so I went, obviously, "He's up there!" And of course they made a beeline for the long-haired one with the beard—poor old Richard Branson—because everyone knew it was long-haired people that were causing the trouble in them days. How fucking hilarious is that? It has been suggested that Malcolm pointed me out to the police, but if he did he did so unsuccessfully. I was long gone.

In them days, the police were so backwards. They were brutal, there's no two ways about it. They were schoolyard bullies—hobnail boots and truncheons on anybody who questioned anything about them. But when they raided us at the boatyard when the boat docked, they had no idea what a Sex Pistol looked like. There wasn't anyone in charge of this fiasco, so they came down looking for the long-haired ones, but also at the same time an awful lot of it has to play into the fact that I was so young—a young kid with septic spots and a dinner jacket and a twenty-inch waist. "How's he a hooligan? Don't make me laugh, get the older ones!" Instead, my brother Jimmy got arrested, and a whole bunch of other people.

That day it seemed there was an amnesty for all except if you were a Sex Pistol. Quite astounding. We can't even go up and down the River Thames on a boat? We're in the middle of a river, and someone has decided that's offended them? Quite frankly, a society that learns to mind its own business is one I'm looking forward to. No grassers here, know what I mean? Take care of our own. Eliminate bullies because we take care of our own. No need for it. No child molesters in a proper manor.

That night, I wandered around the Embankment for a while

with Sid and Vince, a close friend of Sid's, just looking for something to do. There really wasn't much to do, because there was no money to do it with. So I wandered my way back to my humble abode. Walked a great deal, I remember that, and I was really annoyed because I don't like walking when I'm drunk. Because it's kind of like not walking at all.

With all the outrage and headlines rolling on, it was a week or two later that I got attacked outside a pub round the back of Highbury Quadrant, near Wessex Studio, where we were recording the album. Dodgy area. *Phwoar*, you get many mixtures of crews down that way. And—typical me—I picked the pub of anxiety, times ten. I knew what I was doing when I picked it, I knew it would be kind of semi-trouble, but I didn't think it would be quite like the machetes and Gurkha blades it turned out to be.

Every now and again there are people that just want to kill you. Whether they're right or wrong about their determination: *don't doubt their determination.*

To my misfortune, I found myself accompanied by daft fellas that weren't prepared to defend themselves—Chris Thomas and his engineer, Bill Price. I mean, Chris Thomas, what the hell does he know about life? Not much. How can you make good records with two deaf ears? Actually I'll take that back: with one deaf ear and one tuneless ear.

Random acts of violence are what I was well used to from growing up in that part of London. Order of the day. I warned them it was about to go down, and they wouldn't listen. We were up against a good firm of people, and we could've handled them, but those two just separated, and straight into me these lads came. It was a little bit more upscale than the usual, because when people are coming at you with knives and machetes, it's a whole new level. One of them had this sword thing—I don't know what the correct term is but it was a long blade. I've seen it in jungle films, but I was understandably not expecting it in North London. That's all right, I'm still walking.

I got a stiletto blade in the wrist—the exact same hand where I got stabbed when I was out with Paul Cook in early Pistols days, actually. That left hand has really suffered the wear and tear.

On the plus side, that night I was wearing an amazingly uncomfortable pair of leather pants. They were extra-thick and very difficult to sit down in and they hurt behind the knee. That must have been one big old heifer that they made them out of. Really, industrially strong—if I'd had anything else on, I'd be Ian Dury, or Legless Lazarus. I'd definitely be walking with a gimp or a limp. Then again, if I'd had a lighter pair on I might've been able to move quicker. Let's not be too grateful to fashion, shall we?

Life's like that. Situations can come up, and you have to be able to sink or swim by the second, and act very quickly. Sometimes things can be life-threatening or financially threatening or whatever the situation—you have to be fully armored. So when you go out publicly you can't really allow yourself to be in any kind of inebriated mentality because you're making yourself a potential victim. There are people out there who prey on that and they usually lurk in collectives.

The average Joe on the street may be fine, but I quickly realized that I couldn't be caught like that anymore. I was never going to allow myself to be turned over by a mob of mugs again. From time to time you're going to run into the real top-level lads, but they tend to be all right—they've got nothing to prove, they're not out for a trophy. That's the problem with it.

Back then, I suppose I was the prime target of the moment—and still am, in many ways, that's never gone away and I have to be aware of that. It's jealousy, ultimately. Jealous of what? God, if only they knew! Being Johnny Rotten was never easy. To maintain the integrity that I think I have is a daily grind.

From then on, I knew that I couldn't be scurrying off to the off-license out of my mind, wanting a top-up. I get spotted, so I have to stock up when I'm sober. Lesson in life!

That summer, I have to say, it royally messed up my social life.

I couldn't go out to see bands in the way I naturally would have. I couldn't go anywhere alone. It was impossible. The animosity would come not even from members of the audience at a gig, but the bouncers at the front door. It would start from there. And the club owner harboring a resentment or an attitude or a belief system that went against my own.

At the same time, there was sheer glee—because you have to take whatever enjoyments you can out of life. And the fact that I knew I was managing to annoy everybody all at the same time was all the reward I needed. Wow! From Piss-Stains Prentiss, my old English teacher, I'd gone on to much bigger things.

Malcolm sorted me out to live in some posh bird's flat in Chelsea, to hide out from press intrusion. To this day, when you hear about pop stars who've got involved in scandals and they go undercover, they're always in some bird in Chelsea's apartment—it seems to be where everybody ends up. There must be an entire genre of Chelsea birds with apartments, waiting to hide out apparently troublesome pop stars. Hilarious! But I'm sure that from Malcolm's angle, it all fed into his big idea of me as the "man of mystery."

So there I was, isolated in this upper-class neighborhood, and the most I could come up with for entertainment was the curry house at the end of the road. And it being Chelsea on a Saturday night, even there would be some squawky scene. Working-class Johnny Lydon aka Rotten wanted to go back to his familiar haunts, but he couldn't because of the public smear of the newspapers, so he couldn't do nothing. Instead, it was isolation.

Friends and acquaintances would come over but that would be about it, and very quickly I'd be boring them, because I just couldn't go out with them. At first they didn't understand, then it was obviously "There's no point in going over there, he ain't coming out—and we don't want him to!" And this is from my best friends.

You have to understand: the animosity against me was so volatile, so "up there," so beyond anything I think had ever been seen, in the postwar "tabloid" era or whatever, and it was very hard to overcome that. At the same time I knew deep down inside I'd warranted this kind of attention. I'd earned the passport, I'd earned the wings.

I didn't last long in Chelsea and soon I was hopping all over the gaff—here there and everywhere, from earlier-on squats to rented apartments. Just awful, terrible times.

Trying to share an apartment with Sid was the biggest mistake, because he was hanging out with Nancy at the time and that was unbearable. That one was in Sutherland Avenue, Maida Vale. It couldn't have been a worse place, because it was very drug culture-y—as in, heroin-y—around there. You wouldn't think it, looking at all the nice middle-class rows of houses, but that's exactly where all the darkness truly resides.

Where would I be without a witch hunt? After all these years, I've grown to like them, and love them, and indeed find them very comforting. My bottom line is I don't do anything to hurt anybody, that's not what I'm up to. So whatever your excuses or reasons for trying to lock me up, well, you can do that, you can imprison my physicality, but you cannot imprison my mind. CANNOT.

Dealing with the press, and all the "foul-mouthed yob" stuff, we had the perfect reply: "Pretty Vacant." Is that what you think I am? Okay, have a bang on this number! Well, I'm not pretty, and I certainly ain't vacant, so what do you make of me now? Not so much attack to defend, but it perfectly tackled the agenda that was being put up against us.

And it was a very powerful agenda. We had Rupert Murdoch and Robert Maxwell, the whole murky scene of the British tabloid press, really on our case in such an enormous way, basically preaching hate and contempt for us, and we had to really swerve healthily around these things that we were being drawn into. We had to pick our battles a little more carefully in order to survive.

We were being involved in all manner of TV and radio discussions that were just going apeshit wrong. There was one interview, and in it I was asked—the implication was—"Is Malcolm the mastermind behind you?" and I said, "Malcolm's the fifth member of the band, we're all equal." Which of course Malcolm took as a compliment. It wasn't meant as such—I was trying to diminish their opinion of him. I had to tread politically all the way through this, and at the same time be a ferocious voice of accuracy, but not get pinned down into a corner where they could imply that I was just trivial. You keep the engine burning.

There was a time there where punk was really exciting. X-Ray Spex, the Adverts, the Raincoats, the Slits—those bands had different approaches that were fascinating to me. They had the feminine influence, which is interesting musically. It was different social learnings going on, different sharings of thoughts, which would normally have been closed to music. Fellas and girls in the same bands, it was an amazing thing. They came across as level—it wasn't just "Now sing something pretty over the top." They were full-on equals, very entertaining, and it opened up so many possibilities in the songwriting. What a great time! And that side of it wasn't competitive, none of us were competing against each other. To me that was punk properly developing into something really awe-inspiring.

For some reason, however, the Clash started setting themselves up as our competition. There was a headline in *Melody Maker* from Joe Strummer: "We're going to be bigger than the Sex Pistols!" It was infuriating, and I talked to him about it. When direct quotes come out that I think are divisive, then something needs to be said because I don't want to see any of us divided. What attitude is that? None of us are doing this for that kind of chart competition, competing for places. When we start those internal wars amongst each other then that opens the floodgates for the arseholes, and whatever you're doing creatively, you've got to keep a hold on the arseholes.

There was some rubbish war going on between Bernie and Malcolm, friends falling out, and Bernie was trying to use his band as a weapon to get back at Malcolm. All very stupid stuff, and there I was, a young person watching adults behave like that. What made it all the worse was that certain members of the Clash were actually responding to that rally.

Bernie was feeding them a lot of politics and Joe would come round to the various houses I lived—even one I had way out in Edmonton—and he'd always have a Marxist book in his hands, and he'd be studying it and writing down notes. Then: "Oh, the *Six O'Clock News* is on, I've gotta watch it!" Rather than be able to take the BBC with a pinch of salt and be able to read between the lines, he'd grab the headlines and "be inspired." That's what fueled "sten guns in Knightsbridge" and all of that nonsense.

It's not my way. I liked Joe, and I liked Paul, and Mick Jones was such a happy-go-lucky fella, but Bernie was feeding them all that college-union, "declare war on society" stuff. If you wanted a good night out, to meet interesting people, backstage at a Clash gig wasn't the place. That was full of studious learners: "Yes, hmmm, yes, I'm with this program. Yeah . . ." DULL! FUCKING DULL!

Joe had always been so friendly, but as soon as he took the Clash too seriously he became *unfriendly* and indeed got involved in squabbles with some of my friends. He began to lack a sense of humor about himself. He took himself way too serious as purveyor of some kind of weird socialism, and was definitely out to grab himself a crown. He went too far, in the same way that Hemingway would overdo it. Or, you know that Rodin statue of the man frowning? The "Oi've bin finking" statue, as I call it? That's what he self-consciously turned into. His was a conceit that was very repulsive to me. *Phwoar*, what a pose. But one thing I've learned: we're all just only humans, we've all got our warts.

The Clash had a very middle-class approach to everything, and their audience shared it, and the smug journalists loved them, and of course they set the scene for all the deadbeats, all the bands that just wanted to do everything at a hundred miles an hour and scream and shout. That lot were of no interest to me.

Through them, punk grew into a standardized uniform, with the charge led by the mass media. The *Daily Mirror* would put out articles: "How to dress like a punk." Then those kinds of arseholes would turn up, and the whole thing just turned upside down. Many of the bands that came along then thought that the whole idea was to try to out-Rotten Rotten. And so violence crept in, and before long you had the Sham 69s, propagating violence through ballet. Just arseholes. Dumb, moronic, smashing-their-heads-off-walls-to-show-how-tough-they-were fools. They weren't listening to nothing. They were incapable of learning, or growing with a thing, or seeing any hope or prospects for the future.

We were saying "No future" in "God Save The Queen," because you had to express that point in order to have a future. No, this lot really didn't want one. It's like runaway horses. Once the stampede starts, how do you get them back in, into the herd? And indeed, why should I? If that's really what that lot wanted, then fine. Go, charge! And when you're over the next hill and I can't see you anymore, all the better for the rest of us.

My mum and dad were very supportive of me through all the vilification that was poured on me, but it was very hard on them. The negative reviews would really upset them. They weren't great lovers of the noise I was making but they knew the negative press about me being a bad person wasn't right.

Mum always wanted to know what my world was—it was a bit of a mystery to her—and so I showed her that it wasn't deep dark and secretive and a wrong 'un. I took her to see some gigs—Alice Cooper, Gary Glitter, anything that was going on across at the Finsbury Park Rainbow—and yippee-aye-oh, she was well up for it. She was just proud that I, with no apparent talent, could somehow manage to find my way into that world. Because there was no indication for them until the first day I started in the Pistols that I had any interest in being in a band or writing songs or singing at all.

I'd be more than happy to bring my friends around, oddball collection that they were. It used to always make my mum and dad laugh—my dad, in a deeply sarcastic way—but I always had strange friends. Probably they were the only ones that would talk to me. They were all in their own ways in similar positions, socially. None of us could really find a niche in society.

I'd have to ring my parents up and beg them—beg them—not to do interviews, and warn them that they'd be chiseled. "Oh, wi *feel* wi need to stand op fer yi!" "Please, don't! You're just gonna make it worse." They did one interview with the *Islington Gazette* which was a particular hatchet job. They gave them all these photos of me when I was young, and none of them were returned. Terrible, and a very spiteful article.

I instinctively wanted to protect my family from the public circus. With the Pistols I was thrown in there, way at the deep end, and I didn't think my family had the tools to cope with it. It's very hard to come to grips with the world that you used to view as important and relevant and caring and real—i.e., journalism—when it actually turned out to be a savage vindictive opinionated bag of bile.

Thanks to the press, we were involved in this huge quagmire of misinterpretation and Chinese whispers via the media, which start out as a tiny lie, spread into an enormous one, then explode into an atomic bomb with no reality in it at all. And it's very hard to hunt it down and correct it.

The media doesn't exactly chastise itself for getting things wrong. And where do you go? Who do you ring up and go, "Oi, you can't write that about me! It's a fib, it's a lie!" "Oh well, you have to prove that." "With what money?" There was no money in our pockets at that time. We weren't earning, couldn't play shows, couldn't do anything.

All of this created a wedge between us, because these situations separate you, they don't bring you together at all. I think it's a device that the media are not aware of, but it definitely has the end result of destroying you—*if you're weak enough.* We were weak, but not too weak, because we kept the ship afloat somehow. But again, Malcolm went into hiding, not talking. You couldn't find him; there'd be a load of people wrapped around him, all his old college cronies: "Malcolm's got a headache at the moment," or "Malcolm can't come to the phone, he's busy." On and on and on.

I never quite figured out what drove Malcolm. A very fertile mind, but prone to being poisonous from time to time. Self-defeating, actually. He'd create wonderful situations but he'd fold back on them. He'd light a bomb, but he wouldn't want it to explode, so he ain't no Guy Fawkes. He'd like the idea of Guy Fawkes more than the reality.

After Grundy, he became emasculated, he had his balls cut off, and he did that to himself. He lived in fear rather than fearlessness. Too much education, so the intellectual process just ends up in self-doubt, because you overthink a situation to the point where you're killing the joy of it and you're killing the instinct. We do things in life to put ourselves into a position where our instincts can take over. Not with Malcolm. Instinctively he was correct, but then the intellectual process would negate that. I think first, then I act. Or, in Malcolm's case, he withdrew.

Malcolm wasn't an out-and-out crook or a thief, but what he'd see as important to spend money on wouldn't necessarily be what I'd be agreeing to. Malcolm's leanings were always artistic and madcap, whereas for me the motivation was "I need somewhere to live, give me the money!" Malcolm's argument would be "Well, if I put it in my name, the transaction will go through ever so much quicker!" A great deal of problems arose for Steve and Paul, because their flat was in Malcolm's name. I could see that problem coming.

I think there was a shrewdness on his part—like how adults tend to manipulate children. Even after all this advance money, we were still only on fifty quid a week. Everything was now all being saved up to be invested in a Sex Pistols movie. This was a project I bitterly resented, because he was keeping it to himself. It was *his* project, it was all his ideas. Let's just say, that's what somebody that has an Andy Warhol complex can bring to the table. He was always very impressed with Andy Warhol's "everybody has their fifteen minutes of fame" thing, and he fancied that for himself.

One collaborator he tried out for this masterpiece was "titsploitation" director Russ Meyer. When I met him, I didn't like him at all. An overbearing oaf—pig-ignorant and obviously a perv. There was a very odd thing between him and Malcolm. I knew that these were two people that could never get on. Not ever. There was no common ground. Meyer was very brazen in his approach to sexuality, and I suppose Malcolm was trying to scoop up the droplets from that. Malcolm would emblazon himself with other people's efforts, so I suppose he fancied himself at that point as a ladies' man also. Russ would look at Malcolm and go, "You look like a lady, man!"

I actually put forward Graham Chapman from Monty Python for the job, because I'd seen his antics in a pub in Archway. He did a little trick with this small dog, where he'd lie flat on the floor and pour a bit of cider on his genitals, and the dog of course would lick them. If Malcolm was talking about making a film about us, then I thought that's the sort of person who should be directing it. But he was not going to be tolerating Malcolm's phoniness. Malcolm eventually would have to run away and hide from people like that, because you have to put your wares on the table at some point. Anyone can talk up a storm, but you have to declare how big your guns are, and Malcolm's weren't sizeable.

Finally, I got Malcolm to get me a place of my own at Gunter Grove, near the World's End in Chelsea. I was fed up with all this moving about, and I knew damn well it weren't going to last forever with a manager like Malcolm. At any point they could pull the rug from underneath me. I wanted something for myself to be fully grounded with. I didn't care where, but Gunter Grove was cheap. I think it was Stevie Winwood who actually owned it, and Island Records were selling it off for him. I got in there, and the best parties I ever had were there.

At the same time, the authorities were desperate to pin something on me, so that's when the police raids started, and one way or another they never really stopped.

I genuinely felt I was bearing the brunt of this ferocity. There was no backup from my band, and definitely none from the management, so who the hell was backing me up? Some bored kid in a council house, who I had no way of communicating with?

In fact, I actually took to responding to fan letters around that time. I used to do that quite a lot because that was my only real outlet. I never went on to meet any of my correspondents, as far as I know or remember, but I do know that a lot of the letters were "thank you—thank you for making me able to think for myself," along those lines, which is bloody heartwarming stuff, and then two hours later your front door's kicked open by the police! I've always had time for people that wanted to communicate with me in that way. Always.

Me having my own flat annoyed Steve and Paul very much. They'd be like, "Who do you think you are, hurr hurr?" I'd reply, "Well, I'm writing the songs, aren't I? And who the fuck's that cunt, Malcolm? That's *your* mate, *your* King's Road luvvie-duvvie. He ain't mine, he don't like me, he don't like anything I do, yet you're all profiteering off it. Quite frankly, if 'Anarchy In The UK' or 'God Save The Queen' were with any other set of lyrics, they would not have been what they were. They wouldn't have had any direction, they would have been pointless, dull, rubbish. Just another trash pop band."

Their "you can't sing" attitude stemmed from the fact that "boy band" was ultimately what they all wanted. But to be brutally honest about it, if I can't sing then what were Paul and Steve—and Glen—doing? They wanted celebrity, I suppose. I was naturally attracted to the exact opposite of that, and by default got them to where they wanted to be.

But because it wasn't something they quite understood, the rebellious spirit, etc., it was a problem and it would always be a problem and it never got any better. And it was never resolved, it was never openly discussed. The basic attitude given to me was one of disgust for everything I did. And out of that environment, of course, I quite naturally chose Sid. Have some of this punishing number, baby! It was the better route. Poor old Sid, my mate, it destroyed him, and it breaks my heart to say it because he was pushed in at the deep end, but at the same time he made the band better. It was never about nicely played melodies, and why the hell should it be? Nothing in rebellion is about gentle melodies. It just isn't.

Still, there was that yobby image that was being cast out there about me, and I'm not like that. I'm a quiet, contemplative kind of soul, the deep thinker, and oddly enough very rational. That wasn't what was being put out there into the publicity machine, and that was a pity. I tried to do my bit to correct that when I did radio shows and played music I liked. There was one on Capital Radio in London, with Tommy Vance, which got a lot of attention. I played Can, Beefheart, Culture, Neil Young, Peter Hammill, Dr. Alimantado—and all I got from Malcolm was "How dare you? You're ruining punk!" "What? *Excuse me?*"

That was the beginning of the end, where we were virtually no longer speaking—that issue—because for me it was an opportunity to play all the music I loved and adored, and explain the reasons why and what it is I'm doing right now and where I am in the scheme of things in the world. And he was just furious, because Malcolm's presumption was that punk was the New York Dolls, Iggy Pop, the Ramones—but the Ramones didn't exist for me at that point, because I had Status Quo! And the Flamin' Groovies were never top of my chart parade.

It was just him trying to tailor us like we were some new silly T-shirt he'd come up with. Control freakism. Did you think I'm some kind of packaged hamster you just purchased and put a sequined neck-choker on?! You daft cunt, telling me what I do and do not like. Fuck off! I was really angry—*really* angry. We've got to learn to stop thinking in terms of categories as a species.

This is that and that's that. No, there's cross-pollination all the time. And I don't believe in six degrees of separation, I believe in a continuum.

That's where a split began and quite a serious one—a land-mass separation of what is punk and what isn't punk. I'm sorry, I'm on the correct side of this, fighting this backwards thinking of trying to justify yourself by trashy aversions to things. And that's no shame to Iggy or the New York Dolls, who I love and adore—they fit very nicely in with my Todd Rundgren. It's people who experiment in life that interest me. Not just "Wham bam, thank you, here's a pile of trash, and look how junked-up I am."

I didn't want the junkie image thing to creep in, and of course Sid bought well into that and wanted to live the New York lifestyle—so there was Malcolm's hook on that one. Malcolm was very enamored with New York.

We finally persuaded him to book us some gigs in England for August. We ended up having to advertise ourselves under assumed names like the Tax Exiles, Acne Rabble, and S.P.O.T.S., which stood for Sex Pistols On Tour Secretly.

Having to go out incognito was at the same time ridiculous and challengingly refreshing. It turned into the world's worst-kept secret. But it kept the authorities off our backs. Whether or not we were banned outright is a moot point—maybe it was all part of Malcolm's alleged masterplan. I don't know what it was that some of these local councils thought we'd be getting up to, all the riots that would ensue, but it never really happened in that way. The only negatives that we ever really faced were from apparent "music lovers," hahaha.

On that tour—which was only half a dozen gigs, so it never really felt like touring—it became a close thing with local audiences. Very warming, but it seemed that every time anything was working in a really good comfortable way here, for us and the audience, Malcolm would find a way of sabotaging it, like he was scared of it actually being successful.

He got very scared of the long arm of the law who were eyeing us with malevolence, and he backed away into the fiasco of his movie—called *Who Killed Bambi?*, at that point—as a lighthearted escape from the reality of what it was we were actually all about. He was also very fearful of dealing with me in any verbal confrontation because he knew damn well that I had the artillery.

In many ways it became a power play. A very odd situation was unfolding: I was being blamed by Steve and Paul for bringing "that arsehole Sid" into the band. Malcolm stirred all that up and got them two very angry—and by such a scene isolating me—and then tried to create a friction between me and Sid. Because Sid and Malcolm oddly enough *were* communicating. So he was playing both sides, Malcolm, with me losing out in all of these scenarios. And those situations then developed into the nonsense that they became.

From the beginning, Malcolm didn't make a good job of the whole interpersonal thing within the band. He really should've been ashamed of himself. It was now all spiraling out of control, but he would throw in spiteful digs and rumors that would cause all manner of trouble. No two of us were ever told the same thing. It wasn't great fun to be yelling at each other, and when we unraveled what he'd said to each of us, we'd realize the divisions all traced back to him. Then all of us would go, "Right, let's get him to 'fess up" or whatever, and that's when, of course, he'd be behind locked doors.

I remember Steve once smiling and happily saying to his face what a cunt he was, which of course Malcolm would smile at and take as some kind of achievement. So that's how *their* relationship was working. You weren't going to be able to move these unbudgeable nonsenses.

I just got on with being myself, until at some point after trying to share an apartment with Sid, I started thinking, "Well, do I really need to be in this band?" The stuff I was writing and thinking seemed to be beyond this now. I had bigger ambitions than just being involved in this domestic drama which offered no reprieve.

Never Mind The Bollocks, Here's The Sex Pistols, the album, when it finally came out that October, was a nice end result, I have to concede. It showed to me that Steve had great capabilities; he could be taking that guitar to all kinds of different, new, exciting and original places. It was like a guitar army, rather than just a messy noise, going, "That'll do." He had a good learning curve on him, Steve. That's how I viewed it at the time. Recording had gone on forever, and it'd almost turned into a guitar roadshow, but my God, I was liking it. The blending together of all those different takes had made a delicious end product—although it left the band with something of a problem if you wanted to reproduce that live.

Richard Branson did a great thing to promote *Never Mind The Bollocks*. He filled the Virgin record stores with these *Never Mind The Bollocks* posters, the yellow posters with the blackmail lettering—particularly in Oxford Street in London, because he had two shops there, one at one end and one at the other—so we just blurted from the whole window all the way round. Fantastic.

The same posters went up in stores up north, but the north had a different attitude to it—in particular, Nottingham, where they decided to take a local store to court for their supposedly offensive window display. They were to be tried for the "Indecent Advertising Act of 1889." So we had to go to court. Well, we didn't *have* to go to court, but I volunteered. And I wanted Malcolm to go too. We were gonna go and stand up for our right to use the word "bollocks," which to my mind, reading the Oxford dictionary, is a perfectly feasible Anglo-Saxon word for "testicles." Malcolm backed out, of course, and so I was driven up there by some Virgin representatives, because they understood the importance of it. We hired QC John Mortimer, writer of TV's *Rumpole of the Bailey*, to prove "bollocks" was actually derived from a nickname for clergymen.

I'm against the banning of any word, so I was more than

happy to sit in the front row in the courtroom to hear what this judge had to say, telling me what word I could and couldn't use. I was absolutely dying to get on a stand and give a speech. I had prepared one, I'd really, *really* worked on this, I hadn't drunk for days, I kept myself really sober—but they didn't give me the chance, because the judge went, "We must reluctantly find you not guilty." So then we whizzed straight off to see some friends of the record store people, who ran the local radio station, and had great conversations. I got another great chance to play my favorite records to back up the court-case victory. And of course I immediately went into "Where's Malcolm? What a wanker, so I'd like to dedicate Cliff Richard's 'Devil Woman' to him." I was enjoying this "ruining punk" lark.

We drove home later in one of Branson's buddies' Aston Martin at high speed. Fantastic day. Virgin were backing me, they were supporting this. For me it was a personal tragedy that not one of the band wanted to be there, or the management, and I really seriously felt from that moment on this was never ever going to be a unified group. Because they were lacking the courage of commitment. By not turning up they were completely devaluing the Sex Pistols.

In early December we were all set to play our biggest ever gig in and around London, at Brunel University. Unfortunately, it turned into an ill-conceived nonsense, thanks to guess who. We had no equipment that anybody could be hearing us on, just appalling, and Sid's drug nonsense made the whole thing vile and difficult and painful to go through. The cheapskatedness of it.

There were hundreds of people in Brunel that night, and hundreds outside, they came from everywhere—so we should've at least had a good sound system. I'm not blaming everything on that, but it's one thing for the band not to be able to hear each other, quite another to have to strain your ears out in the audience. An unforgivable lack of consideration. But Malcolm wanted to create a scene of chaos. Bullshit, he just didn't want to spend the money. He wouldn't learn: you've got to give a lot to get a little.

The only respite was when we played two benefit gigs on Christmas Day for striking firemen and their kids in Huddersfield—a matinee show for the kids and an evening show for the adults, which turned out to be the last gigs we ever did in England. It was great to do it for them because they were all broke and nobody gave a damn about them. These people weren't going to have a proper Christmas, so we laid it all on, flooded the place with cake and presents for the kids.

Here we were, the alleged most toughest band in the world, and at the kids' matinee show we'd have to play to seven-year-olds! There's an awful lot you have to leave at the door to do that. To start with, I was thinking, "How on earth am I going to sing 'Anarchy' here, with any sense of realism?" Well, kids totally knock you into place with that. They're going, "You're just one of us, John, a big stoopid kid."

Then the cake started flying, and it went into absolute insane mental brilliance. Absolute slapstick. It showed our lighter side. It was *Carry On Sex Pistols*, with Steve as Sid James. Kids can be such a good bounce back to reality. It knocked the stuffing out of Sid too. He was trying to be the hard rocker bloke, but how can you be tough with a Christmas cake in your face? It reminded us that it all had got a little too serious.

As a band, that was probably the closest we'd ever been, but it had come to the point where Malcolm just wanted the band to cease to exist. We just wanted him to go away, but he carried on with that poisonous behind-the-scenes stuff, and it became a total no-hoper. We were right on the brink of falling apart, but not before . . .

Can you imagine what it was like for us Sexy Pissups to have the opportunity to tour America? It wasn't like nowadays, where any fool can stump up the air fare. Most people could never afford plane tickets back then—ever-never-ever—and certainly not the likes of us. *Phwooar,* off to see John Wayne Land, yippee-aye-oh! *And be paid for the privilege*—absolutely astounding! It's the major benefit of being in a band: you really do get the opportunity to do stuff you could never have dreamed of. Sure opens your mind, I can tell you. Whatever happened, we owed it to ourselves to cash in our chips on this one.

America, to us, was *Kojak*, *Ironside*, and dare I mention *Starsky and Hutch*, a show I only remember for the car. America was all big-arse cars, just like in the films. They've actually downscaled in that respect these days, so you would think there would be more room on the freeways—no, there's just more cars.

American rock, though, was in desperate need of a shot in the arm. It was just West Coast banality. Mellow, drippy blancmange like the Eagles—*aaargh!*

I love me musics. I like to know the be-all and end-all of ALL of it. In fact, sometimes I'm more fond of the things I hate, they're oddly more rewarding. But the Grateful Dead were so moribund and boring.

The idea of us "conquering America" was fantastically hilarious. Before we could get out there, however, I was having proper serious problems getting a visa, thanks to my speed conviction. The only thing going in my favor was that, one stupid night not long before we were due out there, I went to a club with some of my escort friends—Linda Ashby and her crowd—and I was stretched out on the staircase when someone tried to rob the cash register. They tried to run up the stairs, but they tripped over my foot, fell backwards, knocked themselves out, and suddenly I was accoladed in the press for stopping a robbery.

Suddenly I was the hero of the hour. Ouch! At the time, I was rather spooked by it. It was nothing I wanted mentioned at all, and denied all responsibility of it. I thought, "Look, my friends aren't going to like me for this one." But it boded well for getting the visa to America. Eventually, the authoritarian figure who interviewed me at the American Embassy said, "Well, you've done things for society." Wow, is that how you viewed it? But it paid off, it got me my work permit. Although that soon came to be regrettable.

Malcolm, in his wisdom, had decided that we wouldn't play in the big cities on the coasts, like New York and L.A., but instead would play to "real people" in the South. Now, Malcolm, there was a man who didn't understand the working class. So, on a preposterous schedule, we crisscrossed America in early January 1978, amid the ice and the snow, on what was basically a school bus.

Many aspiring stars would've been broken by the experience, but not I! The sheer joy and privilege of being able to look out of that bus window and see America whizz by me was utterly enthralling, and in particular because this was the South. It reminded me of the cowboy movies my dad made me watch as a kid. You could relate to the names of the towns, just through the television. Bloody hell, it was great. After all the intrigues, I felt like a kid in those moments. The landscape of America impressed me no end, and I fell in love with the country. Regardless of the piss-poor situation that was unfolding before us, there was still that joy there.

At the same time, I knew I couldn't do anything about Sid. It was an impossibility that you couldn't unravel because Malcolm was using it as a tool to unravel the whole thing anyway. He had to destroy what he couldn't control in his megalomaniac way, which in the current light of day was very childish of him. But that's the absolute truth of him, that's how he was. He was a very jealous man, and if anybody came up with an idea that was something he wished was his own, rather than celebrate it and taking it on board, he'd work against it.

Little did I know that he was secretly taking singing lessons. He looked at Johnny Rotten and he thought, "I can do that!" I just wish he'd have seriously gone into it, because then everyone would've seen the lack of talent.

Although everything was falling apart between all of us, and the shows were horrendous, I tried to hook up with Steve. In fact, we spent a great night once. Steve had this shoe-box full of marijuana. "Fancy trying this?" Funnily enough, I did, and it was hilarious, but we had to deal with the problem of Sid who was two doors down hanging out with a black drag queen. It was ugly and foolish and not a Sex Pistols-y thing at all. It was more a sad interpretation of Lou Reed, and wallowing in the problem of "Is there any heroin around?"

It was hard on all of us, fighting from our own corners, but never grouping together properly. We realized that Sid wasn't the problem, we all were. We just couldn't get on with each other, and that was that. It was pointless trying to continue it at that point, because the outside influences were just continuously poisonous.

The media were crawling all over the tour. Sid did an interview with *High Times*, who were following us about. Now, *High Times* was a drug culture magazine, but it was rumored that it had CIA connections. So really what it was doing was finding out what you do, who you do it with, who's selling it to you, where it comes from, and where it goes. And Sid didn't have the capacity to understand that you don't get involved with those outside agendas. Plus, he'd be willingly waffling away his nonsense, his drug delirium, to these folk who were more than happy to print it, and then we'd end up with the reputation as a smackhead band.

I didn't write these songs for it to be that way, and when Steve first started this band I'm sure that's not what he had in mind either. From innocent kids on the blag, so to speak, to an arsehole out of his brain on heroin getting conned. And, oh yes, of course he had his money stolen by this black drag queen, who actually beat him up too. So it was like, "Ooooh, this is just terrible."

At the gigs, he'd be trying to out-Rotten Rotten. He'd compete with me onstage, and attempt to stand in front of me, make out that he was really tough and hard. He'd fight with people at the front, and the sad thing is I know why, because I felt the pain in him. He was doing that really as a subterfuge, to cover up for his own feelings of inadequacy. He knew he couldn't make the grade, it just wasn't there, he didn't have it, and so self-destruction became an emblem he could stand behind, because that's the easy way out, isn't it?

It was terrible, to watch the demise of a very close friend; it breaks your heart. But at that precise moment in time, I was plain furious that he just wasn't getting it. "Hello, matey, you're in the most privileged position in the world, people will be dying for this *power*, shall we say, and to throw it away and make yourself and everyone around you look like an idiot . . ." He was a controlled robot. You've got to learn in life, you've got to learn it quick and keep it for the rest of your life: pull your own strings, and have no puppetmaster—and—habits—are—puppetmasters.

What broke my heart was that some people were watching him and actually thought that was the groovy end of heroin. Sid's behavior becomes an act of criminality against humanity, for me. His example is one of self-destruction. How is that appealing? And then you've got a media ready to package that, because it takes away from the political content of them songs. Suddenly there's not a real serious social message, there's just a drug addict.

I had made somewhat good amends with Steve in the middle of all this. Of course his angle was "That Sid's got to go." "No, that's not gonna resolve it!" From there on, Steve and Paul started flying between gigs, and booking themselves into different hotels—it became ridiculous. I couldn't, for days on end, in the middle of a tour, speak to other members of the band, because "Malcolm" was hiding them from me—really childish, silly stuff, that you wouldn't think five-year-old girls in the playground would get up to.

During a soundcheck in San Antonio, I wanted to try out a new song I'd written, "Religion," and they just wouldn't cope with it. Didn't wanna know. Fine. Leave it like that, then. It was all too silly for words, when you look back on it—pulling faces, ignoring me. Sid was up for fumbling around with it, but that wasn't what I wanted. He didn't have the chops to get with it. So I kept "Religion" on hold, and used it later when I started my new band, PiL. It was probably for the best that, at that time, no one was talking about making a new album—it never really came up as a thing to do. I'm really glad, because it would've been another major stress point—how much can you take?

I don't know if I really wanted to do it like that, in that same way, and I obviously didn't, because even to this day, I can't go back on that sound and try to repeat it. It's boring, and it would be wrong, and it wouldn't work. For me, anyway. It's like this: I know that's where the money would've been, to put out a Part Two that's just like Part One. But no, money don't come into it. If I feel that that's challenging my sense of creativity, my creativity wins, regardless of the financial problems that it can create for me. I will continuously take the risk because I don't see it as a risk; I see it as the very point and purpose of what it is I'm doing. You only get one opportunity in life and I got it with the Pistols and I intend to use it well. Use it absolutely well, use it to its ultimate extreme. Sink or swim.

Malcolm, apparently, wanted to get the loony cult murderer Charles Manson to produce our second album from prison. I'd be reading things like that but no one would have the front to say it to my face. There were many, many rumors like that. Like, after the American tour, we were buggering off to Brazil to work with Ronnie Biggs, the Great Train Robber. That was the whoremongering of it, and none of that would ever be acceptable to me. It was just looking for cheap headlines and watering down anything serious or good that's going on in this. Behind all that way of thinking—it's glaring to me and it should be to anybody reading this—that it was Malcolm's resentment. He had no control, and so he was trying to take it back into a world of silliness, where he would have a place. A world of cop-out.

So, by that last gig in San Francisco, I'd lost interest really. I'd become incapable of caring about writing another song for this outfit. I felt like, "That's it, there's the full stop. I've achieved as much as I can in this environment." So that's how it ended up with me saying, "Ever get the feeling you've been cheated?" We were a betrayal of what we started out as.

From my point of view, at that last Winterland gig, who gave a fuck about me? Well, they gave even less of a fuck about Sid, other than they wanted to use him later—and Sid was eminently usable at that point. Sid hooked up with Malc, basically for drug money, and of course I didn't want anything to do with that. Malcolm was leading him into "Yes, Sid, we'll get you what you want," and then, when push came to shove a year or so later, abandoned him completely.

It was very clear to me, there was nowhere for me to be in this band, or with any of these people. And so when Steve and Paul snuck off to Brazil without even mentioning it to me, it was perfectly fine—I expected it. It was a relief, actually, but then a puzzlement—"Where's my ticket home? Why no money? What, my hotel bill here hasn't been paid? WHAT?!" And then I ring up the record company, and the answer was "Oh, we've been told that Johnny Rotten's gone to England, so we don't acknowledge you." Ridiculous. Insane.

The only person that paid any attention was Joe Stevens, who was a friend of Malcolm's, and a photographer. I told him what was going on, and he was just puzzled and horrified by it all. Eventually he paid for a plane ticket for me to go and stay with him in New York and clear my head. What a fantastic thing to do. What a fantastic fella Joe Stevens is.

It took about a week to get in touch with Bob Regehr at our American record company, Warner Brothers. He thought it was insane behavior on Malcolm's part too. He came and met me in a hotel, to help sort me out, and right there, somebody laid a writ on him for some reason. Bonkers! I've no idea what the situation was, but obviously it really bonded the pair of us—which really paid off for me a couple of years later.

I did an interview with the *New York Post* to give my side of the story. I didn't want to, but it had to be done. I wasn't in a right frame of mind to deal with it at that precise point, but as Joe Stevens pointed out, "Look what this fuck Malcolm is saying about you," and it had to be responded to.

It was really wicked, really spiteful. He was just trying to make sure I had no chance of an ongoing career, trying to stick nails in my coffin kind of thing, rather than just leave it alone and say, "Well, there it is, a parting of the ways . . ." No, and that really reinforced my attitude about "Right, I'm going to get you back, fucker! Full steam ahead—as soon as I get back to England, straight to a lawyer . . . I want this fixed."

To my mind they'd wrecked everything that was brilliant and glorious about the Sex Pistols, which was unity, and they tore the arse out of that through selfish shit. And it all ends up in what? Celebrating a train robber? At my cost, my expense?

Then I have to run a lawsuit against them going, "Hello, don't I count? Remember me? I wrote the songs!"—at least the lyrics, and quite frankly, being real honest with myself and everybody else, I don't think anybody ever bought a Sex Pistols record because of the lead guitar solo or the drums or the bass—although I couldn't have done those lyrics without those three things. But I never got the respect and love that I think us as a band truly should have had for each other. So eat shit and die, you cunts. That's my polite way of saying, we could've been good.

WHO CENSORS THE CENSOR? #1

JUDGE NOT LEST YE BE JUDGED

Apparently there has been an old audio interview of me from 1978 doing the rounds lately online, where I'm talking about Jimmy Savile and basically saying, "Everyone knows he's a child molester, but we're not allowed to say." I don't remember the interview. I've been told it was by Vivien Goldman, but I was speaking dangerously out of hand, way before all of this became public knowledge. You have to tell it like it is, and how you really see it, and say what you have to endure behind the scenes.

People were calling me a filthy disgusting Sex Pistol or whatever. But what the fuck is *that*? Do you *not know* what is going on with *that*, that institutionalized, decrepit pervert? He gets his OBE, then later he becomes *Sir* Jimmy Savile, but I don't think there was any doubt at all about what he was really getting up to. In fact, I don't think enough of it's come out. Everybody knew. It was common knowledge, but unspoken.

From a very early age, looking at him on *Top of the Pops*, you knew that he was just a wrong 'un. And he was always having a smirk, and "letting you know." You could see it in the eyes what he was doing; you could read the body language.

So that was also how I knew about Savile—his eyes. I could tell he was deceitful, and harboring something dark and ugly, and he was smirking about it in the knowledge, but not declaring. The full-on audacity, it used to drive me nuts. That's what I do, I watch a person's eyes, and I know what's going on with them from that. For me, the best actors or actresses, they do all the telling in their eyes. Katharine Hepburn, Peter O'Toole, Charlotte Rampling. You get so much depth in what they're up to. They can go beyond the words. It's almost musical.

Radio DJs in Britain in the early to mid-'70s had become godlike. If it wasn't young children they were abusing, they were definitely abusing something, because their power became overwhelming and dictatorial. And they would propagate themselves there on BBC Radio as the purveyors of good taste, and careers could hinge on their negative impact.

It certainly took an awful long time for the BBC to spin a Sex Pistols record, and I doubt whether they have to any large extent to this day. Ever since, I've suffered all manner of rigorous avoidance, all from these purveyors of good taste, who at the same time are up to horribly corrupting things. You had no option but to stand up against it and get banned outright forever, or try to toe the line, which of course I couldn't do.

6

GETTING RID OF THE ALBATROSS

At the time people in England wanted to live in tiny little boxy rooms that would be easy to heat in the winter. My front room at Gunter Grove, on the other hand, was a wide expanse, practically like a ballroom, with a kitchen out the back, and two tiny bedrooms up top. That's exactly all I wanted. I think the main room had probably been used as an office space by whoever Steve Winwood had had in there, before I got my hands on it. I had other ideas.

I could plonk a record player on the counter, and a TV with a coat hanger for an aerial between the two windows, which had a semi-balcony outside, and play music as loud as I wanted, and have as many people as I could possibly fit in there, which often seemed about three to five hundred, haha. In the summer we'd drink up on the rooftop—that was always a favorite with the chaps.

The word was out: "You can just turn up at Johnny's, he'll talk to the walls if he has to, but he *will* talk!" I'm a very quiet, solitary kind of person, but I'm very open to significant leaps of faith in any direction. And I do love interesting company.

I ran Gunter Grove almost like an open house, with a thriving community spirit about it. I had an awful lot of visitors, all of them very interesting. What a great collection of people! Every couple of weeks I'd throw a party on a Friday, and just let all my mates know, and the label people at Virgin, because I had a lot of friends there by now. So there'd be a lot of music-y people, but ones you would never expect to meet or have anything to do with. There would be disco bands, and unexpected guests like Joan Armatrading, the folkie singer-songwriter, who I found to be great fun.

There were also film people, authors. I even had the lead singer of the Bay City Rollers turn up with, of all people, John Barry, the composer. What a great evening! Then, later, in the middle of that powwow, Stomu Yamashta, the Japanese composer, dropped in. It was like an otherworldly experience, four very different people, but with a commonality of making music. Thrilling conversations!

What a marvelous eye-opener to me. After the misery of the preceding months with the Pistols, it began to make everything seem possible again, as I chatted away to people in all walks of life, musical or otherwise. For me, it was about getting a varied outlook on life. If you do not open yourself up to different outlooks, you are doomed. Doomed to keep repeating the same failures as all the other idiots.

Originally, I just had this maisonette apartment, the upper two floors of an end-of terrace townhouse. But the neighbor that lived underneath couldn't bear the noise, and so not long after I moved in, he sold me his place, the bottom half. I actually had some trouble raising that money, in the light of my court action against Malcy and the boys.

I never lived there completely alone. I'm not that kind of person, I don't like loneliness. I *do* like variety, so an endless parade of people came through for varying lengths of time. To start with, I had Dave Crowe, whom I'd known and kept up with since William of York school days, and another friend from Finsbury Park, Paul "Youngie" Young—the one whose gangster-moll mum ironed his bondage trousers. Paul's a very cheeky chappie—

a snappy dresser, and a complete ladies' man. Oh, the girls love that fella, and he's as smooth as butter with it. He's the kind of bloke, if you want to pick up girls, that's who you hang out with—just stand next to him.

It was great to share with Paul and Dave initially, because we knew each other so well. I'd spend the whole week with Dave and John Gray putting together mixtapes for the weekend. "Aaaw, what records shall we have?" And "How do we back-to-back these?" Dave had a reel-to-reel tape machine, so rather than going all turntable-y with it, and being stuck on that all night, we'd prerecord it all, and use the double echo that was on his Revox for the joinings between tracks. There'd be bits of *Ben-Hur* thrown in there, and all sorts of nonsense.

One of our favorite tunes in those early weeks of 1978 was the dancehall reggae hit, "Uptown Top Ranking" by Althea & Donna—"In a mi khaki suit an t'ing!"—what wonderfulness that song is! One night that backing track came on because we'd put the dub version in one of our reel-to-reel mixes. Unbeknownst to us, Althea and Donna were actually in the room, and they jumped up and started singing.

All the time it would be like that, it would be proper joy, loving the music, loving the social scene, and loving the fascinating interest everybody had for everybody else. Proper times. Precursor of what Rave was trying to do, I suppose. All agendas catered for, except of course for the jealous and spiteful who were never welcome. Hard to suss them ones out, though, from time to time. You make a mistake, and you think, "Ah well, give 'em a chance." That's my way, I'll give anybody a chance, and if they step wrong with it, well, there's the door.

For a while there, it got a little like, Johnny versus the rest of the residents in the vicinity, what with all the noise complaints. The neighbors next door never heard anything through the wall until builders came in and thinned the wall down for some weird reason. Then, suddenly, the noise went straight through. British builders—I tell you, you want to watch them. They were so bad, removing bricks in an adjoining wall, and they removed them to the point where I could see them at work from my upstairs bedroom. "That's too much, come on! How are you going to paper over that hole? Put the bricks back!" I'm positive it wasn't an intoxicated vision, by the way—it *was* reality.

Another complaint was hilarious: it was some Italians that lived opposite my backyard. They came over and said, "We don't mind loud music, but what we really can't stand is that reggae!" You've gotta laugh. So it was a matter of taste. And of course that didn't create issues. If people tell me what's really getting to them, I'm very open with that. I agreed not to be playing that stuff at 4 a.m.

The curiosity of reggae for me was always that it's not an aggressive music, the lilting rhythms were just beautiful, but my God, the dialogue was *diff'raaant*—totally 'ardcore. That juxtaposition is alarmingly loud: incredibly sad songs of pain and suffering, or revolution even, inside happy melodies. A very effective way of getting a message across.

A song like "Born For A Purpose" by Dr. Alimantado was just life-altering for me. The lyrics, I thought, were genius, and particularly fantastic if you feel you have no reason for living. Like, don't determine my life! "*Whoah!* Hello, Jamaica! You've got some good brains going on there." For me it was utterly one of those moments when you hear a song, and it's an affirmation of the highest order. Like, "Aaah, we're out there together—people who care and really consider what it is they do, and who know that they're on this planet to do something positive."

Everybody in the reggae world came to Gunter Grove at one time or another, because we certainly had the kind of sound system they'd enjoy, but mainly because it was a house where you weren't predetermined or judged. Welcome, one and all, and no troubles, ever.

In the early part of '78, I actually spent a month in Jamaica. I used to get on with Simon Draper, who was second-in-command to Richard Branson at Virgin. He was an ex–South African army fella, but he really cared, he was genuine. He knew when the Pistols fell apart that I could end up in all kinds of problems, so he gave me something to do—go to Kingston, indulge in the latest sounds emanating there, and help sign acts for Virgin's new reggae subsidiary label, Front Line. The label, I thought, was a great idea, and I felt a massive respect to Simon and Richard for what they were doing for me on a personal level.

When the opportunity came up, I said, "Look, I'm not going there alone. I know there'll be other Virgin people, but I need a working crew." The crew were people that deserved to be there, and not much really to do with signing anyone. I wasn't expecting them to take on that workload; that wasn't their role. I wanted people directly related with the place. So I asked Don Letts, the dreadlocked DJ from the Roxy club, and Dennis Morris, a photographer who took a lot of pictures of the Pistols towards the end.

Don and Dennis seemed right to me, because they had family there, and that was important. I was thinking from the heart, not from the selfish perspective of a gang of Brits abroad going to Spain. It wasn't going to be a holiday of oik-iness. I could have filled airplanes full, but I wasn't going to be like that. I wasn't going to do this to abuse Virgin either or take it as an easy ride. I took it very serious. For all three of us, it felt like a musical pilgrimage.

Branson met us at the airport in a 1940s Rolls-Royce with a flat-top roof. It was just amazing, driving through Kingston in that absurdly pompous, Raj-of-India car. Jamaicans being what they are, they're very loud—dey let ya know a t'ing or two, mon! It felt like walls of abuse, cynicism, and wit being hurled at us. So that was my introduction to Jamaica, and I'm eternally grateful.

For that, I worked kind of semi-hard for Branson, to make sure he was paying attention to the right bands. It was thrilling to be visiting the different studios and hearing the different styles. And going to the record stores, and just getting into it big time, and getting to really love many of the people.

The open friendliness of what Rasta was offering then was

astounding. It was, I thought, a good clue to what a proper new world order could be. At that time, as everyone knows from their Bob Marley records, they were enduring the worst of the worst out there, with all the poverty and political violence, but they'd be waltzing through it with a smile, which was quite inexplicably excellent of them.

Look at the troubles they've had to endure just to be Jamaica, from slavery onwards. My God, them fellas have been through the wringer. With all the wrangling between the People's National Party (PNP) and the Jamaica Labour Party (JLP) (or the JVC, as I called 'em), a civil war was going on. Very chaotic. When a Jamaican says "Peace!" it carries some weight to me because they really, genuinely mean it. They really do know how to endure warfare. Jamaicans are not cowards, they're brave fellas, brave men, women, and children.

We set up shop at the Sheraton Hotel in Uptown Kingston, and soon found that most of the musicians already knew we were there and were queuing up to see us. That was breaking many social taboos in Jamaica at the time because there was a very negative thing floating around, where Rasta was associated with dirt and filth and laziness and trampishness.

There was a great record out then called "Ain't No 40 Leg Pon Di Dread" by George Nooks, which was answering this horrible urban myth that was spreading around, about how a Rasta was found dead on the beach, and they found centipedes in his dreadlocks. They obviously weren't in there before he died, but we all know how headlines can take over, and reality is turfed out the window for sensationalism.

So, there in Jamaica, everybody presumed there was all manner of filthy beasts lurking in the hairdo, and so Rasta was always supposed to keep his locks hidden. You could end up in jail for flashing them. A yell went up, for instance, when Don jumped in the hotel swimming pool, and his hair was floating on the top, while the rest of him was four foot underwater. When these fellas came to see us, though, we weren't having any of that. "You can take your hat off . . . Look, I'll take my coat off too."

I'd gone out there with next-to-no clothes. I had a trench coat, two T-shirts, a pair of brothel creepers, a wide-brimmed hat—a Lee van Cleef number—and a blue tartan bondage jacket and not much else. I wasn't expecting that kind of heat.

As an aside, the bondage jacket came from a suit I'd had made by Vivienne for the American tour. I'd insisted that there be some crotch space in them pants this time! And I didn't want the zip to come from the back to the front, I wanted a man-zip, *at the front*. And I didn't want a silly little towel, I wanted a kilt, and a bolero jacket. Unfortunately, by the time I got to Jamaica, the bondage pants had rotted out, quite literally, from touring. Not being in any frame of mind to do any washing, all that was left really was the jacket. That jacket I quite happily sent to the Metropolitan Museum of Art in New York a couple of years ago, so it's done its world tours.

Going to the beach in JA, that was a nightmare. I was very body conscious, I still am. I don't like to strip down and run into the sea unless I can do it real quick, fully clothed. I was aware of how white I was, and how that would draw howls of laughter, because Jamaicans will tell you straight away. But I couldn't stand sweating on the beach any longer, so I just did it—took my clothes off and went swimming, and the howling actually was good. It's friendly, if you catch it right. They knew I felt foolish, and looked it, and I became very red, very quickly, because of the sun. And this, for them, was their winter.

We were only supposed to stay in Kingston for two weeks, initially, but we decided to stay on another fortnight, and dug into our own pockets. Don and Dennis took us all to meet some of their respective families, so it became a family-orientated thing, not just a cold business venture.

One of the worst things that happened was at Dennis Morris' aunt's place—or was it his grandmother's? We went round there and, you know, people don't have money, so when they put a bowl of soup or stew in front of you, you show respect and you eat it. Of course, it was loaded with scotch bonnet—a type of chilli pepper, which must be the hottest thing I've ever known in my life. It was un-be-*liev*-able. In 110-degree heat! So the trench coat had to go. Then a girlfriend of a saxophonist called Dirty Harry came round, and she went, "Jahn, ya got nah cloaathes," and she bought me this gray top, so I wore that for the rest of the time.

It felt really good to give Don and Dennis the opportunity to meet their relatives. It did Don a world of good—he truly found himself in all of this, even though he had to endure the wrath of all the native Rastas. They'd be going, "Rastaman eat *lobster*?!" That's part of the Rasta trip—no shellfish. Fair play, Don stood up and went, "Yeah!" He wasn't adhering religiously to a dogma that didn't make too much sense.

His grandparents, on the other hand, didn't approve of him being Rasta. It's always hurtful to meet relatives in those situations. At the same time it was something that needed to be done, and it's very difficult to say goodbye when you leave—and there I was in my ridiculous outfit trying to blend into the background.

The Jamaicans are so funny. They'd say to me in the street, "Are you a gunman? From May-hee-ko?" But then U-Roy, the DJ originator, was like someone who'd just stepped out of a Clint Eastwood movie. A great fella, mad as a hatter, but full-on with the Rasta thing, and I had lots of puzzlements. There was a hammock in the yard and I asked about it, and it turned out that's where his missus had to sleep at night, because when you're on your period, you're not allowed inside the house. That's, y'know, grrr, kind of a deal-breaker for me, innit? I'd see parts of that aspect of Rasta in other people we'd meet—the woman would have to walk behind the man by some distance.

I thought you can't have those divisions and then rail against inequality in your music, because you're enforcing it on other people yourself. So I saw a lot of problems with Rasta. To my mind they've got to bring the women and children in on an equal level with them. Room for improvement! And dangerous bugger that I am, I had to tell 'em!

Then we'd go into these very precarious ghetto areas and meet people like Tapper Zukie. He'd be very proud to show you his gang and their guns, and waltz up and down the street waving them. I'm like, "Oh my God, what am I doing here, there could be a gun war at any second?" That's Jamaica, and these are very dangerous situations we got ourselves into. Open-minded foolishness really was the only passport in and out of them. Sometimes, if you're too aware of your surroundings, the good things might not happen for you, never mind the bad.

We went to quite a few sound systems. Typical Jamaica, though—these things don't start kicking up until about midnight, and you're so stoned by then, to the point of barely being able to speak or even stand. I'm sorry, but that pipe will knock you out. It's a hard thing to become accustomed to, and being the white boy in the house, it's shoved in your face straight away. There's your manhood challenge. I'd be mixing that up of course with Heinekens, and at that time beer was frowned on in reggae circles, but still it earns you brownie points—two things to put you flat on your back, both at once. Well done, John!

My favorite people were probably the Congos, a vocal group who recorded one of their greatest albums at Lee "Scratch" Perry's Black Ark studios. I loved them and their families, and just their generosity in life. It was inspiring to be honest and really frank with them. All judgment went out of the window, as soon as you'd sit and talk with these fellas, truly classic examples of passive resistance at its finest. Meant no harm to no one—superb, my kind of people.

We actually went to the Black Ark. I even tried to record with Lee Perry, but I couldn't get to grips with him. Too many distractions, too nervous, and too stoned. I was trying to fiddle about with the Pistols song "Submission," to do a Jamaica-inspired version of that. You're trying to lay down a vocal, and there's just too much going on. There was a guitarist there called Chinna, and he had one of them wah-wah mouthpieces you put on a guitar. That was his big toy, and it was so distracting. It was a sound I didn't like.

On a lot of Perry's mixes from that era, you hear this creaking noise in the background. I found out it was the studio door, the hinges were rusty, and it was just people coming in and out, which they seemed to be doing most of the time. His equipment—or lack of—was very, very primitive but achieving outstanding results. Admirable. And it was all one-take.

Perry went mad not long after I was there and burned down his studio. He'd supposedly had an argument over money with Island Records, and apparently painted over all the masters with green paint, so that they couldn't take them off him. Rather than having his music stolen, he destroyed it.

When I was there, the bigger the bong, the more active he became. Don't know how that works for them fellas. I still don't know to this day how you can smoke the weed at all without destroying your voice. It takes the top end right out of me, and all that flows out is a scratchy noise. I don't have the pipes for that.

To be very honest, it came as a surprise when I was asked to Perry's studio. I didn't get my chops together. I wasn't prepared for it. Ouch. Tail-between-the-legs time.

The funniest moments were probably with the bass and drum supremos, Sly & Robbie. They were fully affiliated with Island, so it wasn't really about work, it was just social. They really are like the Nile Rodgers of Jamaica. They ribbed Don no end. They ribbed me too. "C'mon mon, what you wearing dat big coat faaar?" By the end, I was down to flip-flops, shorts, and a T-shirt, because that's how it should've been right at the start. I was terribly self-aware of just how white and pale my skin was. I was looking like a cross between a concentration camp victim and Dracula.

In terms of the music I picked up, I got completely drawn into a whole area there called "Dread In The Arena," which was the big thing at the time, and all the offshoot records that were coming out using that theme. It was fantastic, with lots of Johnny Clarke dub versions. When Front Line eventually put out Johnny Clarke records they didn't have that aspect, and that's the very aspect I would have wanted them to have focused on—the pure dread of what dub was. It's all about tripping—*mind tripping.* Just free up yourself. You don't have to dance correctly or know any of the right moves, you do what you like. Just as long as you know how to enjoy yourself. Fantastic. That's what dub was.

I grew up with reggae music; it was always around me as a kid, so it was fantastic to really feel it as it should be felt, actually in Jamaica. I loved it there. Jamaica became part of me.

Gunter Grove became my backdrop to get myself properly sorted out, and get a new group off the ground. It was at one of my parties there that I met Gloria Knight, who was actually a writer for the *Sunday Mirror*, and was married to her editor. We kind of met through someone else in a roundabout social way, and it just seemed deeply hilarious to be hanging around with anyone that had any connection at all with what I viewed as the gossip rags. They'd been making my life a misery for the last year or so! It was like this: Oooh, this shouldn't work. But it did. Good came out of it, and bad came with it. She suggested that I needed a lawyer.

There was some weird stuff going on around us while we were in Jamaica. McLaren had got some of his people out there, sneaking around trying to film me for his idiotic movie *The Great Rock 'n' Roll Swindle*. Also, Branson had Devo, an electronic band from Ohio who were signed to Virgin, out there, or at least some of them. He may or may not have been trying to get me to be their singer. I certainly don't think he ever asked me. Sometimes inebriations lead to foolishness. I know we caught one of them spying on us. He'd come down onto our balcony, he was looking through the curtains, and Dennis frightened him off. That was really my only connection with them.

The idea of joining a band like Devo, replacing the lead singer, would be an absolute no-no to me. It would make me very angry. That was Malcolm's idea with the Pistols towards the end, replacing me. You can't replace the singer. It will always be the band for me—the singer. That's the direction and the persona and the energy of the thing, particularly with Devo.

I was constantly aware of people trying to codge off me, like parasites, using me to prop up the tuppence worth of talent or involvement that they'd never really had in the first place.

Most importantly, Malcolm was trying to claim my name, Johnny Rotten, as his property, and that had to be stopped. How can you try and steal someone's nickname? *Huh?* On what grounds? There's no doubts about it, I'll see you in court, boy. Which I wouldn't have bothered with previously. I would've just let it go, and moved on. But that kind of spite to try and fuck me out of my own life, my own name, my own career. Very wicked.

So Gloria put me in touch with Brian Carr, a solicitor who specialized in the entertainment business. Not long after that, she and I seriously fell apart, when her paper went and wrote a story that I was a recovering heroin addict. My God, talk about getting it all upside down. I could have revealed all sorts of her social errors but I'm not one for that kiss-and-tell kind of gossip.

But Brian Carr, the solicitor, was hilarious. When I first met him, he looked like Abraham Lincoln; he had that same beardy thing, black semi-wavy hair, and sapphire blue eyes. He was a weird-looking man. When he talked there was always spittle rolling out of his bottom lip onto his beard. So I found him to be "Oh God, he's so unlikeable, this could work!" And indeed, he was very good there for a few years; he's the chap who really got the proper barristers to sort out the case against the Pistols' estate and got everything set up the way I wanted it.

I didn't want to walk off with all the loot or anything. I made sure that when it came to a settlement, we surviving members would get equal shares of the spoils. Even though I was harboring a really serious resentment for the way Steve and Paul had behaved against me, I didn't want blood money or dirty money, as I would view it. I just wanted what was mine, what Malcolm tried to take away from me.

Funnily enough, as we settled in at Gunter Grove, we became aware that Peter Grant, Led Zeppelin's manager, lived across the road—and Brian Eno lived at the top, in a converted church. There was immediately a rumor that Peter Grant offered to manage me—again, one of those fabulous press rumors—although, after Malcolm, what a delicious rumor! I was feeding off it a little—you're always considering the options.

Peter Grant earned a wonderful reputation when Led Zeppelin first went to America as this full-on hard fella. No back-down from him! That's all well and fine, but I don't want bullies in my ranks. I don't want anyone to turn around and go, "Do this, or else. . . !" I'm never gonna have that, because Johnny's an alpha male. It couldn't've worked, and wouldn't've worked, ever.

The only way that I knew Peter Grant owned the house opposite was through something Dave Crowe said. We had a pet pussycat, which Dave named Satan. It was a young kitten that was abandoned, and Dave told me that it was "abandoned by that bastard, Led Zeppelin's manager from across the road. He kicked him out and I saved Satan."

He saved Satan all right, but Dave went back into his cubby hole, and the cat litter tray was up in my place, right next to the kitchen. So I had to be the one to clean up the cat litter, even though it was his pet. Poor little Satan, it was a tiny little thing, jet black, a mini cat—it must've had a growth deficiency. It looked kitten-like, even in adulthood, but it never really bonded with me. It would go down the staircase and meow outside Dave's hatch. By now, Dave had moved into the downstairs apartment, which was self-contained, and we never converted it fully, we just knocked through a small hole with a hatch, to get up and down from one part to the other. But Dave would never open the hatch, and that poor cat was abandoned.

Meanwhile, I was trying to put together my new group. Jah Wobble was still one of my best mates, and he'd often been picking up Sid's bass to have a go—probably more than Sid did. Wobble was very much still a novice with it, but that's not what matters: I wanted him in.

In subsequent years, Wobble's been making out that what we were trying to do was some kind of dub band—and this is from a bass player who was barely learning at the time. He's looking through seriously warped lenses—maybe he's trying to hoodwink his way into the current crop of "whitey does reggae" bands.

That certainly wasn't our angle at the time. At least, it wasn't mine. I wanted something completely new and refreshing from what very quickly became the format of the Sex Pistols. We were only together for a brief time but it did become a format—a format of writing which bored the pants off me.

Being open-minded to all kinds of music was Lesson One in punk, but that didn't seem to be understood by many of the alleged punk bands that followed on after, who seemed to be waving this idea of a punk manifesto. I'm sorry, but I never did this for the narrow-minded. I was horrified by the cliché that punk was turning itself into.

I didn't—and still don't—have too many punk records in my collection, because I never really liked them. Buzzcocks, Magazine, X-Ray Spex, the Adverts, the Raincoats—those, I liked. They were skirmishing on the outside of it rather than the typical slam-dunk bands that drove me nuts, because they all sounded the same, all chasing the same carthorse. I'm not impressed by macho bullshit bravado. It doesn't have any content and it's not actually aimed at anything other than trying to show off your masculinity. Failed!

You had all these males-only bands trying to out-threaten each other. To me that's the lowest common denominator. There were so many of them all doing the exact same thing, all of them completely stupid, not understanding Rule Number One: there are no rules. And yet this lot rigidly adhered to rules and regulations. They became the new Boo Nazis.

For my part in it all, I'd opened up an entire new different genre and way of viewing music, and what happened when the door was opened? In walked all the flotsam and jetsam, who were very proud of being stupid.

I was coming from the standpoint of sharing my life's experiences, not to go into isolation, as punk was doing to itself. It narrowed its outlook—for me, propagated by poor old Joe Strummer. In his mind, he was leading this political punk thing, with a vision of us all standing there like Solidarność, waving banners, and that's a load of bollocks. If you're not doing this for the poor old biddy that lives next door and can't afford the heating in the winter, then you don't count at all. Studded leather jackets for everyone is not a creed I can endorse.

So my mindset was "I can't take no more of that." I was envisaging a consortium of like-minded loonies prepared to jump into the next universe without any tools, and find our way that way, and be an exciting possibility. And indeed it was, because we were not playing to any set standard of musical clichés. In that way, the band can't be revisioned in hindsight into any one person's schemes. We were schemeless. It was a free-form adventure and things like musical inadequacies didn't matter at all. Not to me. I had a stomping ground I could have stuck very safely to and just done Johnny Rottenisms, and that no doubt would have worked, but I wasn't interested in it. Sorry, but I'm a big risk-taker, me.

I'm the one that rang them up. I'm the one that asked them in. No one came to me with any ideas in that way. They were all skirting around unemployed, and so—bingo!

With no minor difficulty, I tracked down Keith Levene, who I'd known from hanging around with Sid at the Hampstead squat, back when amphetamine was the new buzz of the day. He'd since been in the Clash, at the very beginning. I knew he'd worked hard inside that band, but I also knew he didn't fit in with them. Their manifesto was too limiting. He'd come backstage to a Pistols gig one time, and told me how deeply unhappy he was working with them. His attitude was "Look, I do all the work, I write all the songs, I get no respect. That awful rubbish, listen to them. *Aaargh!*"

So I kept that in mind. Whenever you ran into Keith, he never had a good word to say about anybody. That thrilled me no end—I'd never known such a professional misery. When you're young, you can find that entertaining about someone. But once you get into your twenties, it's not so entertaining, because he hasn't learned from it. I look at myself in all of this too. I used to love the word "dismal." "What do you think of that?" "It's dismal! I'm bored!" I don't think I ever meant any of that, I was merely perfecting the art of dissatisfied youth.

After the Clash, Keith had been in a band I'd put together called the Flowers Of Romance. It was a good collection of people, just mates floating in and out, having fun, and so I gave them the name. I liked Marco Pirroni from Adam & The Ants anyway, I knew him from hanging around, and Chrissie Hynde, and it was a good idea for them to be forming a collective, and see if that went anywhere. Keith and Sid and Viv Albertine all passed through, and how it ended up is utterly beyond me. There was another name, the Moors Murderers, which may or may not have been a different band. It was a vague, unimportant thing, but showed good faith in breaking the tethers of punk cliché. And that was definitely where Keith was coming from.

Keith is acerbic. Basically, he's a bottle of vinegar, so you chuck that into the crisp bag and you're gonna get all kinds of flavors. Keith's musical background is interesting. He has flirtations of Wishbone Ash lurking in his previous history; he learned to play from that kind of music. He told us that he'd had guitar lessons from Steve Howe of Yes. I would hate to find out that was all a fabulous lie. It kind of made sense: Keith had a different insight there, to what was currently going around. He was outside of punk clichés.

Once we got together, Keith's playing astounded me. The idea was around back then that after Jimi Hendrix no one could ever play the guitar again. There was no point, the instrument was finished. But to my mind Mr. Levene's playing absolutely proved that to be not true. I thought it was very creative and very different, kind of discordant but at the same time always resolved itself musically. Very trancelike. It wouldn't skip a beat, but it would soar off in many different directions without ever losing its focus. I found that intensely riveting and very, very inspiring. A sense of, he played like a rhythm guitarist, but he took the rhythms to such extremes.

So the landscape was a lot broader than some people's little nail varnish appliers would lead you to believe. There was room for expansion, in an incredible way. All we needed was a drummer. We auditioned a bunch of them, but Jim Walker was the best by a mile. He'd come over from Canada, just to get himself into a punk band—well, my God, he picked the best one in the world, didn't he? Coming from abroad, he was an unknown quantity, but he absolutely stunned us all. I thought, "Wow, the inflections are really grabbing me and exciting me. Cor, they've got my corpuscles bouncing!" Disco, African, a bit of everything really—almost a Ginger Baker kind of approach.

Jim had a very open mind and he didn't come at you like a muso. He was excited by the craziness of it, and indeed as it came to pass, he was way bonkers himself. Way out there! He didn't have anywhere to stay, so I gave him a room in the basement at Gunter Grove, and gave him money to get furniture, and he spent it all on a moose head. When I saw his room when I finally went into it, there was nothing in it but newspaper on the floor, and the moose head on the wall. He had no interest in comforts of any kind. I don't know how he slept in it or what he *did* in there.

The PiL house—which is what Gunter Grove had become, because Keith moved in an' all—was very much centered around what's on TV and what's on the record player. Jim said he didn't need to be upstairs with us—his room was in the downstairs part—because he could hear the bass rumbling through the floorboards. So he'd be down there in the dark. Very odd. As I keep saying, I'm attracted to oddities in life. Him buggering off to London on his own from Canada reminds me of coming out of hospital and having to blend back in at school. I appreciated his sense of adventure.

We had no real concept in advance of how we wanted to sound, other than "We're going to do something different here," because none of us wanted to imitate our pasts, and would've been uncomfortable doing that. The sound really formulated itself from the first rehearsals.

Very early on, we wrote the song "Public Image," which was the freest moment, like escaping the trap of the Pistols at a stroke. The writing and envisioning happened down by London Bridge, in a rehearsal room south of the river. Wobble was getting the lilt of the bassline, Walker was just exceptional, cracking away at the groove, and Keith was really bang on it and really enjoying what we were doing. We were formulating a different approach and doing it quite naturally and things just fitted so well together, and the words just flowed.

I was so proud of each member's contribution, and they gave me a great space in there to shapeshift my voice, to try something different and go with the sentiment of what it was we were all trying to put together here. I wanted to declare where we stood in the world, and "Don't be judging me by the publicity machine and all that nonsense that I had to put up with in the Pistols." I was about taking a completely different step aside from that, and I knew there would be consequences. I knew there would be resentment from the alleged punk world because I wasn't staying inside the confinements of the box. But that's their fault, not mine. Punk to me doesn't accept them kind of authoritarian approaches.

This bit was important: "I'm not the same as when I began/And I will not be treated as property." It's just saying, "Who are you telling me what is and what isn't? You can either pay attention or you can get stuck in that hole in the ground that you've all buried yourselves into. Well, pull the soil over the top. Goodbye."

It was a great, great song and, just in case any member wasn't aware of it, "Public Image" belongs to me.

Only joking! I wasn't referring directly to the band. It's about: Johnny Rotten, that's me, don't try and take him away from me, and don't rewrite his history.

It was very cheeky of me to begin with "Public Image," the name of the band. I'd taken it from a beautiful book called *The Public Image* by Muriel Spark—her wot wrote *The Prime of Miss Jean Brodie*. A very small book, but it's a great storyline, about how the publicity machine turns an average actress into a monstrous diva and she wrecks everyone around her. I didn't want that happening with me or my imagery.

In Public Image Ltd (aka PiL) I wanted to keep the Johnny Rotten side of things well out of it. I'd moved fully and very comfortably into the persona of John Lydon, who didn't need scandal to flog a record. It wasn't so much that we would earn our rights wherever we were in terms of sales, but by the quality and the content.

The "Limited" part of the name was about limiting our public image, to not allow the scandal rags access to us. To keep our private lives private, thank you. Keep a definite distance away from a scandal-mongering publicity machine. Which would've been what Malcolm was treasuring and I found to be detrimental. It's bad for your health that angle, it really is.

The double entendre was very deliberate, though, about setting up as a limited company, which I did with Brian Carr's help. We wanted to be completely free of all attachments and dictations, running ourselves. The idea was to try and break into all areas where you could possibly earn money, but to offer quality goods. And to break away from fear of the corporate, to have our own version of what we wanted the idea of "corporate" to be. A cooperative, meaning all working in it, all together, all doing different things but all for the ultimate good of everyone—a kingdom without a king, a republic without a president, based logically on a premise that common sense would prevail.

The idea was to broaden out and work with creative people in other areas. The first move in that direction was calling in Dennis Morris, who was involved in artwork as well as photography, to collaborate with us on coming up with a PiL company logo. I took my inspiration from the chemical company ICI's logo, but we made it look like an aspirin pill, funnily enough. It was something I remember from my childhood, passing the ICI building on Millbank in London, which had a big round logo on the front. I was always impressed by that, the power behind that cold corporate imagery. It would be a novel approach for freedom fighters.

Virgin had first option on signing us to release music, due to my previous contract with the Pistols. They took up on that and at the time it was, "Oh, thank God!" Going out and hunting down a new label, I didn't have the energy for that. It would have set everything back a couple of years, because you were going to have to tour an awful lot before you got your reputation back, in order to secure a proper contract. And because my direction was so changing from the Pistols, that would've been an uphill battle all the way. Any offers that were out there would've been based on us delivering *Never Mind The Bollocks, Part Two*.

I only found out after signing the PiL agreement that that's what Virgin really wanted too. That might be what they wanted but I gave them what I thought they *needed*, and they weren't shy of a few hits because of it.

We signed an eight-album deal, and got an advance of seventy-five grand. There were mounting legal costs from the court case with *that* lot, and it brought with it all manner of accounting and tax bills outstanding, so financially I was heading for a really bad place as a future in all of this. But we prevailed. It was a matter of scraping by. We just got on with it and kept it as cheap and cheerful as possible. In setting up the company with Brian Carr, I insisted that everybody get a weekly wage. I just thought that was the right thing to do. Most bands don't do that because it's a real economic burden. I thought that would keep us together better. You're working, and you've got money in your pocket. I don't think you should have to deal with what happened with the Pistols—trying to barely make it on twenty quid a gig.

Before we went any further, I had to go to America to tie up a PiL deal with my new mate Bob Regehr at Warner Brothers. I asked if I could bring someone with me, and they were kind of surprised when I turned up with my mum. She'd always wanted to go to America, and she'd been very ill—I'm like that, when anybody's ill I'm Doctor John. It's my way—I grew up that way because of the necessity of having to look after my brothers.

She needed a break from the pressure of having been diagnosed with stomach cancer. At the time the doctors were plying her with all sorts of awful downer medicine. It was making her walk into walls and things. They meant well, to take her mind off the situation, but it just made her lose the plot, so I took her away from that and she had a great time.

I borrowed the money to take her on to Canada, and meet her sister in Toronto. So there was a bit of a family thing going on there, and all very important. I gained so much from it too. I needed to find my roots again before I burned myself out fizzling off into the wonderful world of ego and vacuous pop star celebrity, or infamy, or whatever you want to call it. You can get lost in it and you can get out of touch with who you really are and what you're doing all this for. The glare of a flashbulb, it's very much like a deer-in-the-headlights effect. When the press surrounds you it's easy to fall in love with that moment and think of yourself as more important and relevant than you actually are.

It was very necessary—I didn't realize how much so, and for both of us.

• • •

Levene and Wobble, from day one, were at war with each other. Even at the very first rehearsal I was dealing with Keith's contempt for Wobble not knowing how to play. I'd be backing Wobble on that. "Well, he'll find out soon enough, won't he? We're all learners here; no room for another Glen looking down his nose at everybody. Stop that!"

Keith's a very spiteful person, and very difficult to understand—or indeed, eventually, to tolerate for too long. Gosh, I must be such a nice person because I managed to. There were situations there where Wobble just wanted to kill him. Just kill him. Murder him. Tear him apart.

Keith was very bright and constantly setting himself challenges, which always impressed me. He was always fiddling around with some box of tricks to try and advance himself and be useful. What I didn't realize was that he hadn't grown up at all, since I'd first met him. He would express himself in cowardly snipy ways: backbite-y, under-the-breath comments, the sourpuss face, the act of deliberately making himself look uncomfortable in a nice friendly environment. He'd come up and sit in the middle of the room and try to make everyone feel unpleasant. That kind of childishness, looking for attention. What a calamity, because all he got was laughter and ridicule.

Between me, you, and everybody around at the time, what the boy was really doing was heroin. He was poisoned by chemicals, or a chemical imbalance in the brain, but it made that rat-arsed, snarly, contemptuous cunt unbearable to put up with. It was otherwise inexplicable that he would behave from time to time so completely like a spoilt child.

He was living in the back half of the downstairs part of Gunter Grove, in his subversive ratty way—and all for nothing, no rent, and there was no "thank you" for that either. He wasn't in control of himself but that didn't mean we had to put up with that and endure that. It was piss-poor ugly. Wobble had many points to make. Once he twigged what Keith's problem was, he wouldn't let him alone. There was no saint in any of this.

I'd have to be constantly in between that situation, to the point where I missed an awkwardness between Jim Walker and Wobble, which was an equally wrong situation developing there—Jim Walker feeling bullied. It wasn't working out quite how I felt it was supposed to be, where everybody was appreciating each other.

So it was out of the frying pan, and *totally* into the fire. I'd picked all these people, and kind of left it leaderless. I thought, "People will work things out." Unfortunately, it doesn't go like that in real life. At some point, you have to stand up and say, "Stop that!" Then the lines are drawn and a parting of the ways is usually the end result. It was the same as when I was young, dealing with the daycare kids, or my little brothers.

Initially, the anger issues fed well into the songs. The creativity was thriving off those personal vendettas and animosities—much like my first lot! But that isn't how I wanted it to be. You can't change it. When people really, really resent each other it's just got to stop. It's just a matter of timing. And all the way through this, the record company was consistently on my back, saying, "You should get rid of Blah-Blah . . . Blah-Blah isn't good for you . . . you should work with . . ." "No, I want to stick with my friends and see if we can work this through, and I don't want no record company telling me to just get rid of people willy-nilly, not allowing me to find a sense of loyalty in the home turf." Ouch! Punishing, hard times.

In August, before the release of "Public Image" as a single, we were due to make our first TV appearance on a new Saturday evening music show on ITV called *Revolver*. Keith had decided to travel up to the studios in Birmingham independently, with his new friend Jeannette Lee. Jeannette had just broken up with Don Letts—they used to run Acme Attractions together, a shop on King's Road, not far from Malcolm and Vivienne's.

Apparently, they "didn't want to be with the crowd," so off they went, and I got a bit angry about that, because they didn't even say, "See you later," or anything like that. Somebody shouted, "Let's go to Camber Sands instead!" So that's what we did. We didn't turn up at this show at all, just went off and had a cold blustery day by the sea in Camber Sands. I'd never seen the place before, so it was kind of thrilling. We had a real laugh and a hoot running around the sand dunes, being nutters.

For my mind, someone needed to be taught a lesson there about good manners. Separating themselves from the band in that way—well, here we are, we're the band, so fuck you! You want to travel separate, we're going somewhere else. It turned into one of the best drinking parties I've ever been to, with myself, Wobble, Jim Walker, and my two mates John Stevens and Youngie. The pressure was gone of sitting there worrying about doing a silly TV pop show. Every now and again, you've got to say that what really matters is the well-being of not only yourself but the proper people around you, and at that time that was Wobble. He was my mate, and he took offense to this an' all, and he had every right to. With Keith's neglect of us, it was all "his show," he wanted nothing to do with us. How's that ever going to work out on live TV?

Frankly, if you're walking into an imminent disaster, best knock it on the head. I'm possibly making excuses. But I'm not, I'm telling it as it really is. And I'm very happy to know, Keith came back finally, with his tail between his legs. When you behave like a spoilt brat with us, that's how we're going to treat you. We're not gonna tell you we're not gonna turn up, we're just not gonna turn up! Rather than resulting in fisticuffs, which I'll always stop, this was the better behavioral plan—a passive resistance to what was fast becoming a serious problem in PiL, which was Keith's ego.

When the single "Public Image" came out in October, it opened up a lot of people's minds about my capabilities. I wasn't just a one-hit saucy boy wonder, but then for some mysterious reason Virgin decided to hold back our album. What they really wanted, of course, was just the Sex Pistols Mark Two, and that was never gonna happen. They said they didn't know if there was a market or a niche for it out there. For a while there were actually leaked bootleg cassettes of the album doing the rounds at various record shops, and that forced their hand to release it as soon as possible.

In all seriousness, I deny all responsibility for the bootlegs. It really jeopardized our initial kickstart, because they ended up releasing the album a week or two before Christmas—the best place to lose a completely different approach to music. At that time of year, people want happy holly songs and greatest-hits compilations. Anything else gets buried.

The first album cover is a satire of all those "serious" magazines, like *Time,* which filled the racks at the newsagent's, faces glowering out at you. The album itself is to me sensationally acidic. Cutting and biting. I suppose it was the anger of having to deal with Pistols fallout, and the court case, and what was already going on in PiL. They were abrasive songs, because they were *scathing* and necessary in their production. Stuff like "Theme," with Keith swishing razorblades all through it—fantastic!

I'd become incapable of caring about writing a song for the Pistols and felt like, I've achieved as much as I can in this environment, but it didn't stop me writing songs. It wasn't "That's it, the genius has expired!" I don't analyze what I do in that precautionary way. There's no brakes on my engine, it's just full-on until the day I die.

"Religion" was the only one that predated PiL—the one I'd tried to work up with Steve and Paul in America. It's very much the last time I ever took a Sex Pistols kind of approach in a song. It met with a negative there, so I moved it into PiL and shapeshifted it into a far more enterprising piece of work. I'd experienced the Catholic Church from an early age, so I knew where I was coming from on this.

In PiL we could separate the music from the voice, and we did it with "Religion," between the left and right speakers, so that you had the option of both together, or one at a time. There's all that echo on my voice, so it sounds like I'm sermonizing. Well, isn't that what they do to us? They preach *at* us. Yes, it's theatrical, but those are the tools used against a congregation. Sometimes you just turn the gun around. Point the cannon in the other direction and see how they enjoy that. Hellfire and brimstone.

Jesus Christ and me, we know what these priests would be yelling and screaming at us from the pulpit when we were young, basically telling us we were filthy no-good heathens and we were all going to die and rot in hell. And that's just about where most of them sermons ended. Then a few hymns sung out of tune and that's it. That was a waste of a Sunday.

All demons come from religion, and the Catholic religion is the most serious demon of them all. "Annalisa" is about a young girl who died when her parents allowed the Catholic Church to perform an exorcism on her in a small town in Germany. That poor girl was fucked by the stupid environment she was in—a small town full of small-minded people—and I believe her problems really were that she was coming of age as a teenager and she was coming to grips with all the things that teenagers go through—that sense of individuality and rebellion and indeed sexuality. Her overtly Catholic parents were trying to stifle that and, after trying to punish her in countless ways, they starved her to death. "Purge the demons."

The harbinger of death was their religion. If you're force-fed a diet of this kind of stuff, you will convince yourself that someone is possessed. Rather than deal with reality, people prefer to believe in gods, ghosts, and demons. I'm so antireligion. It's foolishness and a bluff and a scam and a con, and it leads to great tragedy. I can't see any good at all in it. They've got to stop pulling the wool over people's eyes. Those in power like that wool. It stops you thinking. It keeps you in a state of permanent mindless acceptance. That's the opposite of me, opposite to my nature.

It's gut-wrenching and absolutely heartbreaking, the subject matters I started going into in these songs, but it's all worthy of study and investment actually because it ultimately makes me a better person, being able to see things from other people's point of view, while it being also, at the same time, my own story.

People seem to think a lot of the rest of the album was just venting spleen about the Pistols' demise. Not so, although I can see how it could be conceived that way. "Low Life" was finding something inside myself, a possibility that I didn't like, a lust for attention, and I wanted to remove that permanently. It's almost scream therapy, although I didn't know that at the time, I just thought of it as self-analysis. An argument with myself, just looking for what's right.

"Attack," on the other hand—well, that's Malcolm. "You who guarded all the loot . . . You who buried me alive"—yep, I think that's very clear. There's no stone unturned in that one.

We were dipping into all manner of studios to get the album done, always late at night, when the hire rates were the cheapest. There was one place in a basement in Chinatown called Gooseberry, which a lot of the soul, reggae, and pub rock bands used, because it was cheap and cheerful and very dirty. You crawled down the staircase in this basement and there it was, the same old filthy brown carpet and stink of stale beer. A perfect place to just get on with it. You weren't there for the décor. That's where we did "Fodderstompf"—the annoying one that goes, "We only wanted to be loved." We made that up on the spot there. We realized when we got into record mode that we didn't have as many songs as we thought we had, and anything would do. Anybody who had an idea would spur everybody else on.

With the album coming out right before Christmas, we had very little scope to promote the album live, to try and rescue it from its hopeless scheduling. There were a couple of gigs in Europe, but when we were looking for a booking in London, I wanted to play the Rainbow in Finsbury Park so bad—home turf. The promoter said, "Well, there's only two days available, and that's Christmas Day and Boxing Day, and as you know you can't play then." "*Is that right?*" Those were the magic words; that was the genie out of the bottle.

Hard as it is to believe, you couldn't buy anything at Christmas, or on a Sunday, back then. You worked during the week, you get home late—Saturday, if you went to a football game, it meant you wouldn't get anything in—so you had no opportunity to buy anything or do anything or go anywhere or find any other alternative. It was an entrapment. Sunday was the Lord's Day and it was a sin for people to sell you anything. And Christmas—doubly sinful! What? Says who?

I'd made it very clear, hadn't I, through the Pistols, that I weren't gonna be made to follow rules that I thought were written by fools, for fools. You can be as religious as you like—no one is forcing you to buy a Crunchie bar on a Sunday, but don't tell me I don't have the right to do so. So we opened up the agenda, socially, for "Why can't shops be open on Christmas Day? Why are we restricted to medieval law? Or religious doctrine?"

A lot of the sweet shops and off-licenses were, and still are, run by immigrants whose religion isn't necessarily Christianity, and they showed us the way and stood up to archaic laws. It was good for them and it was good for us. Some people would go racist with it, but I'd go, "No, I want their religion please, I'll have some of that—just long enough to get myself a Twix."

So, we managed to challenge all the laws of the time that you couldn't perform or work on Christmas Day and Boxing Day. And it changed everything in England. The Lord did not come down on us.

Our debut gig, however, was in Brussels—a riot, apparently, after Wobble kicked a security guy in the head. Of course, it was the bouncers getting out of control. Everything's fine if they just stay away but they tend to try to show off and use their physical stature for standing in front of the band and thinking that makes them look good. They're wrecking the gig; this isn't the "bouncer show." If it was, go sell your own tickets.

I've always hated bullies. Anyone who comes on like that and tries to tell us what to do, it's gonna go down very badly for them. So, in that case, not quite remembering all the details, I'll back Wobble to the hilt. It's our stage, and if you're not meant to be there, suffer the consequences.

The following night in Paris, Wobble was felled by a pig's head thrown from the crowd. Pigs' heads are fairly cheap in that part of the world, aren't they? The French swear by roasted pig's head. My only complaint was, where was the apple in its mouth?

From the off, it's quite amazing what we had to put up with. All the way, it was resentment: one, for being so different; two, for not being the Pistols; and three, wanting to attack the Pistols thing—all in the wrong place and time, and not really understanding Mr. Rotten at all. That's the way it was turning, punk was turning into that voice of ignorance. The true content and message was strip-mined right out of it, more's the pity.

Come Christmas Day and Boxing Day at the Rainbow, we filled the place, but there was a real problem on Boxing Day with some West Ham football oiks who came down and tried to disrupt the affair. "Hello, this is Finsbury Park, this is solid Arsenal, mate! If you're looking for it, you're gonna get it!"

It was the first night a pre-Rambo John Stevens ever officially ran the security for us, along with my brother Jimmy. You know, you use the local lads. We all kind of knew there'd be some firm coming down to try and wreck it. But more fool them for not appreciating what it is we were getting together here. But at least they had a day out. Even the enemies can't complain—they were entertained!

So that went off, and that's not anything actually to do with the PiL gig, but by the time that reaches the newspaper headline area, it becomes like a PiL riot.

We went onstage very late. "Sorry, it's Christmas, fuck you!" We'd brought in—and this is why it took so long to get ready—a set of subwoofers that actually played bass so deep that you didn't hear it, you *felt* it, and if you stood too close to that, the prospect of shitting your pants was highly likely. Subsonic woofers! Fantastic! The stage was humming. You could feel it through every part of your body. It was brilliant, beautiful, where sound became really threatening. And it wasn't the sound, it was the vibrations.

It was all illegal, we found out later, because it can make you physically ill. It was creating such a feedback onstage that we were feeling ill before we even started. It went, *Bvvvvvvv*! It was feeding back and looping off the bass strings leaned up against the amp. Wobble didn't have the sense to turn the thing off. Or he had the sense to leave the thing on, one of the two. I suppose there was an element of Can running through my skull at that time, when the bass tones from their PA made the stage collapse at the Roundhouse. In a band like PiL, you have to find out what the consequences of extreme music are.

In the back of the venue, all through the dressing room area, the windowpanes were rattling, window frames were coming loose. So there it was, we had a health and safety problem before we even began our first number! Oh my God, there's gonna be death in the house! Great! And on we trot, and it was a very insane gig. Above all, the Pistols fans couldn't understand a damn thing, because they'd never heard a noise like this before. So it was very much like the early Pistols all over again: "What is that?!"

Wobble was new to playing the bass. The concept of turning it off when you go off to wait for the encore didn't occur to him. The feedback was ferocious. The rattling was severe. If we'd gone on another hour, the building would've fallen down around us. There were complaints, but you'd have to be a sourpuss to do that. What else were you going to do? Watch *Ken Dodd's Christmas Special*?

7

OPENING PANDORA'S BOX WITH A HAMMER AND CHISEL

When Sid Vicious got arrested for murdering Nancy Spungen in the Chelsea Hotel in New York, Malcolm was just panic-panic-panic. It was the tail end of 1978, and the whole sordid tale was splurged over the tabloids' front pages. Of all people, I heard it was actually Mick Jagger who intervened, and put lawyers in to try to help Sid out. That was a huge "wow" for me, because me and Sid, we weren't the best of mates at this particular point—we had problems, mainly about his drug nonsense.

One of the last times I heard from Sid, he turned up with Nancy late at night at Gunter Grove. He wanted money so he could buy drugs. When we wouldn't open up, he thought he could bang the door off its hinges. Well, I'm sorry, but we had the police to do that for us, and they were very good at it. Eventually, Paul Young ran down the stairs and chased Sid and Nancy off with an ax—he wasn't taking any chances because he knew that Sid always carried a knife. The only reason there was an ax on hand, by the way, was because Paul was a carpenter.

I don't like heroin addicts, but I do like my friends. I wanted to do the best I could for Sid, but Malcolm kept me away from him. I only found out through my lawyer, Brian Carr, that, "Yes, Mick

Jagger got his lawyers involved in it, and they're looking to protect him." So, I asked, "What has Malcolm done?" "Nothing." I don't think Malcolm lifted a finger. He just didn't know what to do.

Sid got himself into a terrible situation, revolving around owing money to some serious drug dealers. Now I know this one thing in life: heroin dealers cannot afford to fuck about, and if you fuck them about, they—will—fuck—you—*out.*

That's what Nancy Spungen introduced him into when they went to New York. So Sid's idea of a cool and trendy lifestyle soon became a depressing problem of "Where's the next fix?" And that's how my friend, that silly boy from Hackney, ended up lost and confused in a strange land. You've really got to know what you're doing in this world, at all times, and you've *really* got to know what drug dealers expect from you.

Listen, peoples, do not ever get yourself in the position of owing money to those kind of dealers, because there's a lot of reputation on the line for them. Sidney should've known that. I know who was out for him, and if you're talking strictly in terms of moral, principle, right, and value, I had no right to stop them, because he'd overstepped the line. That's an absolute New York fact.

New York was harder than most towns at that particular time. A lot of it was Mafia-run, no two ways about it. Don't be telling those guys, "Oh, fuck off, I've run out of money. Huh huh." Then there's gonna be a big sweep-up, there's gonna be a settlement.

The basic truth is, Sid was not "street smart." You have to survive with "street smart," and you cannot survive *without* street smart. But do not ever get yourself in the position of owing large amounts of money for something so self-indulgent, because you have to be dealt with at that point, because that ridicules the crew supplying the deal, and they're never gonna accept that.

When he was first arrested, it was so tragic and sad: Sid's defense was "I don't know what happened"—with a knife stuck in Nancy. Oh, come on. Oh, for Jesus Christ's sake. Figure it. Seriously. Over your head. Warned you. Sidney wasn't a smart fella. I warned Sidney time and time again. Don't get in over your head, not with nothing. Play to your level.

Nancy was killed, and that poor foolish boy was left holding the knife, not knowing what's going on. To me there's no mystery in it at all. You owe money, that's what you're gonna get. And there ain't no police going to hunt it down any different.

The boy's life was over, and there he was in Rikers Island jail, in New York, with really not much option. As soon as he got out on bail—bang!—he banged up another number in the vein, and goodbye. He comes out, meets his mum, and dies of an OD—he allegedly committed suicide on an overdose of his mother's own making. Fucking fantastic, huh? What a great lifestyle. Don't look for no mystery in it. This is what you get because this is what you want. Get the PiL song now?

Sid's passing was a pain in my life—a serious one. I wrote songs about him for quite some time after that. They're all in there somewhere. He just couldn't see the wood for the trees. Yet again it goes back to education. Education is not necessarily what the schools teach you, it's about acquiring this way of having an insight, and being able to gather information correctly. And Sidney lacked that potential. I always felt like I was protecting Sid, always. Wherever I'd take him, I knew—aaargh!—he's gonna cause a problem. But that's all right!

Once he stepped out of line, and was left to his own devices, oh my God, it became stupid, stupid, *stupid*, no value, principle, system, or logicality in it. Don't mess around with drug dealers, right? They mean it. They have to. Just say no.

While all this was going on, just the other side of Christmas, there were PiL crises coming to a head at Gunter Grove, between Wobble and Jim Walker. I never understood what the rows were about—well, I do, sort of—but there was some kind of bullying going on. I suppose Wobble was feeling inadequate about his lack of playing ability, and therefore somebody else had to suffer.

Jim quit. He had very quickly seen enough, and suddenly, after less than a year, we were without a drummer. I thought that Jim would go on to do other great things in music. But no, he went off to Israel to work on a kibbutz, and madnesses like that. He's not actually Jewish, so it was an even bigger move. I believe he does things in film these days.

We tried out a number of replacements, but none of them seemed to stick. One was pure disco, another pure reggae, and neither could adjust to anything outside of that particular format. Another was just not getting the vibe that was there between us.

For a while there, myself and Rambo were jokingly telling people *he* was going to be our new drummer. Rambo went along with it and it might've worked, but he'd have to have learned how to play within a month. That was way too much pressure for any human being. I'm glad, and I think he is too, that it didn't happen, because we found a way to work together that was much more beneficial to both of us, further down the line.

So, it was just a case of whoever responded to the advert. *Exchange & Mart* was our favorite read of the time. If you tried through the music papers, it would always be the wrong kind of arses. They'd turn up with some silly imagery, rather than content from them as human beings.

We messed around with Richard Dudanski for a bit. He'd been in Joe Strummer's 101'ers, but he wasn't really up to it. He was too soft and gentle to really cope with our lack of fear. Poor old Dudanski was a bit of a hippie, but then he wasn't, because he was balding. Hair loss removed his hippiedom.

Most of them felt ill at ease. They'd be noting the tensions going on between Wobble and Keith, me and Keith, Wobble and me, between all three of us at once—a very hard thing suddenly to be in the middle of. I do understand their position: Sid must have felt much the same when he joined the Pistols. You're walking into the lions' den, and the lions all know each other. Ouch! Heavy, heavy judgmental scenes!

Oddly enough, even though it was down to just the three of us, without a steady drummer, it became more settled and more confident, now that we'd got the first album out. We'd got into a pattern of recording in bits and pieces, all over the shop. We were never in any one place long enough. We'd be a week here, a day there.

There'd be a lot of nighttime sessions at the Town House in Goldhawk Road, Shepherd's Bush, and those were always very last-minute. The Jam were using it a lot at the time. When they'd finish in the evening, we'd get the tip-off and could go over and use the facilities, so long as we didn't touch the mixing board. The Jam themselves were certainly never in on it. It was a cut-price universe.

So, everything we did was what they call "monitor mixes," rather than going through a big desk. Once you get into the bigger technology there's all manner of mollycoddling of sound going on, where the brittle edge is stripped and impossible to replace. That's what gave those early PiL records their thrilling sound—they've got that raw energy of a band playing live in the room.

The big treat for us was when Virgin would pack us off to the Manor, their residential studio in Oxfordshire. This place was a different universe again, and proper palatial by our standards. They'd book you in for a few days, and you'd be the only band in there. The whole joy of it was that it was "Do what you want." There were twelve bedrooms, so there'd be an entourage included—bring yer mates! There'd be a pile of people speeding in the living room, and it was endless food and endless drink. They changed that later, but in them early days, everything was on an open checkbook—they hadn't got around to tightening the purse strings.

There were fireplaces everywhere, so the best time to book it was always in the colder months for that good old roaring-fire vibe. The dinners were enormous roast things, traditional English. It wasn't exactly a boar on a spit, but it was that kind of presentation. Roast potatoes and a proper roast beef done traditional—semi-raw in the middle—was thrilling to me. I gained so much weight.

They had satellite TV, which nobody had at that stage in Britain. All of us would be thinking, "Great, we can sit in that room and just snuggle up round the fireplace and watch endless TV!" But—*grrr!*—it seemed to be the same channels repeated ad infinitum in Italian and Spanish! And channels with nothing but adverts, and that was it.

But I loved the Manor. It was an absurd, ridiculous "harking back to previous centuries" kind of place. I always felt like, "Hey, here's my chance to be lord of something," and I'm damn well sure that was in everyone else's mind at all times.

Mostly, we'd work the night shift. I'm sorry, I can't be thinking of running into a studio at 10 a.m. My brain starts to kick into its agenda at around 8 p.m. By 10 p.m., I'd be fully focused, and everyone else probably wanted to go to bed. Still, whenever we got these opportunities, the eagerness to get in and record, and the enjoyment and the thrill of operating machinery and pressing buttons and screaming and playing things, was absolutely the driving force of it. That wasn't lost, regardless of all the surrounding ugliness. That was the thrill, the joy, the point and purpose of being in a band, and you can't take that away.

We knew the stuff we started coming out with would annoy the record company, but my belief was I've only got one opportunity in life to do and say what it is I truly feel, and I'm not prepared to back off from that, and I'm prepared to suffer the financial and business consequences, because I think in the long run the work will prevail. It would've been semi-useless to be going out there to write a plausibly huge commercial success, which is what the label would've wanted and enjoyed. I'm Johnny fucking Rotten, you know, that's still me, whatever Malcolm thinks, and *I do what I want.*

For me the success was to do something completely unexpected, and yet a natural progression. The album that became *Metal Box* was not contrived. Contrived would've been to have written an instant hit. Somehow, with all the conflicting lifestyles and personal situations developing, and the pressure from Virgin, we managed to make a really cohesive album. It sounds like it might as well have been recorded all at once, from start to finish. It's a stunning beautiful tapestry of high anxiety.

The idea of it was, it would numb you, absolutely flatten your resistance, just wear you out with its omnipresence. I think we got there.

I was fiddling about a great amount of the time on a Yamaha keyboard, which I just loved. It was a cheap nasty thing, but it was one of them earlier ones that had swirly and swishy orchestra sounds, if you twiddled with some of the knobs. I just loved that. In fact I used to play that thing to the point where my wrists would develop these cysts, these lumps, which were very worrying at the time—I didn't know what it was. You know when you keep doing a thing over and over again relentlessly—day in, day out—how the lactic acid builds up in your hands? It may have been Jeannette Lee, who was lurking around with Keith, who said, "Oh yeah, I heard if you hit the cyst with a heavy book, it will burst and go away." The only heavy book I could think of was the Bible. And one good whack—oh yeah, it went away, but the pain! Ouch!

So I realized keyboard-playing wasn't going to be my future. But it gave me a great deal of fun and intrigue—how to suss out a song in a completely different way, so I could weave in and out, snakelike almost, with the vocals. I loved that.

We had no real rehearsals for *Metal Box*, we just didn't have the money, so I'd be writing in my head, and thinking of different formats for myself. A looser agenda, less of a hold on the reins vocally, and pushing myself into really challenging ways of singing and presenting it. And of course when we came to recording, it was about compromising the vocal presence, because I wanted the music to be so powerful, to declare us on a way interesting level playing field, where you don't really need the vocals to be so upfront, which was the pop way at that time.

You have to strain a bit to hear the vocal on some of the *Metal Box* tracks, but that's the point. It will creep into your psyche slowly but surely, where you're almost unaware, or you're lulled into a false sense of security—or a false sense of *insecurity.* Either way, it's getting into your mindset and it's affecting your perceptions of the world around you. At least that was my ambition: making you think bigger things.

We were largely responsible for our own productions. It was the art of balance to get the kind of heavy bass we wanted. This didn't just involve reggae: a lot of '60s mod music was very heavy-bass-y—the Yardbirds, the Animals, they had that deep sound in there—and funk and disco. But to get the kind of bass we wanted, something had to be sacrificed and so smart-arse me sacrificed the vocals. We dropped the vocal a notch to push the bass up, because it was about the aural tapestry of the whole thing, a tour de force on all levels. You didn't need the vocals to be way out front and all the music to be pulled backwards. We wanted the complete impression to be overwhelmingly exciting.

Often, there were obstacles to achieving that, such as whenever Virgin sent down any big-name producer to "help us." There was one who'd worked with the Rolling Stones, and that was a serious problem. We'd gone up to the Manor to do one or two songs, and he was just arguing that you can't have that much bass on a record. Ridiculous. "Yes, you can. This is how. *This—is—how!*" The man kept on arguing, so I got up on the desk and walked across it in my steel toe–capped shoes and broke every button. "I'm not here for you to tell me what to do!"

Trying to wrap us around in tried, tested, and proven formats was the wrong thing to do. We weren't in the mood to put up with that, or be wrongly motivated and misled. That would be record-company intrusions creeping in there. And all done in very nice "I thought it would be good if we brought in Blah-Blah."

"Okay, I'll give that a go." Twenty minutes later—"There you go, I broke his studio."

Playing-wise, Keith was on fire. Quite apart from his guitar, he was very into collecting oddball pieces of electronic equipment of a musical leaning, which many a time paid off really well. Other times, it was like, "What are you turning up with? This is impossible. What do you want us to do with *that*?"

At the time we were fairly obviously an analog band, but one day he dragged in a Fairlight, the digital sampling synthesizer—fair enough, but this is before there was anything that could ever possibly work with it. It was just a keyboard-looking thing with a computer, but we had nothing that would sync up with it. It was ahead of its time in many ways, and I do believe it ended up with Kate Bush and her Heathcliff.

For years, them things were just too bonkers. Now I understand what they can do, but we were not computered up at that time. And indeed the technology wasn't there. This instrument of sorts was a control board really, and was too far ahead of its time and had no relationship to us. That would happen quite a lot with Keith: he wouldn't understand how it couldn't actually work.

If situations were left to Keith, he would've spent every penny on some mad boffin's dream of PiL making music for the third universe by investing in electronic telepathy. These'd be the kinds of conversations he'd want to be indulging in. All well and interesting but that's all they are—theories. There's no reality for that to work in.

Getting ridiculed for these things would of course send him off into a real sourpuss one. Keith was a very difficult person to sit down and talk about average things with. That stuff is really important because it forms a bond, a friendship, that allows you to then move into the serious stuff. With Keith, it was always hypertension. Which is thrilling also, but frustrating when you're trying to curb the excesses.

Let's face it, I'm not innocent here, and I've pointed it out to myself: if left to my own devices 100 percent, it would be chaos. I'm always one for teamwork because of that. I know I'll go way over the top and out the other side of it. I'll be the first to say, I wasn't excessive there, but I'm damn sure all the people around me think I *was*, and have told me so. Rambo will often say, "John, you don't know when to stop, you'll tear the arse out of anything." It's true. All of us need our friends to be watchdogs.

Despite the tensions, there were all kinds of jolly romps going on, particularly at the Manor. I'm talking about pranks, most of them lighthearted, but sometimes a bit over the top, and anyone was a target. I believe it may have been Karl Burns, who was a drummer for the Fall, who came down and crashed out one night on LSD, only to wake and find his bed was on fire. Actually, was that him? Hard to be sure, but I think so.

This kind of nonsense would go on all the time. We'd be partying in the front room, and if you were foolish enough to think you could fall asleep in that environment, you were setting yourself up. It was very much a "last man standing" environment and that would be the thrill and the challenge of it, to see how long you could endure. And I'm not sorry to say—I'm actually very happy to say, because I loved amphetamine at the time—that was an amphetamine-fueled world, from time to time. Not every night, or when I was singing, but you'd look forward to a special night when you could really—again, I know, a cliché—let your pants down. But you daren't do that because you'd suffer the consequences!

Against this jolly backdrop, I was losing my mum to stomach cancer—the worst and most painful one of the lot. I spent as much time as possible with her at the hospital, but not as much as I should've.

One time I went down there, the local priest came round, this freakish Jesuit monk. He'd just come back from Africa—one of them people that touch you and you're "healed." To bring this into my mum was offensive enough, but then to turn around and say that the reason it wasn't working was because I was challenging him—that was despicable. I was really, really upset. I don't like being victimized by conmen. Every analyst, psychiatrist, spirit toucher, ghost hunter, psychic, or priest on this earth is there to do you wrong.

The greatest crime is that when Mum was actually dying, she wanted a priest at her bedside, and of course he wouldn't come. It's all about money. How do you drag a priest away from the pub, and the young chaps? It was all very, very painful.

Mum had always been loving in a very quiet way. There wasn't much said, but that's all you need from your parents, the right kind of attention. Before she went, she asked me to write her a song, which became "Death Disco." I only got to play her a very rough version. She knew what I was up to. I had to curtail it a bit, because what I wrote is very directly about death, so I wanted her to feel it was more about the challenge of an illness. A rough demo of it, with indistinct lyrics, would be slightly milder than the full clarity of "You're dying—urgh!"

It's only by researching these areas of your psyche that you're going to free yourself up. Don't separate music from anxiety and pain, and thereby you'll find a solution. I've never come to grips with death, but through music I kind of found a way of dealing with it. I'm questioning myself very seriously in songs like that. It's borderline mental breakdown. It's me howling in bitter agony. Grief, grief, grief, but at the same time you've got to give joy for those you've loved. Not wallow in the self-pity of it, but rather celebrate the good things about them when they were alive.

When we released "Death Disco," it caused great confusion. Was it a dance record? What *was* it? It certainly didn't mean "death to disco," as some people interpreted it. In fact, when Morrissey came out with "kill the DJ," I thought he was making a misguided reference. Me? I'd loved my nights down at the Lacy Lady in Ilford, and all the music that went with it, but you can't be laying down a bog-standard typical disco pattern. It doesn't mean you need to imitate or duplicate. You advance or destructuralize or whatever it is you need to do, in order to adapt the journey to the content. And Johnny don't sing in Michael Jackson stanzas.

I was very pleased that "Death Disco" launched all manner of intrigues about what it was we were up to. I was fascinated when I heard *Record Mirror* put us at Number Eight on their dance chart. I thought, "Cor, what are the soul boys going to think when they hear *that*?" But it really did make it onto turntables in clubs, no doubt helped by the fact that we put out various different mixes of it.

To our punk followers, it was saying, "Look, why are you lurking in the shadows, boys and girls, get out there under that glitterball. Here's your opportunity! And get a load of it, you're dancing to the death of my mother, you bastards!" That's a lot of hardcore tasking going on.

It didn't get on the radio, of course. It wasn't on Tony Blackburn, or even John Peel—for all his apparent open-mindedness and celebrating the wonderful world of music, it was quite a narrow agenda that man had. In many ways, as my Jamaican friends would say, "Dis free I up." We didn't have to consider playlists.

Still there'd be one or two at Virgin going, "Why don't you just write a hit song?" "Hello! I'm writing songs, I don't know whether they're hits or not, and I don't care!" Everything I've done has always had a very good turnover and that's just the fact of it. The problems arose when either there weren't enough pressed or there wasn't enough backing. Those are the two major issues of why records don't sell. If you don't tell people they're available, they might as well not exist. That's the record company's job and the way record companies were changing at that time, it became less and less their job. The idea of outsourcing—promotion would be purchased independently of the record label by the artist—was becoming the order of the day.

So there was less and less support, and I was cast out as

eccentric—a lone wolf—not in touch with the psyche of the population. Well, sod off! Is there any reason I should be? I don't write to patronize, I write to deal with issues or problems that directly affect me, and thereby, to my way of thinking, affect everybody.

Put it this way, I sucked eggs for no one. I knew damn well it wouldn't go down too well when I went on *Juke Box Jury*. This was an old BBC TV format of mindless celebrities reviewing the latest pop releases.

On the one hand, I was still dealing with the punk hanger-on sorts, who got stuck in the original foothold and weren't prepared to take the next step, and they tended to be the most bitter. "That's not punk, you're selling out!" "Oh, for fuck-off sake! I gave you the boots, you're wearing them, but now learn to walk in them. In fact, actually, change them—here's a new pair!"

At the other extreme, the Beeb were probably hoping for another F-bomb outrage, but I wasn't going to give them that. I willingly went on there, but only after me going, "No! Grrr!" and whoever was around going, "No, you should, it'll be good." I was also aware that by doing it I was going to annoy certain members of the band who'd think I was hogging the limelight. There's that element, always. But I'm very good, I think, in them open forums. I think I excel in them. You're not going to get what you expect—you should be expecting better, and that's indeed what I give.

I went on in a rather fetching red silk suit. I had two of them, one in red and one in green, both made by a designer friend of ours, Kenny MacDonald, who used to have a tailoring store on the King's Road. I really liked his approach to clothing. He was a Jamaican-parentage fella, a serious nutcase, hard and tough, and his clothing was very original, and very mental. All of our early visual statements were coming from Kenny: the gray check suit that I'm wearing in the "Public Image" video, for instance, that was one of his. He created the red fur coat I used to wear, that everybody thought was a dressing gown. He also made me one in white where I looked like a polar bear. He used to make very bizarre cuts of tweeds and things.

Anyway, this red silk number was a fairly sensible cut, but in wack-arse fabrics. Very, very good fun. I liked Kenny a lot, but I heard he's been in jail for many years. We have patience.

The presenter of *Juke Box Jury* at that time was Noel Edmonds—Mr. Condescension, obsequious to the last, a man whose daily condescending was truly on its way down the staircase into the basement. What he does is a con, descending.

Joan Collins was on that week—I mean, who the hell is she to tell you what's what about music? Also there was Elaine Paige, who was sitting next to me, and whom I got on with famously, oddly enough. People like her are not my enemy, but the pretentious pop lot are. She sings in musicals. She was great fun. She was just "You're so funny, but right!" I'm a sucker for that chat-up line. She grasped that where I was coming from was actually a sense of fun in all of it, rather than trying to be deadly serious about the frivolity of pop music.

You were meant to hold up a little disc that said "Hit!" or "Miss!" for each record they played. One time I just held it sideways on—it's neither a hit nor a miss, *it just is what it is*. Then that stupid bloody host with the beard was expecting me to like a particular record because it was made by my alleged fellow punks—a Siouxsie Banshee thing. "No, that isn't how it works, I'm not a cliché and I won't be pigeonholed into doing the expected just because that's what they want of me. I will express my true feelings and you can like it or lump it." For my mind that's the presumption of why I was invited on to that show, you'd get an honest response.

At the end, you were supposed to do a handshake thing, and wave to the camera—this was all set up beforehand. "No! That's false, that's Joan Collins' universe, not mine." I wouldn't get involved with the sociability side of it. I ostracized myself and walked off. Maybe sometimes I go a little bit too far, ahead of the norm, but slowly but surely it does have an influence—*Juke Box Jury* got shelved after that series. Good work, John!

Perhaps it was no surprise, then, that our next Public Image Ltd single, "Memories," didn't do well commercially. The song itself jumps backwards and forwards from one texture into another, we thought, in a very poignant way—from brittle to warm. I loved it, but it wasn't a hit because it was nearly five minutes long and we knew it wouldn't get any airplay because of that. Keith and I were in total agreement on this: neither of us knew where you'd cut it, or what you'd cut out, and what would be the point? It was the length of it that got the true emotions across. You can't cut out the last chapter of a mystery novel just because it has an extra twenty-five pages. But what that does, though, is it alienates you to playlists. Which honestly has never really been a problem for me.

The tracks getting longer was something that evolved quite naturally: "There isn't any point in stopping this, because we haven't run out of ideas yet." It wasn't any great analytical study about it: "Oh, I think we should do a ten-minute one!"

"Albatross," the one about Malcolm's cowardice, ended up at that length, if only because that's all we could fit on the record. It deserved that length. You let the song dictate the pace and the time, rather than you trying to master it and control it and make it all note-perfect. I find those kind of approaches to be stifling, a contamination.

It's not how I feel a human being actually physically and mentally works. When we sit down to relax, or agitate ourselves, and/or whatever, we use music almost as a backdrop to our own ramblings, our own thoughts. We were giving you something there to work at that with, where you don't have to be focusing on counting the beats, and "What are the dance steps?" It's more "What are the mindsets?" Every time you play a thing like that back, it's different, there's another angle you come by, so you're not trapped by the precision of it all.

I loved the recording area in the Manor. Once I got in there and got over all my fears and phobias, I would be there for hour after hour after hour. What we'd do, we'd break it up into sections, each person would have their own time in there to sort themselves out, to work alone, and then combine the efforts. That was interesting. You'd go in and you'd hear what someone was trying out or fumbling about with or rehearsing on their own.

For me, I couldn't get myself interested in going in until it was *way* late at night. Everybody felt that too, because it would be sunny out, and this was a fascinating old castle-y type structure, and there's acres of fields to run around in and things to suss out, and there's a three-mile hike to a pub and that's good fun. And the old stonework, and "lording it up at the Manor," and all the food and drink—I mean, you just can't help yourself.

On the down side, there was the screaming from Keith, and knowing that Wobble couldn't control his temper beyond a certain point and was highly capable of mangling Keith—who in fairness was a squealing ferret. Wobble would go into a freeze, like a silence, and I knew what could possibly be coming from that, and I knew I had to stop it. I have this constant attitude that violence can't solve anything. You cannot have it in your workplace. If anybody oversteps that line—that's not a rule, that's an absolute value—then they've destroyed themselves.

Other times, there was a great sense of fun. For instance: ashtrays on piano strings! I've had a deep love of harpsichord music since I was young—stuff you'd see on TV, some old Bach or Beethoven, done on the original instruments. It would absolutely thrill me, that sound. There you are in the studio and there's a grand piano, and I'm out looking for metal ashtrays to get that twangy metallic buzz going. What a racket, much more thrilling! And, on the piano keys—elbows only!

Keith, Wobble, and I were all on the same page with this kind of experimentation. All our animosities were at personal levels, which was a shame. All these discordant resonances, these sonorous tones that drift in and out, are just wonderful. They're almost tune-destroying, but soul-wise very beneficial. You need lots of places, not only for the performer, but the listener to spark free thought from. Of course, none of this would be considered proper work in the studio. But you get great results because you're really intrigued by sound, at least I am—totally. That's the one thing of being an avid record collector. What you do is you build up this repertoire inside your psyche of different sound elements. You're not imitating or copying, you're advancing it. Or, as many musos would tell you, *destroying it.*

Lyrically, I was going to a very different place too. While Joe Strummer would be busy watching the news trying to assimilate political headlines, I'd be listening to the story of a young girl raped, and I'd find the humanity in that to be much more interesting—to try and get across the great grief that girl went through. She had been captured by two men, blindfolded, bundled into the boot of a car, driven out to the country, and raped. If she hadn't run away she might even have been murdered, and she remembered very little because the pain of all of this was so overwhelming.

All she did remember was a tune played on the cassette deck in the car, and that's how they caught the men responsible—when they traced the car, the cassette was still in the tape deck. It was the tune she remembered. They never mentioned in the media what the tune was, but I actually found out that it was a Bee Gees song. Having a great love and affinity for the Bee Gees I found that even more interesting. So, hence the line in the song: "And the cassette played poptones."

And all this to perhaps our most hypnotic and uplifting track, with Keith in full flow. What a great way to draw the listener in to grasp that poor girl's pain, and have a sense of empathy for her. Every time I perform "Poptones," that story is totally going on inside me. I'm completely that person, that victim. Same with "Annalisa." It might be foolish of me but it's what I do. It's an empathy for all victims. It's not a great thing to do that, to put yourself in the position of a victim, it's very sad and deeply hurtful. But nowhere near as severe as the pain that the actual victim went through.

It was about dealing with different human emotions, the ones that normally get pushed aside. If it be Pandora's Box I'm opening, I'm doing it with a hammer and a chisel. I broke that padlock, I broke that fear of the unknown.

"The Suit," on the other hand, was about my mate Paul Young who went and borrowed my suit without telling me for some date he had with a girl from Totteridge Park. He put it back afterwards and of course it was all smelled-up. And I loved that suit! I don't mind sharing clothes, but it's a bit much when it's your best suit that you're saving for the right moment, and it's got stains on it—and not all beer stains—and there you are like you've been rolling in the hay. He could have dry-cleaned it, you know! Cheeky monkey, haha.

"No Birds Do Sing" is a reference to that Keats poem, "*La Belle Dame Sans Merci*"—"O what can ail thee, knight-at-arms,/ Alone and palely loitering?" etc., etc. It's a haunting, ghostly kind of a poem, so I thought I'd apply that love-forlorn thing to life in suburbia. It's quite an irritating song, but you try living in Tring.

I don't think anybody, up to that point, had attacked synths quite the way we did, to create these tense atmospheres. "Careering" is about the troubles in the North of Ireland. The ethnic divides across the nonsense of religion, which later translated into gang warfare. I can't support any group that believes that, by the killing of another, it improves their cause. For me, once you've murdered someone, you have no cause at all.

Musically it had to be so stark to achieve its aim, which is to make the listener aware. In all that Catholic–Protestant nonsense, for me, there's "bacteria" on "both sides of the river"—I'm not gonna pick one load over another. I'm not going to be arguing with people over which "bacteria" is better than the other. Indeed, all religiously fueled or politically fueled situations are contaminants. There's a manipulation that goes on there that the followers aren't quite aware of, and they should be. You've got to know when you're being used.

That song caused me a lot of problems in Ireland, particularly in the South, where they thought I had no right to be dissing the IRA. Great shame. It's the same with the song "Religion": I was put down for making a direct attack on the Catholic Church, but I think you need to. There can't be any subjects in life that are sacrosanct and untouchable, because these are the very things that lead to all the troubles. Armed gangs of murderers will never ever get my vote. And they'll never ever lead to a world of peace and happiness.

On "Radio 4," though, I couldn't fit in vocally, so I didn't sing at all. I cut my bit out, to give the band some space. I was more than happy and pleased to break away from the demand that you've got to be singing on everything. I'm doing lots of other different things, not just singing. There's all kinds of recording decisions and producer stuff and adding little bits myself here and there and whatever, so it's a jolly good mix-and-match.

We wanted to break every rule, every discipline. It was like opening up a toy store to four toy-starved children. We wanted the deepest growl from the bass end, so it would almost shatter your ears. Other places, what you might think is some kind of electronic gadgetry, is actually just the TV recorded onto a two-track, and sped up and down. It's all fairly abstract, but there's always an underpinning of danceability—although, as many of my friends told me, "You need three legs to dance to that one, John!"

Every track was just, work it out quickly and bang it down. Often one of us would have to muck in on drums—Keith stood in on "Poptones," and Wobble on "Careering." I loved all that. I loved the fear. It was like a runaway truck speeding down a hill with no brakes. We were just hoping that there was a rise at the end.

• • •

Not long after the Pistols broke up in 1978, I did an audition for the Who's movie *Quadrophenia*, because Pete Townshend asked me to. He wanted me to try out for the lead role, the one that the English actor, that ratty character with the black hair, Phil Daniels, eventually landed. Come the time, the Who's manager didn't like me, and they didn't think that I'd be able to carry it through a movie. They were probably, frankly, dead right, because I'd have needed some coaching or education as to how a film was put together at that point, and I just wasn't prepared to listen to anyone about anything.

Throughout my career, Pete Townshend has always shown a favorable, helpful point of view. Our paths first crossed in the early days of the Pistols, when we were demoing at the Who's studio. Mr. Townshend found out who was using it and said, "We won't charge you." So I've got nothing but complete respect for him. He also did some favors for my brother Jimmy's band, 4" Be 2".

He's one of those characters who's not understood completely, but in the music/band side of things, many people, if they bothered to step forward, would tell you that he's done everything he can to help you, while keeping his name out of it. He'll make studios available, he'll talk to you, he'll run through songs with you, and he'll tell you what's missing in there. When he tells you Who stories, you get the feeling that you're actually in the band. It's not "And then I . . ."—he's talking as a band to you. So, meeting Roger Daltrey years later, that was the first thing we had in common. "Oh, you've been with Pete? Oh, blah blah blah blah." So we had a root-canal conversation.

It's impossible to catalogue it, other than you feel there's a protective spirit when Pete's around. It's always favorable and open-minded, and if he doesn't understand what your ideas are, he will not tell you different. He will back you even on that, and that's kind of father-like. I know he's someone I could call if I wanted to, but I'm an independent spirit and I don't want to run it that way. But he would be absolutely without doubt available for anything. He's just that kind of person.

Anyway, while I was auditioning for the movie, I'd get my moments on film, my sketches, delivered to Gunter Grove, and they'd arrive in these great big film canisters. It turned out that Dave Crowe, who was still living at Gunter at that time, knew the company who made these canisters in Britain, because their factory was out in Borehamwood, where he'd lived when he was younger. So, one thing led to another, and we had a groundbreaking new way to present our groundbreaking new music—literally, in a metal box.

It was a great idea to present all these expansive tracks across three 45rpm 12-inch records, but almost our entire budget for *Metal Box* was frittered away on this extravagant packaging. We spent more than we did on recording, and Virgin didn't pay for it—we did.

How the boxes themselves turned out was awkward in the extreme. What we ended up with was like two round sweetie trays put together. They were hard to prise apart, and it was impossible to get the records out. It was appropriate, though, because what you were about to listen to could've been construed as distinctly unpleasant—it was made for those consumers who were prepared to put in a bit of effort. Then, once you were inside, the bass was so deep, it'd kick the needle clean off the record. We were slightly in advance of the hi-fi gear of the era.

When *Metal Box* came out, in November '79, it got good reviews, but I trust those even less than the bad ones. It sold out of its first pressing straight off, but its influence just grew and grew over the years. In interviews, I'd describe it as "mood music"—I knew that one would hurt.

Going out to play gigs around this time was impossible. In September '79, we'd headlined an indoor festival in Leeds called Futurama—a catastrophe of a gig. It was so badly organized, and it made us really angry. Outside the venue, as you drove up, was a bunch of blokes dressed in Nazi outfits, *sieg-heil*ing and handing out National Front publications. What? That's unacceptable. There was no security of any kind, the exact opposite of overbearing security, so there were people just wandering around and unplugging whatever equipment they fancied. While you were up there, you knew stuff was getting nicked out of the dressing room, so you'd constantly have one eye on the back of the hall. Just silly, stupid, unnecessary chaos. Thank God for friends like Rambo, who turned up and paid attention and stopped people doing all that.

Unplugging equipment was a very big and popular thing in them days, people would get onstage and go straight for a lead or whatever, just to try to unplug something and see that as an achievement. These weren't what you would call fans or people that were into the music; these would be football-y type gangs that would go to venues and deliberately try to wreck them. Particularly if you were touring up north and you were from down south. Whoa—*getcha!* That was their mentality. It required a lot of bravery to stand there and fight back, to take that on without resorting to violence. You have to understand the psyche of the audience and control what you do in those situations.

People wanted me to be the bad boy, but they also wanted to pull me down for it. It was just walls of animosity. It was a problem, because every time you stepped out, you had all these prehistoric monsters who wanted to drag you back to the past. It's always going to be that way when you deliberately put yourself out there on the cutting edge of learning. I could take that kind of situation quite happily, though, because sooner or later, those people who are at first against you for being oh-so-different, eventually get it. It took years for the first PiL album and *Metal Box* to be understood. By that time I was two or three albums down the road and, of course, those were then the albums they weren't understanding, so I'm always three, four, or five albums ahead of the learning curve.

There was trouble at these shows, but for my mind there were always mitigating circumstances, which were not considered. Britain was in a steady situation of catastrophe there at the turn of the 1980s, with endless strikes, riots, football violence, and everything. It was a very violent time, and all bands faced this agenda when you did a gig. But it was very convenient for certain newspapers to hang that firmly on me, and that did a lot of damage. A lot of promoters wouldn't touch us because they thought the exaggerated press reports of those incidents would create even further trouble down the line.

It's always down to promoters, because they're the ones that raise the money, but only if they feel your record company's properly behind you. It affects gigs quite seriously. If your record company backs up whatever it is you're up to, those economic problems that promoters can bring in wouldn't arise. Suddenly we had to sign these insurance clauses before we walked onstage, making ridiculous guarantees that somehow if you agree to them you're implicating yourself as the creator of any riot that may happen—by signing that you'll try your best not to inspire a riot, it's an implication that it's in your framework. You know, Jesus, I'm not responsible for what a mass of people do, *I'm* responsible for what *I* do.

Very quickly, we were more or less canceled out of Britain. The impossibility of getting gigs—again!—was taking away our lifeblood. It leads to all kinds of powerfully antagonistic internal confrontations, usually focused—again!—on "I want more money!" What else do you want, when you've got nothing else to do? It's an understandable human reaction to frustration, and that then becomes your maypole and your bone of contention. And trouble comes from that, it's very hard to control. It caused fractionalisms.

So we were in a pretty bad way by the time we went to do PiL's first American tour in April '80. I had virtually nothing to wear, but I heard Wobble had used my mate Kenny MacDonald to make some suits for himself, that I found out PiL had paid for. By chance, as we were leaving for the airport, I'd talked to Kenny. So we got to the hotel, and Wobble opened his suitcase, and my first thing was, "Right, I'm having that, because I paid for it!" I was always mix-and-match, anyway: my suitcase was equally open to anyone I was working with. Wobble never said anything, but I knew it created a real bitter agenda inside of him. He shouldn't have been using my mates to make clothes without telling me, and having it put on my bill.

Regardless of such disagreements, our media profile was fairly confrontational. *High Times* put us on the cover, with Willie Nelson on one side, and Johnny Fucking Rotten on the other. Underneath, it said something like, "The two sides of music," or at least that was the implication. Fantastic! I was thinking, "Willie Nelson isn't the enemy, you're kind of getting this wrong, fellas!" I've always had a keen listen to Willie's lyrics. He's a rebel in his way: he doesn't want anybody telling him how he should or shouldn't be living.

In another interview, I declared, "Rock 'n' roll is shit, it's got to be canceled." From that point onwards, honestly, America declared war on us. "*Whoargh*, how can you say such a thing?" Well, I was right. Rock 'n' roll had become a very lame duck. Expecting anything new in music to fit into that established genre was repulsive to me. It had become a ball and chain. Think outside the box.

Metal Box was being repackaged on both sides of the Atlantic as a "conventional" double album called *Second Edition*. We did a radio ad for the release proclaiming it as "twelve tracks of utter rubbish from Public Image Ltd." Complete and utter rubbish, for your dubious pleasure! I hoped the humor would be picked up on. It was an attempt at friendliness and openness, using irony, which is something that's possibly not understood in America.

We'd signed back with Warner's for America, but we'd had problems with them again. They didn't like the first album very much, and passed on releasing it altogether. They didn't know what to expect. I suppose they were hoping for "That great rock 'n' roll band the Sex Pistols Part Two," but this is the same lot that couldn't accept the Sex Pistols Part One. This meant I was one step further away from being caught up on.

We played about ten dates, mostly in theaters in key cities including New York and L.A., as well as a legendary TV appearance on Dick Clark's *American Bandstand.* It all got off on the wrong foot when we arrived and they suddenly informed us that it would be a mimed thing. Our equipment hadn't arrived in time, apparently, but we soon got even more upset when they said, "Oh no, you couldn't play it live anyway, just mime to the record."

They'd made up some edited versions of "Poptones" and "Careering," and gave us a cassette to check out beforehand. "Oh my God, they've cut it down to *that*? I don't know where the vocals are going to drop. What are we supposed to do?" None of us knew. Just thinking about trying to sing it like the record was . . . *aarghh*! You can fake it with an instrument but you can't as the singer. "Okay, so you've cut out the point and purpose, it's like removing the chorus from the national anthem, just because it makes for an allotted time slot on a TV show. That's arse-backways!"

Just before we went on I said, "Right, let's just freeform it—basically as a cover for me, please!" I started the ball rolling. I made no attempt to mime, moved around the place, and dragged all the audience up on the stage to dance. For all of the problems that caused—such spontaneous behavior didn't fit with their usual cozy format—Dick Clark, the host of the show, who was a massive star of U.S. network TV, became really friendly afterwards—even though Wobble had messed around with his wigs. We found Dick's room backstage in the makeup department, and hanging on hooks were all of these different hairpieces which, you know, got assaulted. But in the end it played out really well because when Dick Clark did a rundown of the greatest ever performances on *American Bandstand*, Public Image were up there in the Top Ten. And he'd been running that show for decades—almost half a century.

I knew that in that world they were all sycophantically groveling and arse-kissing each other and cliqueing it up, and expecting anybody new that comes in to fit into place. That's what music wanted from me, that's what everything has wanted from me, and it's not going to happen. Not ever. I don't need to find a niche in that kind of society. The more they annoy themselves about my kind of personality the better it is for me, because I honestly don't think I'm doing anything wrong here.

On the plane home from the tour, Levene had a massive withdrawal. I don't remember too much about it, but I wasn't very sorry for him. My attitude was, it serves you fucking right. Keith kept saying he wasn't going there, but it was clear he was. Having to deal with liars really upsets me. I'm very forgiving with friends when they lie, I try and understand what made them do it, but a bit too much of it—when it becomes a public display—and I'm furious.

Wobble at that point had gone beyond the point of endurance. His bullying behavior towards the drummer was just not acceptable. With his animosity towards Keith, he was causing too many personal situations. He was involved in too many of these issues all at once to be just merely a coincidence. His girlfriend would turn up at the house and ask where he was, and insist on coming in, in case he was in the house and hiding from her.

He'd become very mercenary. He'd cultivated this aura of Jack-the-Lad, coasting just for the money. That was the vibe he was projecting, and unfortunately a little too loud and proud, and so he got what he asked for. As long as he'd been in the band, he'd never offered anything like a resolve or an answer to anything. Just sit back and snigger and never actually contribute, never. Never wanted to make a commitment that he might be judged on later, if it be right or wrong. Believe me, that wouldn't be a problem in PiL. But no involvement at all *is* a problem in PiL.

Making wrong judgments is not a problem. Making mistakes is not a problem. These are things we can deal with and move on from, but lack of commitment is a serious error. Knowing that he made himself somehow seem above it all, that was unbearable. You cannot be in a band that works like PiL, and disassociate yourself from the problems, and from the writing situations, and hold your hand out at the end of the month expecting a check. That's gonna come up against a brick wall.

Everybody was arguing about money, everybody wanted more. But when there isn't more, what can you do? I had my mate Dave Crowe in there to try and run some kind of accounting, because he was very good at Maths. I didn't even have a bank account, really, or a credit card, or anything at all up until PiL, but then it became necessary. The wages side, I didn't have direct access to, what I wanted was for Dave to control that, so there wouldn't be any suspicions that I was going and pilfering on the sly. Money is the root of all evil. If there's any there to be had, everybody else wants more. And there's not much of a way around that, I'm afraid. That's the original sin.

Unfortunately around this time Dave Crowe stopped working for me because, let's just say, he had his own problems. It was too risky that the financial side was being orchestrated by a chap who would forget to go to the bank on Friday and leave us all broke over the weekend. That's all he had to do: walk 150 yards, because there was a Barclays Bank right around the corner from Gunter Grove, and he just somehow couldn't seem to make that. That made me feel like a fool, because I'd be the one that would have to explain to everybody else why there weren't no money this weekend. There was money in the bank but I couldn't do sod-all about it.

This was before 24-hour ATMs, and there was no alternative. I didn't even have a checkbook. I just didn't see the need for such things. I'd be quite happy on ten quid a day, I still am, but you can't be like that when there's other people expecting their wages, whether earned or not. It is your obligation, and that sense of responsibility is very serious and must be taken serious. So there it goes, I had to let Dave go, and we've never really talked since.

My problem is that I'm intensely loyal to the people I work with. I say three strikes and you're out, but in the end I'm always there for four, five, and six, because I believe that in the long run loyalty gets you better results than "hissy-fest fights and then separate."

With Wobble, too much rubbish got in the way. He made himself uncomfortable to be around. I'd had enough of it. There were too many maneuvers going on that I thought were sly and underhand. While we were recording *Metal Box*, he was secretly taking the tapes of some of our backing tracks to use on a solo album he was making for Virgin. One time, I actually caught him in the act. One of my best mates! We've never properly made up on that. I said, "We have to part our ways. You're my mate but this is the end of the line for you in PiL. It's dead, it's not working. We're not mates in the band, and that's a tragedy. But let's stay mates." And that's how, for me, it was left.

So now it was down to just me and Keith. Everything was far from on an even keel in the PiL camp, but there was something that brought a little joy to our hearts. Pretty much as soon as we got back from America, Malcolm's Sex Pistols movie *The Great Rock 'n' Roll Swindle* was finally hitting the cinemas.

I was very, very happy, because it was excruciatingly bad. Me and Keith were over the moon at how rubbish it was—overlong and full of Malcolm pontifications. There were various Nazi outfits and rubber masks, for no reason at all that I could see. The best scene in it was the opening sequence, with the hangings and the burning of the effigies. That was great. I thought, "Oh my God, this is going somewhere. This has got a real, deep message I'm going to be terrified of." But no, then he threw it away. It just became an exercise in sticking pins in a voodoo doll—i.e., me.

It was ineffectual and very limp-wristed really, because he wasn't replacing me with anything. He wasn't grasping the bigger issues of what that band really was. He was just trying to trivialize it in order to make himself look bigger. The Ten Commandments according to Malcolm: "Oh, and then I thought . . . And then the idea occurred to me . . ."—in that pompous voice! Who do you think's gonna listen to that, Malcolm? Now we know why you locked yourself in your office for so long—because you were perfecting *that*!

It was everything I wanted it to be, because now Joe Public could see what it was I'd escaped from. And, by contrast, what it is I'm actually capable of—please judge me thereon. If you think I should be involved in a world of swindle—well, fuck you!

I felt like he was taking everything backwards into shyster. It wasn't done well, it was trivial, it was mockery with no good intent. A classic example of Malcolm when he was left to his own devices—a disaster. That film, that album and all of it, it wasn't searching for any good in it. It was all on a superficial fairy-dust kind of trip.

Of course, part of me was thinking, "I'm trying to get PiL going here, and by inevitable association they're dragging us back into this idea that everything we do is a con." That caused an enormous amount of damage for me, because people assumed that to be the truth of it. All they had to do was listen to any of the words I wrote, any two sentences I'd strung together, to know that I wasn't doing it for the money.

And then they're auditioning singers to be Rotten—hahaha!—and you know where they auditioned them? At the Rainbow in Finsbury Park, right by the flats I was brought up in. How we laughed.

Overall, we were really chuffed. It was a happy little period for me and Keith. That summer of 1980, however, Richard Branson invited me over to his canal boat in Little Venice in northwest London. He's since built a studio out of it, but he was living on it at the time—no doubt it was a trendy thing to do, while he was waiting for his new castle to be established.

So I went over there, in good faith, but I was really appalled and annoyed and disgusted when I realized that the meeting was all about this agenda of his wanting me to re-form with Steve Jones and Paul Cook. They were now calling themselves the Professionals, and Branson played me a rough cassette of these awful, duh-duh-duh tunes, that they expected me to write some words for. No songwriting, no direction—just terrible.

I still had all the pressures I was going through with the court case. God, my head—how I coped with it, I don't know. And Steve and Paul were still siding with Malcolm, who lest we forget had stolen my name and was trying to end my career.

I became very bitter about this whole meeting, because I'd invested wholeheartedly in PiL. I'd made the right decisions for Public Image when we started. It was the right decision for me. It was comfortably uncomfortable. At this point, we may have been at a crossroads, but it was still the right decision, and I wasn't going to take two steps backwards into that kind of nonsense. That would be so wrong. Tail between the legs.

The answer, obviously, was "*Fucking no!*"

WHO CENSORS THE CENSOR? #2

SWANNY TIMES

It worries me that this book may get too linear. My biggest fear is that I'll come over lecture-y. The written word is a dry thing, without the emphasis on certain words in a sentence. I think musically, and I talk musically, and that's the way I formulate songs. If you read the lyrics on a piece of paper they don't have the same clout as the pronunciation there in the music.

I love oratory. I loved voice projection at Kingsway College, learned how to really read and project the meaning of a thing. It was something I was shy of up until that point, and it became suddenly really interesting. I looked forward to having to stand up and read what I was writing, or read what we were studying in front of the class. A very nerve-wracking thing, but totally enjoyable when you're explaining it properly.

My most entertaining thing to do was always read Shakespeare with an "ooh-arr" yokel's accent. Then it becomes not the language of pompous rhetoric—it becomes real. "Owt, owt, breef carndle! Loife is but a warrking shadow . . . conspoiring wi him outta lord an deff!" Then it sounds like someone in a pub talking, which is what Shakespeare really meant. He didn't want this to be confusing to the masses. By the time the likes of Oxford and Cambridge got hold of it, they changed it into something else.

The same goes for a lot of classical music too. The harpsichord parts are now played on grand pianos, with thumbs. This is something a music teacher taught me once: if you're going to try to be in any way accurate about playing piano pattern lines, you can't use your thumbs, because thumbs were not part of playing the harpsichord. In fact, there was no place to put them. Fascinating. I took in that lesson more than any other thing. The actual playing part was boring. But the thesis and the theory behind it always fascinated me.

The same with art. I could listen to people talk about what they were interpreting a painting to be to teachers, and I found that infinitely more fascinating than sitting there and trying to do an angry brushstroke on the count of three. "All together now. Got your brushes ready? *Anger!* I want to see anger on the page . . ."

Years later, when I went to Cologne in Germany, there was an art exhibition that some of the local Germans wanted to take me to. I went, and there was a Captain Beefheart segment. Really small things, but I really understood the anger, and the Beefheart way in his paintings. To look at the real things rather than on album covers was fantastic.

Phwoooar, I wanted one of them paintings. I also wanted a *Blue Peter* badge, from the BBC TV kids' show. But I never wanted a *Crackerjack* pencil. Everybody wanted a *Blue Peter* badge because it was a great thing. It may well have been plastic, but it had that beautiful three-masted ship printed on it. Plastic was very modern at the time and therefore incredibly exciting. It was like a medieval shield, a tiny little thing, but impressive. White with the blue ship on, that's the one I wanted.

As an aside, this idea of what a singer's voice should be or shouldn't be is revolting to me. *American Idol*, *X Factor*—they all expect singers to do all the trills and all the runs that singing instructors require—the gospel background. What a load of bollocks, man. Why can't you just sing the way you FEEL? It doesn't actually have to be what you would call musical, just how you feel in the moment, communicating something. The concept of tune, or tuneless, to me is bizarre. I know when I hear someone, it doesn't have to be a G Flat Minor, perfect, but it has to be accurate. The emphasis of the words, and the tonality, and the *pain* in the sound that they're procuring, and the message. If those things come across, tuneless doesn't exist.

Where being in tune counts very much, of course, is on boat cruises. That's what *American Idol* is really trying to procure! Boat cruise singers! My God, hahaha! I always enjoyed this story about the Cure, because the singer, Robert Smith—he can't bear airplanes. So the band took the *QE2* to New York and the rumor—I don't know what truth is in it—was that they played on there. I don't know if it's true or not, but I love the idea!

I've never met Robert Smith or talked to him or had anything to do with them. They're complete strangers to me, and in an odd good way, I like that. Every time I've ever got close to people whose music I liked, I've mostly found that I didn't like *them*. Look at the range of music I listen to: it's just about everything, right? I love what I garner in terms of emotion from what I'm hearing in a record. Even when I describe it, I just go, Uuuuuh-huh! I get a lump in my throat, because I love music, I love listening to what people do, but I don't love it when they think music has to be strict, and according to a certain sequence of notes, and perfection of those notes therein.

I actually love vast amounts of classical music. I love Mozart, beyond belief. But I don't think Mozart was too interested in accuracy over emotion. He was a genius apparently, and a crazy fuck. Well, I think it's the crazy fuck that I hear when I hear Mozart, and in particular when I heard him played in a North London pub after my father's funeral. Wow! That was Requiem—dun-dun-nun, dun-dunnn. And you know why it got played? Not because that's such a great tune, it's because it was featured in the film *Barry Lyndon*—my mother was a Barry, my father was a Lydon. They put an "n" too many in his name. But . . . It was fun to see what you would think would be the local hooligans and gangsters, and the Irish contingent, and the huge vast army, that us as a family and neighborhood have collected over the years, listening to that, instead of—what was the current hits at the time? No Doubt, or In Doubt. More like, Doubt-Ful.

I never met Gwen Stefani either. I'm going to contradict myself here, but I'd like to. Don't know if she's all right. You never know, and unfortunately in the world of showbiz, people tend to hang around with people who have equal amounts of money. There's reasons for that, because you don't want people parasiting off you, but money does dictate who your company is, so by default even rock stars run a royalty ring. They're only up there with equal friends who've won Grammys also. You see this at the Grammys year in, year out. The Taylor Swift brigade, you know.

Hold on for one second. Dada-da-nana-nana—tay-kee-laaaah!

Oo, what a lousy drink. Someone bought me a bottle of it the other day, and I got completely sidetracked a moment there. Urgh. Actually, it's mezcal. My God, why don't they just give me the mescaline and be done with it?

You'd think after meningitis and all those hallucinations, a person like me would never dream of going near acid, yet I found acid very tolerable back in the day. At fifteen and sixteen, going to festivals and concerts, I rather liked it. Everyone around me was screaming, "Oh, it'll do this, it'll do that." No! I think from all that stuff that went on when I was young, I've learned to decipher what's real and what's not in my brain. I know when my brain's playing up. I'm now able to go, "Stop that, that's silly!" I use a Monty Python reference there. I used to watch those shows rigidly, I learned so much from comedy. Norman Wisdom, all the one-liners—"There's nothing wrong with *me*!"

In my very young days in Benwell Road, we had a tiny black-and-white TV set. I got used to black-and-white, and couldn't adapt to color very well. We never had color TV at home. I think I bought my family their first color TV. The first one I had in color, with a remote, was a Sony when I moved into Gunter Grove, and it had a remote inasmuch as it had three buttons: on/off, volume, and channel. That was complicated science back then! The trouble was, I never got around to putting a proper aerial in, so there was a coathanger used—the old Irish way! I grew up thinking that TV was automatically always disturbed by someone going to the toilet. I never related it to them blocking the signal; I thought it must be to do with the flushing.

As I said before, I loved *Doctor Who*, but only when the Daleks were in it. The rest was stupid. "That's not real!" That's why I don't like science fiction; I don't think it offers too much. It's a nice exploration of minds, but ultimately the journeys are fairly tedious, because it ends up in that asexuality we call *Star Trek*. It's one step outside of how human beings actually really do evolve, or how communities work, and it's not a step worth taking. Science fiction doesn't seem to understand that. It seems to always come from the point of view of one man's lonely journey, and therefore for me very judgmental. That's how I view Asimov—judgmental! Not great. There's not an "us" thing in it.

Whereas, again, Shakespeare, which you'd think would be bizarre to me, isn't. It's the language. He's using words, and the sound of them sometimes, rather than just the meanings, so the meaning becomes something else entirely; you just follow the poetic beat of it. And the pronunciation, you garner ever so much more information from that than you do by just observing the words on a piece of paper.

I love Shakespeare live, if it's done well. I've seen James Earl Jones do Othello years ago—phwoar, wow! I love the audacity of just walking in off the street to some lesser playhouse, to see something that looked on the billboard like complete crud. Like fringe theater. It's thrilling, the embarrassing closeness to the actors, then realizing what it is they're going through, to be able to pull that off. It's kind of a really good cheer-up for me, because that's what live gigs are. So I understand it from that point of view, and I listen, I suppose, more intently than the regular audience.

The nightmare of everything like that, though, including modern dance, is that the audience let the performers down. One time, me and Nora got taken to see *Swan Lake*, the ballet, by Caroline Coon, the punk journalist—back when she was hanging out with Paul Simonon, from the Clash. Paul's warm, I like him, and that night, the four of us had great laughs. Caroline's line was, "You're so into *Swan Lake*, John—there's even a bit of it in 'Death Disco,' so you're going to love it!" It was her wicked check on me.

So Nora and I went over there, got in the cab, and there we are stood outside the theater—we might not even have known beforehand. Bloody hell, you took us to ballet! It was astounding, but, I tell you what, I got bored really quickly. I couldn't help it, the bar looked more enticing. When you see ballet on TV, as boring as that can be, you just see the leaps and the pointed toes and all of that, and that's barely endurable, but when it's live, with a live orchestra in the pit, you can barely hear the orchestra.

But when forty girls jump up and down on their pointed toes, you can hear that like hobnailed boots at the back of a terrace. It's like an invading army of hooligans. It's really loud; the wooden floor echoes and reverberates. I thought, "Urgh, that's bloody discomforting!" That's the other thing for the spectator: the pain involved in that, it's pretty hard to watch. Nora would tell me stories. Her sister took ballet when she was younger, but she wasn't good enough, because she was slightly bigger-boned, shall we say? She later ran a ballet school in Germany. It's a sad thing. When I saw her sister's feet, just how that big toe is nobbled into something really, really ugly. And the arthritis and the pain in later life. Now, at least, I understand the work ethic.

But the audience at these things is vile and snobbish. When they're too trendy, and too involved in their own arsehole back scene of it, they're missing the point.

When you're talking about missing the point, though, the majority of punks win the prize! They just got involved with the clothing, rather than the content. They certainly missed the politics, didn't they? America's interpretation, from their early days of poetry-reading punk, to absolute violence punk—just awful, both aspects of that were too much, and too ridiculous for me. I don't want the oversimplification of either end of that. I want an absorption of all of it, but a good sense of right and wrong about it. Don't go too far into anything, it gets you wrong.

Listen, I know I'm swinging left, right, and all over the place, but it's all roots to dig into, and that's the correct procedure, ultimately. I can't do this *all* chronologically.

8

JUST BECAUSE YOU'RE PARANOID, IT DOESN'T MEAN THEY'RE NOT OUT TO GET YA

The title of this chapter comes from a poster that Poly Styrene, the singer from X-Ray Spex, gave me while I was living at Gunter Grove. They used to lock her up occasionally and take her off to the madhouse. She'd break out and always make a beeline for my house. They even came for her at Gunter one night, so she bought me that poster, because it was kind of relevant. Double negatives, I love 'em. Apparently, much later, Kurt Cobain turned that idea into a song lyric. Maybe he got it from a picture of my living room at the time.

The sentiment, coming from her, was really charming—wonderful, from a nutter! Nothing wrong with you, Poly! It's the shitstem around her that was wrong. I thought she was borderline genius, really—the songs, everything about her was just hilarious! She may have been inwardly very depressed, but outwardly she was good company, a fun person. She was good fun until the ambulance turned up for her, with the police. And I had a house full of natty dreads at the time. Don Letts and some of his posse had come around too, so they were panicking. They thought it was a police raid. No, no, lads, it's just the nutter squad.

John Gray's mother was a nutter. I liked her, but he felt embarrassed about her. I thought that was a bad way to be, because people are what they are, and I've always found people who were slightly nutty to be highly entertaining and brilliant company. They view life a little bit differently. Two left shoes, maybe, but two left shoes are okay, if that's what you've got. I'm totally inspired by people who come at things in different ways, so lunacy to me is not a thing to run away from. It's enthralling to be in a lunatic's company. If I had a job in a mental institution, it would be the best job ever, to me, the thrill of my life. I wouldn't be able to tell whether I was the patient or the doctor. Sometimes the two are the same thing, you know? I loved that Pete Hammill song: "The Institute Of Mental Health Burning." Burning!

Life at Gunter Grove, however, was starting to get me down. I was bored and fed up with it. I felt entrapped—trapped in my own house. It didn't feel like my house at all. It had become a common room for the flotsam and jetsam of London at that time. Very uncomfortable. I had no way of switching off, other than locking the door and moving out from time to time.

Martin Atkins, who'd joined us at the very end of *Metal Box*—I think he only played on one track and had been playing with us as our live drummer—realized very early on that all this was stifling me. It wasn't claustrophobia. I was getting dungeoned by a situation of my own making. He said, "Look, I live in a really boring flat in Kensal Rise"—or somewhere like that—"Why don't you come and stay at my place for a night or two, just to clean your head of the constant pressure?" But I rejected it. I wish I hadn't, but I did. I was suspicious of doing that on so many levels.

I couldn't give up the feeling that it was *my place.* Once I'd started accepting that my place was the problem, it was a hard thing to come to grips with. It was obvious that my house *was* a problem, and the people living in it—by which I mean other band members.

I'd liked the idea of all us PiL-ites living under one roof. I've always said that when recording a band should all—*all!*—stay together in a completely abstract universe, outside of your regular commute. The end result, though, was that I had no escape at all. I couldn't escape the dilemmas. All the others could go off to wherever. I had nowhere to go to, because that was me, right there. That was everything I had, and I was very proud of the achievement of what I had. But what I had was turning bad.

I'd hardly ever see much of Levene. He'd lock himself in the basement. Weeks could go by. One time there was a really bad smell emanating from down there. I was half expecting a rotten carcass, but it transpired that it was just a garbage can he hadn't taken out. He never bothered himself with domesticities like that. It was all beneath him. But I thought he was dead. An anxious moment, and I obviously over-worried it, so that created a huge row.

Then there was Dave Crowe down there too, beneath his hatch, with similar "lifestyle issues," shall we call them. Maybe that house was haunted—he was in the same room where Jim Walker slept on newspaper with a moose head and no furniture. Strange things came out of that room.

When Dave stopped looking after the administrative side, that created even uglier situations. Jeannette Lee was floating around, through the Don Letts connection, even though they'd now broken up, so it was "Can you help us out here, because we do need some administration occasionally." That worked fine for a very short time, but then she'd start cliqueing with Keith and we'd not see the pair of them for days on end, and nothing got done.

I don't know if you can ever say what Jeannette's role was. I'm sure she'd be mystified by it too. That's the joy and the difficulty of being in PiL, it's a puzzlement as to what your actual role is, because there are no allocated specific technologies. Whoever's available at any particular point to handle a situation must be capable of doing so. Jeannette offered a kind of clarity to us. We couldn't be handling the business side or some of them bloody boring financial meetings, because you were trying to write songs and you just can't cope with it on that level. I went a bit nuts there, trying to run an office and write songs, you can't do it. Answering the phone all day long gives you no time to think outside of that. Structure can be the antithesis to creativity. You need the structure in order to be able to create, but you can't be creating the structure as well.

Jeannette and Keith had a dark relationship, but they were very close to each other, for whatever reason. Who knows? Jeannette could be a real distraction in the workplace. Fellas were mad for her! Some of those fellas were ones we were working with, like Dave Crowe, who fell madly in love with her but kept it all to himself and expected her to know that. He put himself in a world of ridiculous misery there for quite some time. Also, Joe Strummer would be lurking around. It was obvious that a lot of fellas were coming over because they fancied her. That's how life really is, you realize: everybody's after somebody else all the time. It's human nature. It's just sometimes, if the situations in a working relationship get too intrigued, it has to be stopped because then cliques form and separations open up.

Dealing with Keith was a nightmare at the best of times. He insisted that he be on the front cover of *Second Edition*, for instance, but then he didn't like that picture—appropriately enough, they were like distorted mirror images, but he didn't get the fun of it. He wasn't on the same page on the art side at all. He's just not happy with any visual representation of himself. I get that, but you've got to move on and get over your big bad self, and be able to laugh at your own silliness.

I put together the cover for our *Paris Au Printemps* live album, and I put one of my own paintings on the front. If you look at it, that's me at the top, old honky donkey, and Keith and Jeannette underneath, as a pair of poodles. The cover was seen as great fun by everybody except Keith, who absolutely resented his cartoon portrayal. He shouldn't have, I thought it grasped his character rather well.

His heroin problem was becoming a *real* problem. There was a selfishness in the way he went about things, compounded by the drug of choice. There was one particular incident where I tried to get him to go cold turkey, obviously hoping to get him off the stuff. Ever since then, he held a resentment towards me. It seems to be that when you help out an addict in those situations, they don't blame themselves for the position they're in, they blame you. It's deeply unpleasant for the self-made "victim," of course—i.e., Keith—but it's incredibly hard on the helpers—that is, me. When the addict comes out of it, they're just sneering at you. Oh, business as usual. I suppose with a character like Keith, you couldn't really blame it all on the drugs. He's just genuinely nasty, anyway.

We'd do our best—all of us, Jeannette, Dave, everybody that was around in the house—trying to help him out with it, but you wouldn't get much joy out of him.

The world beyond our front door felt no more welcoming to me. Everyone seemed to know that I lived at the house. I still had the tabloid reporters on my back, horribly so. There was always someone flirting about outside Gunter Grove with a camera. To an odd extent, I'd go out and have a chat, and I got to know a lot of them. It kind of became all right then, because they'd be like, "You're just a regular bloke, we all know that, and we're not out to scupper you." And so I could continue my lifestyle somewhat unabated. Fair play, because Gunter weren't no nun's convent.

The fans were a different matter. They'd be carving lyrics into the front door, and scribbling all over the exterior walls. It got weirdly continental. The fan base changed from local punks, to UK punks, to kids from Italy who would demand that they be let in. It's just too much at a certain point. With the SEX shop around the corner—now called Seditionaries—I was obviously on the punk tourist trail.

I actually used to let fans in all the time. "Hello!" "Oh, hello, in you come!" But it got nutty because the emphasis shifted into psychotics, clingers, and just out-and-out intolerable selfish weirdos.

That's what happens. All the open-mindedness in the world really doesn't work the second that one evil cunt who wants to kill you comes in. You let them in and they turn nasty in a heartbeat for no good reason. Unbelievably hard for me to accept and tolerate. Trying to find some passive way of kicking them out of the door is never easy, but you learn from it.

Another worry at Gunter Grove was a warfare that was going on with Jock McDonald, who was a friend of my brother Jimmy's. They had a band together called the 4" Be 2", which was very much an Arsenal hooligan mayhem kind of approach. There were a lot of football firms that started to put records out—the Pistols opened a lot of doors.

Jimmy and Jock just wanted to have a band, and Rambo and Paul Young were in it too. I was very vaguely involved, when I sussed that it wasn't a nasty joke. They were trying at one point to take themselves seriously, and that's when I'll offer any help I can. My dad even got in on it—well, they used his name, saying, "Produced by John Lydon." My dad went, "Well, that was my name before you started using it." "Okay, Dad, that one's yours." Otherwise, I've not much idea what they were up to; I just know that when the taxman came a-knocking, it weren't me what did it.

However, I soon got embroiled with Jock and his nonsense, because he'd fallen out with Paul Young, who was still at Gunter. Jock's brothers were friends of mine, and Jock would turn up with them to try and intimidate me, and they'd go, "No, Jock, we'll beat you up first—John's our mate!" This kind of silly nonsense.

My brother Jimmy is a year younger than me, and we're very close. We're incredibly different, and that's probably what makes us so close. I'm the quiet one, the elder one, the responsible one. But I'm also Johnny Rotten! Jimmy's a saucy fella, he's nonstop comedy, like a stand-up. Everything in life, he finds a laugh in it. I suppose we both do, so it must be a Lydon thing, even though both our parents were so quiet. Maybe we just grew up learning to use words to entertain ourselves, because Mum and Dad hardly spoke. We'd just bugger off and get up to all kinds of hell, and even though I'm the eldest, any troublesome situation we ever found ourselves in was Jimmy's fault. I'm not saying that spitefully, I'm just saying Jimmy has a knack of finding, you know, *catastrophe*. He can't help it. And big brother was there to try and sort it out.

Dublin was a good example—a classic case of, "Another fine mess you've got me into, Stanley!" I was out there in October with the 4" Be 2" and their traveling army of friends and family, all out for the beano.

I got myself into a whole heap of trouble when me and a friend went for a quiet drink in the afternoon. We were in a pub called the Horse and Tram down there by the river next to our hotel. I'd just bought a drink when the landlord seemed to take offense to me. I don't think he liked the way I looked or sounded. I guess I stood out like a sore thumb. Something was said and someone snatched the pint out of my hand. A few of the locals—at least one of whom it later turned out was off-duty Garda—decided to make themselves busy and join in. They were definitely playing a bully game on me—they thought I was some kind of English eejit. One thing led to another, and a bit of a scuffle ensued. There was a lot of shouting and fists began flying. Just not mine! I've said before, I attacked two policemen's fists with my face. That's pretty much the truth of it.

I went back to the hotel to change me clothes and was followed by one of the off-duty coppers, who promptly arrested me for assault. It looked like I was only going to get a slap on the wrist, as I was let out pending charges. But that all changed later. I went off and met the 4" Be 2" at Trinity College and then went back to the hotel for a drink, when the police arrived and took me away. There had been a radio show with my brother Jimmy and the gig in between. Perhaps somebody in the authorities had listened to the radio and had me arrested again after realizing who I was. I don't know. The whole thing was a farce.

I found myself in court the following morning on a charge of "Common Assault." They refused me bail despite the fact one of my friends, Johnny Byrne, had offered to put up the £250 security. He also got me my lawyer. Jock McDonald helped too. £250 was a lot of money back then—thank you, Johnny, it's not been forgotten—but they refused it point-blank. Despite the fact that the guy in the dock before me had been given £50 bail for hitting someone with a hammer at the 4" Be 2" show the previous night. The prosecutors had tried to claim I'd called the bartender an "Irish-pig." My lawyer made a point of telling them I was Irish and both my parents were born in Ireland. My case was adjourned until the Monday and I was taken to Mountjoy—a notorious prison filled with IRA and UDA terrorists and all sorts of psychopaths.

On my arrival, the warders decided to make an example of me. They stripped me, threw me into the yard, and hosed me down. But you know, you can strip me, cover me in flea powder, and laugh at the size of my penis, it doesn't matter. It—does—not—matter. Over the years I've noticed that when these institutions get hold of you, the one thing they're trying to embarrass you about is your nakedness, and your penis. Let me tell you, Johnny's got a perfect penis to laugh at, and he don't care. That's not ever going to be a problem.

Inside there, it was tough—really, really tough and hard—a punishing regime. I tried to have a routine but they made it impossible. The warders would wake me up all night long with their truncheons and make me stand by the bed. With hindsight, what you learn is, when the institution has got a hold of you, then you quickly have to learn to adapt and blend in and try to merge into the shadows. Which is, of course, impossible for me. So ignore my advice, and ultimately just be yourself. That's all I was.

You were allowed an hour of telly, and who came up but yours truly on the news! Then there was a program about the history of music and yours truly was on that too, with all the other inmates surrounding me and looking at me. The embarrassment! I just wanted to crawl under the concrete. The prisoners were fine, though.

Just by being alive, the warders didn't like me and that left some breathing space with my fellow inmates. "God, look what he's got to go through." A lot of the prisoners felt if they stood next to me or chatted with me, that they'd come under the glaring eye of the warders that were trying to make life punishing for me. But I'm like, "Is that the best you've got to offer? You're not gonna make me uncomfortable about being myself, I don't care."

I was back in court on the Monday where the judge sentenced me to three months in prison. Thankfully my lawyer got an appeal in, and this time I made bail. I immediately went back to England, and the very next day started work on the *Flowers Of Romance* album.

When the appeal finally came up, months later in Dublin, I knew my career was on the line. I had to go back. It really was squeaky-bum time. If I'd lost that appeal, my sentence would have been doubled to six months. As you can imagine, I was under massive stress at the time, but the case was thrown out of court in ten minutes. The judge saw right through the contradictions in the two witnesses' statements. They didn't even bother to turn up, at least not till after the case had been dismissed. I was acquitted but not before I was *asked* to make a £100 donation to the "poor box." That's Irish justice for you.

Through 1980, many people had found great entertainment in *Metal Box*'s claustrophobia. All well and good, but then comes the problem of a fan base that wants to hear that sound on everything you do, forever and a day. If that's what you want, I'm not your fella. I don't like doing that.

By now, with Wobble gone, I was into stripping the bass out completely and researching drum sounds, using a collection of loops. I was really angry at the time with Keith Levene's despondency, his seeming unwillingness to put in any effort towards helping in the project, which was of course largely due to his predilections. Most of the time, he was upstairs playing video games. He'd just bought the model of Space Invaders, which came in a little black triangular box, and he'd become completely addicted to that. You couldn't get him away from it. That would be it, morning, noon, and night, just gawping at these dots moving up and down a screen. To my mind it showed a very compulsive behavior. He can't get out of things; he goes too far.

So I got on with it by making drum loops with Martin Atkins. He'd already committed himself to an American tour with his band, Brian Brain, however, so I only had him for a short while. Once he'd gone off, I had to garner what I could out of those tapes, and formulate patterns which I then put vocals to. I had a great time doing that with Nick Launay, a trainee engineer/tape operator at Townhouse studios in Shepherd's Bush. He was all that was available late at night, and thank God for him, a happy coincidence.

When Keith finally came down, oh God, was he snooty about it. He played a little bit and then would just vanish again, so I got on and finished it with Nick. We put together a sound based on drum loops and vocals, and then I started adding things like bits of piano, bass, and saxophone. That's what we had to do to get the thing done. I could hardly call myself a saxophone player, yet there it is on *Flowers Of Romance*. You turn the pressure into a useful tool.

I'm playing just about everything on the album. Keith came in for a very little bit and behaved like a sour pussy. For me, it was at that point my proudest moment, because, for the first time ever in music, I'd done it all without any backing. I'd put things together completely on my own, without having to share the workload. I showed that *I could do it*. I found my way instinctively around various different instruments, and ashtrays on pianos, and all of these things were firing left, right, and center.

I was so full of ideas, because I was free! The prospect of spending six months in Mountjoy—that was a bit of a harsh reality. If the charges were proved, that's what I would've gone down for. And, dare I say, falsely accused. All of that trauma is definitely inside the music, somewhere.

There's a song that particularly refers to the situation called "Francis Massacre." It's about notes that were being sent down to me from prisoners on the higher level, who were asking me to pass them on to people on the outside. But that was impossible, as I was told I would have been searched, so I had to flush them. I wasn't taking any chances. One note from a guy called Francis Moran I made a song about—"Go down for life, Mountjoy is fun." The song was inspired by a combination of all those notes.

"Francis Massacre," in my mind, is directly related to "The Cowboy Song," the B-side to "Public Image," in that it's another yippee-aye-oh clippy-cloppy bunch of noise. I find those things to this day refreshing to play, because I'm aware of the situations I was involved with at the time, and that to me was the best interpretation—screaming angst and cacophonic jagged edges.

"Flowers Of Romance," the song, is at the opposite extreme. I loved that song as much as I love "Sun," which is a thing I did more or less on my own too, for my solo album, *Psycho's Path*. They're *my* anthems, they're in a happy-go-lucky pop format, but to my mind they fit into the same context as, say, T. Rex's "Life's A Gas," which will be forever a guiding light to me. Those are my versions of anthemic festival music, and this all goes back to seeing the Who live at the Oval cricket ground in 1971, with the Faces and Mott the Hoople supporting. Aynsley Dunbar was the DJ between bands—I loved his stuff, he was a great DJ—and he put on "Life's A Gas." This audience thought of themselves as hard-core rockers and, you know, "Boooo!" to anything T. Rex, which they viewed as pop trash and a sellout, but it cut through them. Glorious to hear it over a PA system! It's just an open, happy thing, as is "Flowers." Other people may hear a darkness in there. Well, musically, I'm something of a gypsy, I've got to travel.

"Four Enclosed Walls" has a very Muslim call-to-prayer vibe. What I was doing lyrically was understanding that anything you're blaming on these latter-day martyrs for the Muslim cause, you have somehow to trace back to what the Christian crusaders did centuries before, invading their country with a religious nonsense belief, to explain away the fact that they were out for filthy fucking destruction and thievery. So it's a long, ongoing process. This is how far back terrorism goes, how the crusades can lead to a tragic conclusion. The lines, "I take heed, arise in the West, the new Crusade," are against all religion, because, praise be to Allah, He would be horrified with what the contemporary followers are doing with his message. Listen, Allah's me mate, so's Jesus. Seriously, I won't be swayed from that, because that's what these religions were supposedly offering you, a world of friendship. They've misappropriated it to a world of warfare and abuse and, like all religious wars, that's mutual destruction.

My voice on this record had an amazing hollow sound that came from the stone rooms at the Townhouse and the Manor. The sounds could be so crisp, particularly with drums. I've always had an appreciation for the Led Zeppelin drum sound, so when it came to *Flowers*, that was a yippee and a half, to be able to move into that area. I remember reading somewhere that Zeppelin would record everything separately, and John Bonham would do his drums in his stone cottage. Fantastic. Sometimes the putting together of a song is not necessarily all in the same room at the same time, or even on the same continent.

I suppose I could've just broken up the band at that point, but I thought, "That's all right, I've got tolerance for that—because I've just got out of jail!" My energy zone was way rampant. I thought it was a great record. To this day, it thrills me to pieces when I play it. How bright and fresh sounding it was. Nothing like that had been recorded in that way.

On the album cover, we introduced Jeannette as a member of the band, with the *best* picture of her. She looks like a girl having a cheeky party in Spain, with a rose stuck between her teeth. It's like an English-abroad kind of thing.

As a band, however, we weren't able to go anywhere to publicize the record, which eventually came out in spring 1981. The "hard to get gigs" side to PiL's existence had reached a total stalemate in England. So we started to think about looking much further afield to play, just to get away from this ridiculousness of promoters not wanting to back us because they were fearful of riots. We felt like we were being cut out, being written off as unwanted. So, in order to survive, we had to think outside the box, and keep PiL a thoroughly mobile artillery unit.

We were also trying to develop the idea of PiL as an umbrella organization for multimedia activities, but this was met with resentment whenever we mentioned it publicly. Keith and I had talked about it on a couple of TV interviews in New York the previous summer, with Michael Rose, and then Tom Snyder. These people really gave us some gyp. We were talking about living outside of the chart system, in a world of creativity that didn't have to come cap-in-hand to any of the corporations or institutions, and how we didn't give a tuppenny fuck about not getting a Grammy. The Tom Snyder interview was particularly frosty.

Years later, up came a chance to do Tom Snyder again. I sat down with him and had a really bloody good talk, during the interview and after. I'd call him a definite friend. I really liked him. He started to send me all his old interviews and stuff, and also some really wacky music things, and then a couple of years later he died. So that was a terrible thing, because there were plans to do things with Tom. He actually really "got" the umbrella thing. It's chaps like that, when you have conversations with them, you realize that age is irrelevant; it's the ideas that count. Sometimes you have to have an awful lot of patience until the situation is there for you. Tough tits when you're young, though, the opportunities aren't there.

Keith's understanding of our broadening company ethos was

typically selfish: his sister was getting into knitting, and Keith wanted to introduce her into the umbrella of PiL as someone who would manufacture knitted sweaters. "O-o-oh, GOD! Just, *no*!"

I mean, all of us balked at the prices Vivienne Westwood charged for her mohair sweaters back in the SEX shop days. They were great, everybody wanted one of them, but financially they were beyond most people's reach. My mum made me one once, but without the holes. "You know, Mum, this I couldn't wear in Alaska, it'd be too hot"—really heavily made. She went, "Dem holes ar' silly, who wants a jomper wit' holes?" "Me, Mum!"

But still—"No, Keith, we're not taking up a knitting division . . ." That's where nepotism creeps in, trying to get your family involved. That won't ever work because it causes all kinds of problems. I didn't see us as trying to make cheap imitations of something that somebody was already doing. It's all right if my mum makes me a sweater, but I'm not gonna set up a commercial line. This would be Keith all the time. He'd consistently come out with these wack ideas. That's all well and good, but in the end it gets you feeling like, "Oh my God, not again. Could—you—just—*stop*—*talking*!"

The police raids at Gunter Grove were getting ridiculous. In spring '81 there were three in a three-week period, every Friday, and that was just too much to take. They'd dismantle the place, smash the front door in, tear everything apart, then go, "All right, thanks for that!" and leave. Or drag me down to the station for one reason or another, and then let me go. They'd be dragging me in my pajamas, barefoot. No lift home. Of course, not being fully dressed, I'd have no money on me, so I had to walk back down Fulham Road in my bare feet, pajamas, and red dressing gown. Many thought I was copying Bob Geldof's band's keyboards player, Johnny Fingers, wandering around in me jim-jams. Very embarrassing. It was shouted out, "Oi, Fingers!" Oh, for shame!

Then I'd get back, and of course the front door is off its hinges.

It got to the point where I had to leave a hammer and a fresh set of screws, latches, and nails by the door, ready for the next one. In modern times, you'd have a complaints department to ring up, and get the damage taken care of. In them days, you had to fork out yourself. Luckily, I still had Paul Young living with me. He was a carpenter on the building sites, and would just bang it back up into place.

The only explanation given for the intrusions was that it was a drugs raid. Suspicion of illegal activity. What would be the evidence? "An IRA flag in the back window." "Er, actually, that's the Italian flag." My neighbors had given that to me—the ones that asked me not to play the reggae too loud after midnight—because I had no curtain on the back window. "We can see what you're doing, and we don't want to!" These days, of course, it would be cameras akimbo. Life was different then, people tended to help each other out, so you had a respect. You'd say, "Ouch, sorry, I won't play it like that at 3 a.m. again." Or, in this case, "Yes, I'll put up your flag, so you can't see my botty." I'd be perfectly happy with that. You won't find neighbors bitching bad things about me, because I look out for them and they look out for me. To me, that's a very important part of life.

The police unfortunately had a different attitude. It got to the point where I started to know them. I knew them from hanging around the pubs I'd be in. They'd be there, supposedly undercover—I don't know, maybe waiting for a gun or drug deal to go down. Oh, for God's sake. Let's put it this way: the police in them days had a vicious intent and a suspicious attitude for anybody that was outside of the norm and was an easy target and unprotected by the alleged society at the time—which I definitely was, at least according to the scandalmongering of the newspapers.

The very last raid, the "Johnny Fingers" one, actually happened on a Monday morning, but that was the worst of the lot. Barking angry Alsatians, the whole thing. And do you know what they pulled me on? When they burst the front door open, I came down brandishing one of them antique swords, and I came at them. I didn't see their uniforms. However, that was viewed as an assault on a police officer. So that was the scam, that time. The laws are somewhat different these days and they're more protective of the house owner, and indeed they always should be. To my mind, whatever you do behind your own front door is entirely your own business. Entirely. I could never agree to or justify those kinds of raids. Ever.

This was very early Monday morning, at the crack of dawn, absolutely exhausted, tired, but we had nothing incriminating in the house, save for a teapot full of herbal—which they never found, which was bizarre because they even kicked a speaker over that was behind it. They never twigged! This has a lot to do with Satan, the cat. Satan bopped about, terrified of all the barking, and one of the dogs went for him, knocked the pot of pot over, but didn't really notice it because it only had eyes for the kitten. How odd: they don't even know how to do this stuff properly! Poor old Satan was so terrified of the dogs that he ran away and never returned.

The police attention was too much; it was overwhelming. It meant, for instance, that I couldn't go up to Finsbury Park and hang out with friends, because there'd be *that* following me. They'd be there making everybody feel uncomfortable, and therefore I'd be at fault for bringing that into the manor. It destroys you socially, and for what result? I can't imagine how much money they were spending on putting together a police raid. Surely it doesn't come cheap? And a lot of time wasted, and a lot of this pertaining back to the fact that I had a £40 fine for possession of amphetamine sulfate.

It was very intimidating, and the clear implication to me was that they were trying to run me out of the country. There'd been the three raids in a row, and a couple before that. It became apparent that they were bored with what they were doing. They made it clear to me that they were just following orders—"Don't take it personal, John." We were on first name terms—as I said, a couple of the officers that came in had been following me when I'd go to pubs in Notting Hill. Keith Burton, who was working at Virgin at the time, and actually a few years later became my manager, recognized the police straight off one time I went to a pub near the label's offices. He went, "Oh my God, look, your shadows are here!"

It just felt futile staying in this unprotected environment. There's nothing in the media to back me up, or save me, or declare this to be unjust detention, or unwarranted. The tabloid press only wanted to report bad things about me—and so, time to go.

It was Keith who went to New York first, around that time. It was all to do with this idea that we'd be an umbrella of thoughts. Well, maybe the umbrella was a fishnet. With hindsight, he was obviously following a similar route to Sid—the availability of heroin in New York being famous at the time. Anyway, he'd allegedly gone there on a jolly, and once he was out there he'd found out about this new system of cameras and screens that were being installed at a nightclub called the Ritz. The end result was that we all went out there to work on that too, and we never came back. We lived there for about three years, all told.

The months before I left London for New York almost felt like a holiday. I was trying to gather my thoughts and work out my next move. To that end, I hooked up with Rambo and was going to stay at his place. He'd said, "Gunter Grove is killing you. My parents are away and I'm staying at their house, come over and you can get your head together." He was going to sort me out, because he was aware of the pressure I was under, so we were going to have a laugh together. We had been to Margate on a coach trip with a few of the lads that day. On the way home we'd bought crates of booze and were going to party until the following week. First night, a phone call comes in. I still don't know how they got Rambo's number and I don't know who transferred the call—it must have been someone at Gunter. It's Keith, all bouncy on the line: "Come to New York, we've got a chance to do a live camera display at the Ritz." I went, "I'm round at Rambo's, I can't book a ticket." "Don't worry, it's waiting for you at the airport." So I let John down, because we were gonna have a hoot together, but off I went the next morning.

We still couldn't get gigs anywhere, remember, but the people at this place the Ritz were going to put us on for two nights, as some kind of live music-slash-video production. The idea was to project multiple camera shots live onto one big screen. It was an interesting concept that I thought had heaps of potential, particularly bearing in mind that Jeannette used to carry a camera around in a violin case, and how we were thinking, "Film, film, film!" We realized how important filming was to the Pistols, and yet how little footage there was of the actual events. We wanted everything to be catalogued, but also to think outside the box when it came to live performance with the band—not just us playing in the standard format, but creating other kinds of situations. It could be many other things going on at the same time. Open-mindedness really, and . . . Bingo! A riot started. Or it didn't. It wasn't a riot, it was a fiasco, but an enjoyable one.

The idea was that we'd stand behind the screens with a record playing. We'd make a few noises over the top, with some live drums to bolster the sound. We got a drummer from a music store, a very old fella called Sam Ulano, who had a jazzy sensibility. His kind of music was Frank Sinatra. We could have picked any record to put on that turntable but I was insistent on it being *Flowers Of Romance*. I knew that would annoy Keith no end, because of his dismissive and withdrawn attitude during its actual recording. "You get what you deserve in this band, mate. What—you don't know the guitar parts? That's because there aren't any—you weren't there, you were upstairs playing Space Invaders. Here it is now, deal with it!"

So the album's on the turntable, and Keith's there with his guitar, going, "Brrr twang bang," deliberately being awkward, and the old fella's playing drums to it, and it's fitting in quite nicely, and everybody's got a camera and they're moving around the place, and all this is being projected on a screen in front of us. We're on the stage, so people are seeing the screen rather than us—a screen of loads of different images of each of us simultaneously, split-screen, multiscreen, every combination of cameras you could imagine.

The control boards of the cameras were being manipulated by a very fun American chap called Ed Caraballo. He was converting all these images live for the screen, with flashes of audience and/or whatever. Because I was behind the screen and seeing it all in reverse and up close—and my eyesight is not good—it all looked to me like a Tangerine Dream album cover.

Then—oh dear!—the record skips, because the people leaning on the front of the stage are pulling the canvas mat that we're standing on, jogging the turntable. A front row of elbows is a powerful force—it's almost like water bursting over the dam. And by that pulling, the record goes, "Skip! Skip!" And suddenly: "Boooooooo! It's not a live gig! Fraud!"

It seemed I'd no sooner got off the plane than I was practically doing the gig—I had no concept that this had been advertised as a proper live show over the radio. That wasn't what I'd agreed to. I wouldn't have turned up if I'd thought it was going to be some unrehearsed nonsense masquerading as a gig. I thought it was just a yee-haw, for a crate of lager and a laugh. But for a moment there, it ended up like we were going to get killed. People were chucking bottles, the usual mêlée.

It was absolutely nothing I wasn't used to. I may have goaded the audience a little—I'm Johnny, it's my business. "Silly fucking audience!" I told them. That was the point where it got to the real boos and the hisses. That's an instinctive response. If they felt cheated, then I felt cheated with them. And then oddly enough we're back to "Ever get the feeling you've been cheated?"—the last Sex Pistols gig in San Francisco.

You've got to take control at that point and explain through an aggressive stance that this is not what you've been misled into believing. But at the same time, "Come on, *it is entertaining*—it's worth the money. It is different!" It's an experiment into the future, and now if you look at every single one of the modern pop bands, they have these enormous screen projections going on behind them, not to mention the turntables. That's the idea we were initiating. I'm not saying we invented screen projection, but we invented the cut-up thing of it.

Security just fucked off, and people started invading the dressing room. The only person who buggered off very quickly was Keith. He just basically abandoned it, the very situation he was so proud of, and suddenly it was all smiles again. Jeannette was great fun that night, she hung about. People were saying, "That's the nicest riot we've ever been in!"

I suppose the casual way I approached it all was helpful. "Why don't you all come out to the bar and drink with us?" They went, "That's a very good idea," and did. Then the staff tried to close the nightclub early because they said they didn't want a repeat performance of the earlier catastrophe. They closed the bar about twelve-thirty, one. And then canceled the following night's show because of the so-called riot.

So this alleged fracas was actually pretty hilarious. There was virtually no damage whatsoever to the screens or the cameras. The police were laughing, they even sat down and had a beer with Johnny Rotten. They were just "Hey, are you that guy John, man? You're wild and crazy, that must be really disappointing, that was only a pussy riot!" Maybe I was the precursor to that all-female band from Russia, after all.

The grand delusion and illusion of New York, wrapped around the ready availability of chemical highs and lows, was not at all why I wanted to be in New York. I wanted to be there because it was cheaper, and we could actually get gigs there. They were seemingly readily available, and we could earn money.

For a while after the Ritz fiasco we stayed in small hotels, at cheap prices. We'd share two rooms among the lot of us, but soon we started renting a loft for not a lot of money. This place was on West 19th Street, between 10th and 11th, almost exactly opposite the Roxy nightclub–dash–roller rink. It wasn't Midtown, but it wasn't Downtown either. It was a nondescript area aptly known as the meatpacking district—we were surrounded by container trucks wheeling in and out meaty corpses, 24/7. There was a bar right around the corner called Moran's, where they did brilliantly cheap stuffed clams and the proper imported draft Guinness from Ireland. Delicious. There was even a women's prison behind the loft. In summer, the women would be screaming at us out of the cell windows. The truck parking lots around us were gay cruising haunts. It was quite a place.

It was dirt-cheap and industrial and full of incredibly seedy gay nightclubs that were nothing like anything we used to know in England. These were middle-aged men with beards and no arse in their pants, lots of bending over behind trucks, and wet beards—just full-on filth, pure decadent filth.

Underneath us was a chop shop, where they were welding iron plating on limousines for God knows what reason—never ask questions in New York. I knew it wasn't for the government. It was incredibly noisy, but that was to our benefit too, because they didn't care about the noise we made upstairs.

The loft itself was absolutely amazing: it was 2,500 square foot, with one squalid little bedroom—mine—and one large back bedroom, which wasn't really a bedroom, more an office space, which I gave up to Keith and Jeannette. I took this little hole in between that and the kitchen, because I always wanted to be near the food. When Martin Atkins arrived a couple of months later he took the front room, which wasn't a front room, it was actually a stage. It must've been a venue of some kind at one time or another, because there was a little room in between where a mixing board used to be.

The whole place was absolutely geared towards us creating a Public Image kind of space. But we never got round to it. I wanted to put in a DJ booth and a mixing desk, to turn the place into not only where we slept at night, but a PiL kind of rehearsal–slash–live gig kind of environment. Naaahh. Couldn't get Keith involved with anything. Wouldn't see him for days on end, and always Jeannette going, "Oh, you can't talk to him at the moment, let him get over it." *Aaauugh!* She mollycoddled that fool terribly. It was quite astounding because Jeannette was a hardcore girl. Jeannette would drive men crazy. A big flirt, and yet she had this connection with this idiot, which didn't make any sense at all.

Underneath our block there was a garage, and the idea was that from there we could spring out into all the surrounding states, and play lots of nightclubs and theaters. We did that eventually, but it took a while to "settle in." When we finally got out there and did it, it involved driving there and back, but I liked that because it reminded me of the early Pistols, when we used to do the dates around London, or go on raids up north. The problem now was, we were all going back to the same place. With the Pistols, at least we'd all go off to our own hovels. Here, we were all hoveled together.

Still, it felt great to be there. I imagined that I wouldn't end up in jail quite so frequently, and I could get on and do whatever I wanted to. I had no attitude of connecting with the New York scene, and indeed all of us knew damn well that PiL wouldn't fit in with that. So whatever it was that we were up to it was going to be outside of that. And indeed it was. The people that did come over and wanted to speak to us were mad artists doing fabulously, stupidly, interestingly different things. New York's full of that but it's not a one-level town. There's so much going on in it. At least, there was then.

Jeannette was particularly into opening up the activity of filming out there, because she was very interested in that and therefore we were very interested in her for that. We hoped that whole thing would get going out there, but it didn't. After the Ritz performance, nothing much happened really. All of us are to blame, but we all treated it as a vacation and started to indulge in our own activities, and not jointly, and not sharing those instants, which is not so great.

Jeannette had a very outward character: she could make friends with anyone at any time, and just had a natural way of finding her way into being an important member of a scene, like the nightclub scene. She got to know all the doormen, God knows how. We'd go to clubs because of Jeannette's activity in those areas. I wasn't too enamored with the music scene in the dance clubs in New York. I was not that impressed. The hip-hop thing was taking off and there were radio stations that played New York hip-hop nonstop, but I never really got into it. I thought it was all a little bit samey, with hideous keyboard noises that reminded me of when the Osmonds did "Crazy Horses." That dreadful synth sound.

On the plus side, America had a hundred-odd TV channels, while Britain still only had three or four. How heaven was that? And if you missed anything, it was repeated on another channel later. I viewed it as research. I absorbed myself completely in Americana, in American TV culture. I even found the TV advertising channels to be thrilling, trying to sell you any old bit of plastic. Lessons in how to separate a fool from his money! I found them fantastic, and shocking that people could get so involved with that, and be so desperate as to buy a Glo-Mop at $21—a mop that glows in the dark. Why are you cleaning your kitchen in the dark? You know, there's $21 you could have spent on the electricity bill.

America was way ahead. Everything in New York was late night. It was amazing to us, coming from England at that time, because it really was twenty-four hours. The only thing that would've been equivalent to it in Europe was Berlin.

On the rare occasions I went back to England I'd be aware, from the airport in, how low the houses were. There were no skyscrapers and the whole thing seemed pitiful by comparison to a Manhattan skyline. Shocking. I was well aware that I could've waltzed into the American attitude of, "Oh, you're so quaint over here!" but I understood from an American point of view finally what they meant by "quaint." Our version would be "rinky-dink." But still I loved what I came from, because without that I wouldn't be me.

Out in New York, there was an early reminder of London life, when the Clash came over to play at Bond's on Times Square, for their residency in May–June '81. So they followed me to Jamaica, now the fuckers had followed me to New York! I'm dealing with me PiL problems and in waltz this lot. I can't remember how many nights they played there—was it something like seventeen?—and apparently filled them out every single night. Bearing in mind that their songs didn't have any content, and they really didn't seem to stand for very much at all other than this abstract socialism, they still pulled that off. So Bernie was a good manager after all.

I went two nights running, and I could've gone every single night if I'd wanted to, but my God, it was bad theater to me—exactly the same procedure both times. As a band, they had nothing to offer, character development–wise. Joe just ran up to that mic and screamed, "*Aaaargh!*" in that exasperated strangulated way he had, night after night after night. It was just a pub band—they might as well have been Eddie & the Hot Rods. And yet the masses thronged to it. And so whatever it is that I'm doing in this world, I'm not for the masses, I'm absolutely not.

It seems to be, if you don't have a clear directive other than some vague socialism, you're gonna get problems because an audience will find you not easy to deal with. The Clash were very easy to deal with; they weren't offering very much at all. They never made you have to think about yourself and your lifestyle. In fact they made you feel comfortable. There's the rub. And poor old Johnny Rotten ain't gonna ever make any of you feel comfortable.

There was a lot of cocaine floating around the New York scene.

It was everywhere all the time. You couldn't avoid it, no matter where you went. Just at the local Spanish restaurant there'd be anything available over the counter. It was very open in that way, New York. Apparently a lot of it was Mafia-controlled too, so that was keeping the crime side of it somewhat curtailed. It led to excessive behavioral patterns on everybody's part, and again, cocaine is not a drug for creativity. Not at all. For me, taking coke gave you thirty seconds of high anxiety, and then three hours of the flu—until you did the next line. Then the anxiety was doubled and so was the down, and on and on it goes, until you find you've done so much that you can't come down, you're just in agony. It was not my favorite thing at all, but I've got to say I fiddled about for at least a good year in that area.

Cocaine kind of numbed us out of being creative. It makes you feel guilty, whereas heroin apparently—well, I know so, from watching my friends on heroin—kills the guilt. Cocaine actually accentuates guilt, makes you feel bad about yourself. It's not an escapist drug. It's the world's biggest fucking foolishness. Unless you're living up in the Andes and you need something to give you the energy to go on, I can't find a place for it. Let's just say that, like most things I've ever dabbled with, I've done it to fucking death. I put it in the same region as Southern Comfort—there's something I won't touch ever again. I can from time to time be a creature of excessive stupidity. I'm well aware of the warning signs and yet I'll dive in and just go with it, but overdo it. I tend to lack subtlety. Maybe in later years I'll catch onto that one, the idea of being subtle.

Heroin, on the other hand, that's the real killer; I'd rather deal with the things that agitate the hell out of me, than let it lull me into a false sense of security and then make me desperate for the next hit. That one ain't gonna happen.

A lot of bad drug scenarios went down around us, and created pointless, useless dead-end activity. It wasn't so much centered around being creative as being selfish for self-entertainment. For me there were always drugs about, in all various shapes and sizes, but I had a proclivity for the energy boosters. I was not one for the inactive drugs.

I knew Keith's proclivities, of course, before I asked him to join the band. I was fully open-minded to that, but when you notice that withdrawal symptoms are regularly disrupting your creativity, that's got to stop at a certain point. You hope that they can see it in themselves, and you try to gently maneuver them into a frame of mind where they realize that they're enforcing an incredibly selfish habit upon everybody else, and expecting us to—what?—sit around waiting for them to wake up from this lazy-arse escapism.

People who find me difficult to work with say that I have very lazy behavioral patterns. I do when I'm recharging my batteries, but all in all I can't stand lazy buggers. You've got to be thinking, you've got to be using your brain, because you've only got one life to live and that's my constant driving force. My brain doesn't stop. Dreamscape, we're back at that. It's very much like, well, the body stops but the brain doesn't. It's all part and parcel of what I am.

The closest I've ever seen to what I'm talking about was a live gig I went to in New York around this time by Robin Williams, the late great American comedian. It was nonstop freeform and funny. Utterly hilarious. I didn't want to go, I was alone, but there was a ticket at the box office, and I thought I might as well. I hated that *Mork & Mindy*, but I gave it a chance and he was fantastic live. A lot of the way he was freeforming the jokes and randomly jumping from one situation to another, then somehow or another they all seemed to make sense eventually, is exactly how I feel about myself, how I think I am.

That would be the kinds of things I'd go to see. A lot of "Off Broadway" productions, for instance. Someone who was very good to hang out with was Ken Lockie, when he came to stay with us. He was always hovering around on the outskirts of PiL, but I didn't really want to commit him to the band. I never saw him as a proper PiL kind of person—too quiet. He didn't have that get-up-and-go. A very studied musician, is Ken. He knows his way around all the notes and formats but that's not to me a very useful tool. He was quite a respected musician and friend of Keith's, but as soon as he moved into our apartment, Keith didn't like him anymore, and so I'd end up hanging out with him.

One of the maddest things he took me to was a movie by Rainer Werner Fassbinder, who was trendy at the time. Off we trotted to this cinema on the Upper West Side and it turned out to be all about the gay sex scene in Berlin. It was about one man's life, he was a schoolteacher, and he was explaining the cottaging scene and the toilet scene in Berlin. Grr-r-r-r-raphically! You know the phrase, "I didn't know where to put my face"? Wrong! It was a bit of an eye-opener. I've always known gay people, but it was very challenging for us two to be sitting there watching this. We never talked about it when we left, it was so strange. Then a couple of weeks later, we were in some other social scene with a crowd of people, and somebody said something, and I caught Ken's eye and we both just burst out laughing. All's well that ends well.

I also invited my friend John Gray over for two weeks—he wanted to have a holiday. I got a load of beer in. And guess what we got in to watch on laser disc? *Aguirre, The Wrath of God*, and the other mad art film Werner Herzog did, about the paddle-steamer, *Fitzcarraldo*! Wowzers, what an evening of weirdness! It'll drive you nuts, unless you've got a real good place inside your head for the absorption of that kind of negative humanity. That nitty-gritty kind of thing that Herzog does, he abuses his audience. But not really, you come away and there's a good learning curve in there. I suppose, really, it's the study of temperament.

Little did I know that the art-movie world would soon beckon me to stand in front of the cameras again.

HUGS AND KISSES, BABY! #2

For all the obvious excitement of moving to New York, the whole thing also broke my heart. When I made the decision to move, there was no Nora involved in that, but we never separated mentally. I had obligations towards the band, and it looked like we could get up and running out there. In all of that, I never forgot her. In fact, I couldn't go out with anyone else. I didn't want to, and I'd feel terrible about even considering such a thing. That was the ultimate woman I'd met there, and I wanted that to be forever.

New York offered all manner of temptations, and "no" was the answer to every single one of them. Absolutely no interest. I didn't care if people thought I was a weirdo. I knew what I wanted to be committed to, and what was worthy of commitment.

I'm very, very loyal, me. I hook into one thing, and I wanted that one thing to work. I'd never felt comfortable being a Jack-the-Lad playboy. I can't stand flippant one-night stands. The next day, I'd feel horrible about it, how pointless it was, and what the hell was that all about?

I'm always looking for deeper bonds and relationships and connections. I basically like to study the human beings around me and find out what makes them tick and love them for that, warts and all. That's a far more enjoyable process for me, rather than just ping-ponging about for no particular good reason. You don't learn anything, you get nothing out of it, and before you know it, you find yourself utterly alone, and stupid for it.

You have to learn to give, and that's the big thing, why relationships work—it's a give situation. And the giving is better than the getting. Well, that's my experience. Show me a happy multimillionaire. All these situations are relative. Any human activity is an insight into all human activity. If you're just busy collecting trophies, well, you've actually achieved nothing. Zero. It's just stuff cluttering up your toilet.

I don't want to put down, or sound like I'm viewing negatively, the freedom to operate as you want; that's fine for certain people and indeed that will always be their way. It's not what I can put up with. It's the talking, the chats, the warmth of being close to each other emotionally—that matters much more than random free sex.

I just thought Nora was absolutely adorable, fantastic, unlike anybody around. There was none of that hippie florally nonsense or Biba rubbish going on with her. Nora was directly related to 1940s film noir, trying to find herself in this sense of restrained but sexually attractive gear, and outside of the world of flip-flops and long floral-print dresses.

Nora liked to show a bit of knee in a pencil skirt. From my early childhood, that was the proper look of the skinhead bird. When Nora did wear floral prints, they'd be above the knee, and "happy-go-lucky skipping across summer lawns." You know, that "free of fashion victim" stuff, but then at the same time incredibly well thought-out. The perfect clothing. Clothes are so important, in an unimportant way. Once you understand what you're doing with clothes, you don't have to think about it anymore, you do it naturally, and you wear for the occasion. If it's hot, don't be wearing a studded leather jacket and calling yourself a punk! You're not a punk at that point, you're a plonk. There are moments where you have to wear according to the weather and the location. Adaptability is really what I'm talking about. But I digress here. Sorry, it's the way my brain works.

With me in me bondage gear, and then my Kenny MacDonald suits, I suppose visually you'd have thought, "This is not going to work." But mentally, it did on every level, and all the suspect characters that call themselves my alleged band members and friends and affiliates didn't quite grasp that. The sparks flew. I don't know how you explain sparks, where charisma leans into something just incredibly inviting. There's nothing that can stop that, not on earth, and that's my explanation of love.

It's amazing how people will try to chip away at a relationship so deeply founded in truth. We do not lie to each other, that's the fun of it; we enjoy the truth. I've never been able to share that with anybody, really. In the workplace, maybe, but not in that deeply physical, personal way. It's inexplicable and I suppose it's really the reason that you feel ultimately that you were born. It feels right, it landed right.

The ongoing separation was very difficult, but also at the same time, she knew damn well she could trust me. When Nora finally came out to visit me in New York, there were all kinds of animosities at the loft. Keith was unaccepting, and Jeannette was indifferently weird. It was ugly. I was just trying to keep the two things running. I realized that Nora meant more to me, and indeed Public Image meant more to me, than to be putting up with spoilt nonsense from Keith.

Nora gave me the bolster to get out of the rut. I realized that the scene I was in was a little bit fucking *down*. Every now and again, those that really care for you, they'll give you a bump. I realized how much it'd broken my heart to be apart. I'd thought I was too young for a permanent relationship at that time, but now I knew that's exactly what I really needed.

You have to go through that, you have to find yourself, and you have to get outside of the run of the mill. You know, if I hadn't taken my chances elsewhere, or eyed up the competition, or window-shopped a little bit before I met Nora, it wouldn't have felt quite so healthy to become 100 percent totally committed to her, which is now what I am—and indeed am with everything. I just took a long time to realize that, and commit to it, and "me babby" waited for me.

Once I make that commitment, it's forever. That's how me and Nora are and were. It's quite brilliant how it worked out. I can't imagine living without her, not at all, and it doesn't matter what people tell her about me either; here we are, and here we will be.

9

THERE'S NOWT AS GOOD AS CHANGE

As a member of the Musicians' Union, as a lead singer and a performer, you're automatically an actor, apparently. You're on a roster somewhere, like a pool of thespians waiting to be hired, blissfully unaware. Some of the agencies we were using to push tours had a film agenda too, and so ideas would come about through those kinds of channels. It certainly wasn't anything we were actively chasing but, out of the blue, in came this offer to star in a low-budget Italian film opposite a young Harvey Keitel. It was based on a novel by Hugh Fleetwood called *The Order of Death*. It got very confusing as the film ended up with different names in different countries—*Copkiller* in Europe, *Corrupt* in the States, and *Order of Death* in Britain, which is what I've always called it.

It was a wonderful opportunity to burst into something new, exciting, thrilling, and completely dangerous. Unknown territory to the max! I was terrified, obviously, because it was acting and I hadn't got a clue about how to do it, other than the tryout for *Quadrophenia*, which hadn't landed me the part. So it was a case of "Oh my God, what a hole I've got myself into here." I really had no one to back me up or support me. I had no support system during this period, and the only reason I could do it, and did do it, was because there was nothing going on in the band. Nothing at all.

The decision to do the film was not a difficult one to make. Getting the relevant visas, however, was. After accepting the deal, but before shooting started, I flew out to Rome to meet the producers. Trying to get back into America was a nightmare. They pulled me up at immigration and in my suitcase I had my contract, which had already been signed in New York. Therefore I didn't have the correct work-permit paperwork to go with it. That was a very difficult situation to sort out, but some real good came out of it, because a couple of people that worked on getting me my immigration visas, Bob Tulipan and Maureen Baker, became really good friends.

Maureen Baker is a superstar. She has done all sorts of interesting things over the years; she also used to photograph us when we were in New York, but for a long time there she helped get performers permits to work in other countries. She did all of the Kirov Ballet visas, for instance. That's charming, isn't it? It's an interesting universe. There's so many great people in the world that do so many different things, and if you don't have an open mind, you won't grasp how that can help you improve yourself and your own agendas. That doesn't mean that I'm going to run off and become a ballet dancer. I'm just highly impressed by her motivations, and that in turn inspires me.

So, between May and July 1982, I took time out from the band and buggered off to Italy. Filming was divided between Rome and New York. I had a great time doing it, and I got on very well with a good number of the people involved. I got to work with Harvey Keitel, and I like him a lot. But not only him: there was also the writer of the script and the book, Hugh Fleetwood, and the director, Roberto Faenza.

Watching their approach to putting a thing together was thrilling. I particularly enjoyed Rome—the meetings and the planning of the next day's scenes, and the director's wife cooking pasta Italiano style. Proper Italy! And the rows the Italians get themselves into, and the writer of the book translating what these people were yelling for me—fantastic! Roberto's wife—I thought it was his wife, she may as well have been, they lived together—she was one of the producers, and she was the one cooking, but the rows those two would have! And so actual studying of scripts was absolutely pushed to the back. And "Can you please outline what you expect my character to be?"—that was of no consequence! Personal animosity was the order of the day. I love Italians—they're ready to explode at any point.

Out of all that I had to play this lunatic character, Leo Smith, this spoiled rich brat left to his own devices. He was meant to be from a lonely but wealthy background, and it was sort of about how that can ultimately corrupt you. I could connect to that on a certain level—but also not. Obviously his education taught him that he was superior to others, and that was the crux of my angle, how I tried to make it work. I wasn't getting any help, really, but that's all right. Harvey's best advice to me was "Just get on with it! And take it serious once that camera's on!" Fantastic, thanks. And there I was, struggling, trying to remember dialogue, which is the hardest thing for me. I can't cope with learning dialogue because it's not coming from a deeply personal place. Many actors have told me this: if you have a strong personality, you're gonna find that almost impossible to do. You have to be a blank card.

There was clearly meant to be a *Performance* thing of role reversal going on in the film, like, "Who's actually dominating and manipulating who? It's an alpha-male scenario: who's really calling the shots here?" I thought Mick Jagger in *Performance* was astounding, made all the more so because of that solo song he does, that blues song on guitar, "Memo From Turner." At least he had something to grip on to there, whereas I didn't. The one thing in the back of my mind was, don't end up like Dave Bowie! He's so woodentop on film.

As it turned out, I wasn't rehearsed enough in acting to know how to portray the character. I wasn't able to look the other players in the eye and deliver the lines, because the second I'd catch their eye, the dialogue instantly disappeared from my memory. It would be "Ha-a-arv-ey, what do I do now?"

Of course, I was absolutely intimidated by Harvey. Come on, how could you not be? What a fucking fine actor. But again, at the same time, I was kind of angry with him because he took his role-playing too serious. If we'd go out for dinner together, he'd still be in role. Because his part was a policeman, he'd be looking for his gun in his holster. These things would matter more to him than having fun. I said to him at one dinner party, "Come on, let's have fun!" and he turned around and went, "What is fun?" Seriously! "Bloody hell, wow!" For once in my life, I was left speechless.

He didn't seem to know too much about me, or at least didn't let on that he did. Then, after we'd shot the film in September, he came to the Roseland Ballroom in New York to see me play with PiL, and he was like, "Oh my God, I never knew that's what you did! Wow!" Whatever character he was seeing in me while we were doing the film, he wasn't aware of what Mr. Rotten does when he gets onstage, and how I can let rip, and how I can really get a crowd going . . . I'm wide open onstage, and maybe I wasn't showing that in the making of the film. It's a shame: he could've taught me how to use that energy in an acting kind of way. I've met him since and we're all right with each other.

I was shocked by what good reviews I got for *Order of Death*. The film critic on BBC TV at the time, Barry Norman, said something like, "So far so good, but we have to wait for his next film before we can determine if he's a really good actor, or if he was just playing himself." I most certainly wasn't being myself!

I realized I was outside of my comfort zone, but not in a very interesting way. I didn't like the tension. I can understand the tension of getting ready for a gig—when you actually go on and do it it's an hour and a half of relief. In a film, it's fifteen hours of waiting to do one minute's work, and a couple of side takes at different angles. That is so confusing to the brain. What the hell am I supposed to be projecting here? By the time you do the third angle from the back of your head, you're really getting to grips with the role at that point, so what you've achieved film-wise is all the tension and anger and angst and character development that's ever possible—from the back of your head. It's a different universe they live in, and it's one I can't get to grips with.

I couldn't accept that the film, as it would be seen in cinemas, was uncontrollable by me. What ends up on the cutting-room floor could be my best bits. That's frightening. I know this about myself: I can't work in any environment where I don't have a say in the final creation. I have to be involved in all of it, all the way down the line, and anything that's predetermined by other people's interpretations, it's not gonna work for me. I don't view myself as one of the tools. I'm not a tool! Actors may self-aggrandize and get awards, Oscars, whatever, but really they're no more important than models are to the clothes they're trying to sell. That's all you are: a coathanger of sorts.

So I shut the door on acting, but then a whole bunch of offers came in. Oh my God, you know what, I turned down *Critters*, which was a cheap and nasty knockoff version of *Gremlins*. I was really pleased—crisis averted! How could I do a film like that—fighting these alien fur-balls!

There were heaps of other offers, but let's just say, the acting side of me dwindled. I'd be blatantly rejectful, and I didn't see the potential of a great many offers that came in. I shouldn't have done that. It was a mistake. I've made many mistakes like that over the years. I fancied myself as playing a lead romantic role, in the style of Cary Grant, maybe. This is what I'd be saying to the agents, and—errrrrr, door closed!

Now here's a thing—yippee-aye-oh, my absence for the film inspired Keith to get out of the doldrums. He actually came up with some really good ideas for tunes and songs for the movie. One of them, "The Order Of Death," which had a lot of Keith's work in it, was so good that it became a potential theme song for the film.

Finally we were proving that we were now working in different areas, yet it was still PiL, and it all came back to the same center force. We provided the producers with plenty more music, but they were very wary that Johnny Rotten and his band would take control of the film if it went too far, so none of our material was used in the end. It was a film with the seriously famous Harvey Keitel, and they didn't want an upstart like me who can't act dominating the scenario. Instead, they got Ennio Morricone in to do the soundtrack, who at the time wasn't really respected. People thought he just made trash noise for Italian cowboy movies—laughable, sneered at!—but here we are a few decades later and he's seen as somehow rather genius.

So, anyway: wow, the beast had awoken. Keith was back in action, and there was a short burst of really good and interesting energy. Jealousies had been lurking in there: he thought I was getting too big for the "PiL umbrella" situation, swanning off to star in an Italian movie, and therefore he felt he might be losing some kind of control. Maybe he thought I had just abandoned them, which wouldn't have been my way at all. But I understand that insecurity, because I'd been in that position, where my band abandoned me. I get it, but I don't get it. I thought we were closer friends than that, and he should've been more open. And, in fairness, there was nothing else going on.

Oh, and here's the laugh of all laughs: when I was filming in Rome, Jeannette came over with her friend just to hang out for a couple of days—they just turned up. That was good, because she was grasping the PiL thing with it. It would've been a perfect opportunity to film what it's like on set, but she forgot to bring her camera. Or maybe she *did* bring it, and we just forgot to look at what she did. I think that's closer to the truth.

I want people to understand, always with Jeannette, it was a working relationship. I'm not one for loose-arse affairs, that's not my way. In that respect, I think me and her worked really well together, but looking back on it she was put in the middle there to try and save my friendship with Keith, but it was hopeless, it just didn't work. I can't actually remember where it all ended with her. It was something about her falling out with Keith. There was certainly no big problem between her and me; she even used to come to the gigs for a while thereafter. It just wasn't gonna happen any longer.

Keith's resentments run so ludicrously deep, and they're so pointless, and back then everything just always seemed to end up revolving around his drug problem. As I said at the time, maybe he had too much blood in his drug-stream. If he was having a bad time because he couldn't get a fix, we all had to suffer. And how much of that can you take? I'll put up with anything from anyone if they're creative, but when the creativity lulls it's really hard to endure. To me dependency is a great form of foolishness. The lack of self-discipline and control in it. You should never get into that condition.

I don't want anyone to think I'm being hypocritical here. I'm far from innocent myself but I've never been dependent. The bottom line is, you've got to be in charge, or else what's the point? You *gots* to be in charge.

In November 1982 Keith actually got married in New York—a marriage that lasted about all of two weeks, and then he came crawling back to the loft apartment, alone. It was insane, weird, ludicrous—a situation I never understood at all, even though I was their best man. Her name was Lori Montana, and she was the bass player in a band called Pulsallama. He met her out there. She was a lovely little girl, and she was absolutely innocent and open-minded. Kind of hippie-chick-y. Very odd, for him. It was all "Oh, she's got me clean, everything's gonna be great!" Well, that soon stopped. Game over.

All the time, through this nonsense, I had this thought in the back of my head: that guy's got something incredible to work with. But he refused to deliver those goods after a certain point. I don't know why. I think he doubted himself, and yet I never doubted him. But that information never seemed to get into his psyche, he never understood my backing.

I got the impression, trying to talk to Keith around that time, that he viewed me as a parasite living off his genius. This is how it would be coming across in conversations in the short sharp smirkiness of his approach. That was just too ludicrous by half: sneaking off and wasting studio time and misspending money on *his* ideas, and his ideas were excluding me. In these moments when the light of dawn finally breaks on you, that's painful.

I was not going to give PiL up. I wasn't going to let the likes of Keith and/or any member run off and claim it as theirs. Without meaning to be ridiculous about it—it's like Ted Turner. I understand what CNN was when he started out, and I understand his fury at what it turned into, and his absolute rage that he was pushed off the board of that company. It's a great tragedy—CNN is now a complete farce. I couldn't let that kind of thing happen to PiL, "my own creation."

I did, though, consider making a solo record. Through a friend, Roger Trilling, I'd got very into an experimental scene surrounding the label ECM. A lot of the records would be solo cellists, going, "Pop, bing, twannngggg"—lots of drawn-out, slow, melodious patterns, some of it completely pointless. In the loft, I liked putting these things on and letting it drift in the background while just doing normal everyday things. I found it a very comfortable way of using it, rather than to sit down and realize, "Gosh, this chap's really crawling up his own bunty . . ."

That led to me thinking I needed to put out a solo thing in this kind of direction. Avant-garde? Oh yeah! But not quite in that way, obviously. I was very thrilled that these people had the audacity, and I remember arguing with the musicologists at similar concert places at the time—they'd go, "No, this man studied at the

Royal Philharmonic for forty years." But this is what he came up with—"*Boonnggg!*" And why not? I was absorbing that that's what these people found the most thrilling; they were exploring the tonal quality of abstract random plucking, for instance, or farting down a horn. They were so in love with the sound that musicianship and structure became pointless. A wonderful insight into the workings of the human mind. My record collection, if you come round my house, is very much like, "Okay, I've got a few dance things, but here's the serious stuff . . ." I can clear a room in minutes.

Anyway, my grand scheme fell apart—the solo venture never came to fruition. There was just too much fractionalization going on, so that I couldn't actually sit down and have the time to do it alone.

At the same time, Virgin were trying to make me write a hit single. "C'mon, Johnny, why don't you pen us a nice love song, so we can all make loads of money?" I'm like, "After all I've been up to, who or what is it you think you're talking to?" The core essence of me is, I think, that I write very good pop songs, but I don't write them because I'm asked to or told to, they happen quite instinctively and naturally. To try to interfere with that process in me is never going to work, ever. I do what I want, and it just so happens that what I do is kind of really good from time to time. I don't mean that to sound bigheaded, it's how it is. I'm not gonna disrespect my gift and misuse it by writing cheese.

Thus came the idea of "This Is Not A Love Song." Initially, those energies were dissipated because Keith had the audacity to go off and do something on his own with it in the studio. He'd decided I was a bad singer, but he didn't tell me this to my face. You couldn't actually get him to speak to you directly, not about anything, at this point. He knew damn well I was on to him. As soon as I looked into his reddened snitchy little eyes, I'd go, "What do you want?" And so he'd avoid me.

I'd be hearing secondhand about his conversations in the studio, so I went down there, but he wasn't there, and I said,

"Well, where's the tape? What's he been doing?" I heard it and thought, "Right, we can do something with that." And so, I put words to it—something I don't usually like to do. I like to be in there at the outset so I'm fully involved in the song's evolution, but in this case the song went from there. We had no bass player, so Martin Atkins brought over a mate from London called Pete Jones who played bass. He wasn't great, but it helped, and some gigs came out of that.

We never discussed anything like "musical direction." Anyone I've worked with will tell you I never come at it with an agenda. Still, it got very negative because, with this music we were putting together, Keith didn't respect what he was doing, he thought it was junk. It wasn't. He just couldn't appreciate himself musically. He viewed his guitar lines as throwaway when they were *not*. They were really thought-out pieces. He'd lost the plot. Ultimately all his bile and spite and resentfulness became unamusing. He really meant it. He was like gooseberry jam without the sugar.

I don't know what Keith wanted, but somehow he didn't want me involved and presumed that it was his band, so I had to put the knockers on that. There was no one single thing that ended it, just a collection of incidences up to about May or June 1983. The most pressing issue at the time was, we had some dates booked for July in Japan. There'd been problems getting him a visa, but he knew he couldn't survive the plane journey. He couldn't even survive a van journey from New York to Pennsylvania, which is three or four hours.

To this day, I feel clean with the break from him. Back then, it was almost like a switch that went on, because I was free of the bondage that some of them ex-members had become. I just wanted to go out and earn my wings, exercise my chops, and grapple with the tension and fear that is live performance, and is ultimately to my mind the biggest reward of being in a band.

All I needed was the band. The commitment to Japan was already made, and the tour had to go ahead. I scouted around for whatever possibilities I could find, and ended up with a bunch of fellas that Martin Atkins and our producer Bob Miller found working hotels and bars in New Jersey, playing cover versions. There wasn't anyone else readily available that wasn't coming in at an enormous price that I could match, and along came these fellas with shiny suits and mullets, and I thought, "Wow! This could work!" There was such an image danger that there was an appeal in it and, as fellas, I liked them very, very much.

I went up to see them in Atlantic City, and it must've been the Holiday Inn there, because that's what I was saying at the time: "Look, I've gone and got myself a Holiday Inn band." I wasn't being disrespectful; I just thought this was definitely going to challenge an audience, judgmental fucks that they can be. For me, the only thing that's ever going to be bad for my image is a lousy live performance. I thought, "Well, we're not going to be that lousy," and indeed a great time was had by all.

They were just thrilled to get to a place like Japan, after being stuck in New Jersey. In terms of musicianship they were miles ahead of me, but at the same time miles behind, because they couldn't quite come to grips with the way the songs were anti-structured and deliberately let loose, so we had to shapeshift around what the set would be. Which was fine, because at that time I was into tormenting pop songs—very much so—working within the restraints of verse-chorus to see how I could maneuver that into something exciting.

We even started playing "Anarchy In The UK"—with boys with mullets. Fantastic! It was a matter of "Are people actually listening to the song, or are they judging us by our hairdos?" The Holiday Inn chappies were terrified of approaching "Anarchy," but at the same time they loved the balls of it. The lead guitarist, Joe Guida—there he is in his tight jeans and white sneakers, legs as wide apart as possible, and a mullet, and he's banging out his gruff guitar riff—fucking great! You know? If you're going to write a song like "Anarchy," it has to be understood that it's not just for the fashionably elite, it's for everyone. I was magnanimously sharing that message.

Virgin pressured us into releasing a live album from Japan, but I was appreciative that it would go some way towards paying off our debt to them. It's keeping the business rolling, and keeping the record company interested in you. I don't blame Virgin for everything; I do understand that I'm a very difficult person to get along with. I challenged their sense of economics on almost a daily basis.

It also helped us to get PiL finally established as a live act, and we hit the road through the latter half of 1983. Little did we know we were swimming against the tide of where music was heading at the time. Video production was where all the big money was going, and live performances were neither here nor there to the bigger outfits of the day. Once the wool had been pulled over your eyes and you'd committed to buying anything by those kind of outfits, you didn't know what the possibilities of real bands were. So shambolically ridiculous light shows convinced you that that's what music was all about.

We started working with a set designer called Dave Jackson, who'd orchestrate lighting for us, but our way was very different from everybody else. He put together our "toilet set," where the backdrop was white tiles and urinals. An absolutely clean approach! Shame on my brother Jimmy, because when we ended the gig at Hammersmith Palais in November, he wandered onstage and pissed in one of the urinals. He presumed they were plumbed in!

So I can be accused in my big bad life of having fake urinals. The idea was that out of the antiseptic environment of public bathrooms could come great ideas. The toilet's an incredibly boring thing to have to deal with, getting rid of your waste, but you can let your mind wander and possibilities do arise. It's that, or you play with your pud. I mean, I've had many great song ideas while sitting on the shitter. I won't name them for you, because that would spoil them—I don't want to put the vision of a big macca in anyone's head.

Around that time, we played live on Channel Four's *The Tube*. I had the worst flu. Whenever a TV slot like that comes up, it's usually towards the end of a really hard, rigorous tour and I'm completely drained and flu-ridden. The Eurythmics were on as well, and their guitarist, Dave Stewart, was coming across a bit stroppy—a self-proclaimed genius. All's well that ends well, because Annie Lennox sent over a bit of an apology for his gruffy-ness. She's a lovely person. I've got great respect for her—big time! From "Sweet Dreams" onwards, she had me, hook, line, and sinker. Yippee! I'll lay down and take your pervy rollercoaster, baby!

At our regular gigs, we found that there was a dark, one percent element out there, absolutely out to hurt—really seriously hurt—us or anyone in their path. I don't know what you would call those kinds of people—stalkers, fanatics, psychos—but they make the presumption that you aren't what they believe you should be. They can do you some harm.

At the Paradiso in Amsterdam, somebody came up with a screwdriver and tried to stick it in my back. Luckily it was blunt, but it left a mark there. And of course on came the Hell's Angels security making arses out of it. There's nothing uglier than huge physically overdeveloped oafs running around the stage like Clampetts and things getting confused. The reason I actually got stabbed in the back was because one of the security guards grabbed me. He was trying to protect me, but by holding my hands down I couldn't maneuver, so he accidentally made me an immobile unit—a dartboard. I was furious about that.

It's quite serious when you put together all the attacks that Public Image as a band have had to endure—much more so than the Pistols. Listen, I was moving on, I was advancing my stuff, and these were people that wanted me to stick to my past. It's always been the problem. If you don't like what I'm doing, fine—leave it alone, move on, don't hang around and demand that I go back ten steps and mollycoddle myself in the safety of my past. That's just not going to happen. You've got to bear in mind they paid good money to attend in the first place. I'm always insistent: unless they really push it, don't throw them out, but the more magnanimous you are, the more volatile they become, and then it becomes a free-for-all and it gets misunderstood as "Oh yeah, go to a PiL gig—you can boo the singer and throw things and try to stab him, it's great!"

Wowzers, what kind of a world are these lot living in? None of the fellas that do this kind of stuff are what I'd call hardcore. These are all lonely bedsit bastards that have somehow justified their activities as saving the world from the likes of me, with the wonderful excuse that I've apparently "sold out," whatever that means.

The concept of selling out all seems to have grown from the Who's album, *The Who Sell Out*—their glorious piss-take of advertising. The photography on the cover of the first PiL album was a nod and a wink to that—our version of Roger Daltrey sitting in a tub of Heinz baked beans! That was a very advanced thing to be experimenting with, the idea of anti-promotions. Sometimes some of us in this world of creativity get a little too far advanced for the mental capacities of certain members of the audience. And aye, as Shakespeare would say, there's the rub.

I do find great fun in irony. "This Is Not A Love Song" was, for me, a continuation of "Pretty Vacant." You know, I'm not pretty and I'm not vacant. "This is not" is actually "this is." The idea was to oppose commerciality and greed, and so the juxtaposition inside the song is "Happy to have, not to have not/Big business is very wise, and I'm inside free enterprise," when my sentiment was the exact opposite. By saying one thing, you're actually meaning something else. So "This Is Not A Love Song" is really a love song.

What I'm truly very happy about is that I haven't kowtowed to corporate dictation from Branson and Virgin. I haven't written the songs that they wanted me to. I haven't become the commercial-success arsehole it would've been so easy to turn into under their watch. That way, I wouldn't be the same person. I'd have had a lifespan of two years and made so much money that I wouldn't have to deal with anybody ever again. But that's not interesting. At all. I just can't do that; it has to come from the correct place.

"Love Song" wasn't written deliberately to confuse people, but afterwards I did enjoy the scope of possibilities, and the intrigue that journalists can find in these things. That absolutely thrills me, and I'll always be the first to stand up and go, "Yes, of course!" when really it isn't "Yes, of course!" It's a song, what's the problem with you? It's there to provoke thought. I do like my bits of pop, but the words are there, listen to them and they'll tell you the story. It's just a human being trying to explain his place in the world, and how he interprets his immediate surroundings. And irony will always be there because that's the greatest achievement in the English language, and that's sadly lacking in other cultures. I happen to know for a fact that you cannot translate these songs directly into German, for instance.

Music had suddenly become very corporate. Around that time, Duran Duran launched a single with a video that cost around half a million quid. I might be wrong on the figures, but thereabouts. They got that from nowhere, just straight at you. Videos and big productions were now just the norm. No learning curve in them, but at the same time, I've got to tell you, I loved "Hungry Like The Wolf"!

Years later, I met Simon Le Bon. It was an odd one: the Hard Rock Casino was opening in Las Vegas in 1995, and every musician was invited. I went because my manager at the time had all these free passes and rooms. His name was Eric Gardner and, little did I know, all he wanted to be there for was to gamble. When I went there, even getting into the venue was very difficult. Simon Le Bon spotted me having a problem there, and he went, "Don't you know who he is?" and that was it, I was in. I thought, "Bloody hell, it takes Duran Duran to get Johnny Rotten into a building!"

I liked him as a bloke, and I like a lot of their songs. I like "Girls On Film," and I can't pretend otherwise. I don't have hatred for different forms of music, in fact I've got a great deal of love and openness to everything done by anybody. Christ, I have to: I've got two Alvin Stardust albums.

Anyway, for "Love Song," we shot our video surrounded by the financial hub of L.A. It was, get a couple of thousand dollars, rent a car, get a cheap camera, film it, have fun, and spend all the serious money on having a party afterwards. Always have done. I love making videos when they're on the cheap—to me they're the most fun. I see these hundred thousand dollars pumped into other things, and I don't think they're anyway near as effective.

"Love Song" became a delicious issue with Virgin, because they didn't want to release it as a single. They declared outright it was destined to be a commercial failure, so I found a company in Japan that was interested in releasing it and, although it was going to put me into a dangerous situation with Virgin, well, if they thought it stood no chance of charting, I'd go somewhere where they would be proved different. Of course, it turned into a big hit in the clubs of Japan. So I took that straight back to Virgin—"Whatever it is you said it was, it isn't. It's actually a hit, now you have to release it, or take me to court and be proved wrong, because there's the commercial success right in front of your face." Bingo! It duly went right up the British charts and ended up Top Five. It was also a big hit all over Europe, including a Top Ten in Germany. Prior to Virgin finally releasing it, the Japanese release had been heavily imported into Britain. There was a buzz about "Love Song" before it even came out. We plonked it right in Virgin's lap for them. What more could they ask for?

That's why, over the years, I've always gone back doing different versions and updates on "Love Song"—it's there as a reminder of the powers-that-be telling us that songs like this are not possible. It's a weapon of war, it's as important to me as the Pistols' song "EMI" was.

The water of that whole relationship with Virgin was muddied further when Keith Levene sneakily tried to release an album of the stuff we'd been working on before he left. It was called *Commercial Zone*, it was incomplete, *some* of it was very scrappy and, to my mind, very painful to listen to.

I just thought, "What are Virgin now gonna make of this? Are they gonna come back at me and go, Told you so, that's what you get working with those loonies?" Actually, they helped tear that one down, and quite rightly so. And then it went into trying to get the reel masters, and they were all a mess, so I had to rerecord all the tracks for the album that became *This Is What You Want . . . This Is What You Get*. I felt I had to backtrack to get those songs back into the fold, and not let them be stolen away. I must agree, though, that we didn't manage to grab the intensity of the original demos.

I did it with Martin Atkins and a few other people, mostly at Maison Rouge studio in January and February 1984. That place was virtually under the stands of the old Shed at Chelsea's football ground, Stamford Bridge. I'd walk there from Gunter Grove, which I hadn't sold off yet, but some nights there'd be a Chelsea game on, so in the back of your mind you'd be thinking, "Oh God, here we are walking down the streets with an assortment of bizarre instruments—in the middle of football crowds . . ." And in them days there'd always be "Rotten! You're Arsenal, in't ya!" I never kept it quiet—you are what you are.

Some of the album was done at Pete Townshend's Eel Pie studios. It was the closest we ever came to working together. He wanted to, certainly. The place was in Twickenham, right on the Thames, and it almost flooded out a couple of times, because the river was overflowing. Townshend wanted to be involved, but I kind of shy away from those things—it might distract me and set me on a pretentious course. When the moment's there, if it's right, then do it, but if that's not the right moment, then don't. Let your instincts guide you.

The problem was, Martin and I had learned how to have fun

more than anything else. I loved working with his loops, but by strip-mining what we were recording, we left it threadbare almost. The empty spaces in that record are kind of where the action is. The correct use of emptiness taken to the ultimate extreme.

I wanted to go further into a drum-and-vocal universe, full-on. I had great respect for Martin's drumming—he's beat perfect, the fella—simple, to the point, and that left you lots of room. But he had doubts about himself as a drummer, and didn't want to do it anymore. "Well, what the hell are you gonna play? Flute?" He's like me, he's not very studious when it comes to understanding the ins and outs and intricacies of instruments. We see instruments as accoutrements, not as guiding forces.

There were songs on there called "Where Are You?" and "Solitaire." I suppose it felt quite lonely out there in the wilderness. The album title, *This Is What You Want . . . This Is What You Get*, which comes up as a chant at various points, was an absolute tirade on what I saw was now going on in the '80s. People were being force-fed a diet of vacuous pop rather than content, and you couldn't get content out there at all. If you had anything poignant in your lyrics, MTV would find a reason to cancel you out.

It was a world of amazing isolation. The relationship with Virgin at that point was one where we didn't give a flying fuck about each other. Very hard times. Hello, I'm Johnny—I'm a nurse by nature, and I'm nursing you into the future, and I've never said a word that's wrong, really. I predict well and accurately. I had to dig deep to keep my batteries running, in order for PiL to maintain or sustain any goodness that was in us, because it was a world of "Boo, screw you. Here's the latest video by . . ."—then you'd watch Simon Le Bon on a yacht. And no digs to Simon, he's really an all right fella, but at that time the whole game was about the finances of video backing, and the more money spent on a video seemed to mean the more attention. So, out the window went cause or point or purpose, and in crept "Look what I'm wearing!"

I might have pushed it too far. But guess what? Here comes a hell of a lot more pushing. I don't do this for chart positions, I do this to make the world a better place. I'm arrogant enough to believe that whatever it is I do is actually to the benefit of mankind. I don't really view it any other way. Every decision I make is always based on these principles and values. Am I an anachronism? I definitely began to get the idea that my approach to life was a dinosaur in the mid-'80s because nobody wanted to think about anybody but themselves. What a great pity, to see punk unravel in that way. And pop music to accept any old palaver as long as a big-arsed name producer was attached.

That album cover was a similar thing: the label had lined up this famous photographer—called Norman Seeff, who was living in L.A.—to shoot me and I don't understand how the photos went so wrong. "Oh, you're gonna look great, if this fella takes your picture . . ." Then I go to these sessions and I'm not enjoying them, I don't connect with the fella at all, and the results show that.

In America at that time, we didn't have a label. I had close contacts with a couple of people at Atlantic, and they seemed like great fun to me. One of the sticking points was that they didn't have much of a liking for poor old Martin Atkins, so yet again I was facing a record company that didn't mind me, but they just couldn't bear the people I was working with, for whatever reason. But that's not why me and Martin went separate ways—I never let anybody influence me like that. I don't abandon people, but I do seem to be quite good at losing them. It was a continual revolving door.

By summer 1984 I'd had enough of New York. I'd been there for three years and it was time to move on. So why did I not move back to London? It's difficult to explain, but it's the lack of get-up-and-go. The lack of wanting to change a thing. The acceptance and the general malaise-y attitude of "Why bother? *Yaaawn*. It's not going to make any difference in the long run." "Arrgh! *Yes, it will!*"

But then I get the flipside: "Erm, why did you leave? Why don't you come back?" My response is, "*What?!* What are you yourself doing, to be saying this to me?"

I'd stood up and been counted and noted for being rebellious, and yet I was resented for it, for making people have to think. The media was not protecting me, instead flirting me off as the bad boy that deserves his comeuppance, and that's a very dangerous universe. Quite frankly, I knew that if I didn't get out of Britain, I was going to end up with a very long prison sentence. The endless police raids on Gunter Grove weren't funny. There's no humor in that. They're out to get you, and sooner or later they *will* get you. Back to Poly Styrene's poster: you can't just keep putting your head on the chopping block all the time. I had no allies to back me up. I was well aware that evidence could be tampered with, or faked—so: move on, get out of there.

Plus, there's always the gyppo in me. I've got that gypsy attitude: get up and move. When the facilities have run out, find a new space, recharge the batteries.

Brits tend to go to New York because it's nearest, and it's like another London. It's very exciting when you first live there, but after a while, it wears you down. Everybody's agitated, everything's a drama conducted at a ridiculously high tempo. I'd spent all my early life rushing to everything and if I wasn't careful, I would've rushed to my own grave. And of course, it was a drug-fueled atmosphere in them days. You were practically led by the nose to it, so to speak. New York is a town where it's usually overcast. The days, if you're in the music world, you tend to resent. You live for the night. Well, I'm not a vampire. I enjoyed the lifestyle for a little bit, but not too much, thank you. It was ultimately, for me, soul-destroying.

Bob Tulipan, who'd been helping us on the business and management side, had left the previous year, and by this time we had brought in a new manager, Larry White, who was really instrumental in moving the band to Los Angeles. Larry was a sweetheart. He managed a lot of surfers, who can be a suspicious lot, I discovered. He had a lot of problems representing them, and they didn't understand what the hell he was doing with an outrageous ass like me. But courtesy of Larry, our road crew had soon all become surfers. When we played at the Cornwall Coliseum in St. Austell in November 1983, we were staying in this couple's guesthouse, which had a panoramic view of this Cornish bay—and the waves were about one inch high. The derision these fellas laid on us was hilarious.

At Larry's insistence, everything soon focused on building ourselves a really good office in L.A. In June '84, we got a really cheap rent space way up in the hills, twenty minutes' drive from Pasadena. It was a great little facility called La Granada—a wooden structure with stucco walls, plastic windows, aluminum frames, screens, and an overhanging roof so that the rain didn't wear the walls down—whenever it *does* rain in L.A. The whole thing was basically on a very dangerous precipice so, like many places up there, whenever there is some rain, the mud falls from beneath the house, and slowly but surely the concrete structure is eroded, cracks up, and you fall down the mountain.

I shared the house with Martin Atkins, for the last six months that he was in PiL. We were beginning to have problems with each other, and then my brother Martin came over and stayed, and that caused even more problems. Martin Atkins didn't like Martin. And that worked both ways, by the way. Martin started getting a bit of an attitude about who I was having around. I wanted my younger brother Martin to learn everything he could about how to set things up—he loves the technical side of music, backline, etc.—but it caused a friction. I understood Atkins' point of view, but at the same time, it's my young brother—come on, he's hardly here to replace you.

Instead, my brother ended up working with that famous Swedish guitarist, Yngwie Malmsteen, who I used to call Manigwee—the joke's long lost. It was very good for Martin too, because he was on his own two feet, and he wasn't a problem in the house all the time.

One very major positive of living there was that KROQ, the influential Anglophile radio station, was based in Pasadena, and because of that proximity it was very hard for them not to be playing PiL records on the radio. Yippee, that was a first! From there, PiL got very involved with the L.A. indie band scene, and gigs for us around L.A. and the various surrounding counties soon became very big affairs. No small attendance, full-on packed events.

The concept of Los Angeles is inexplicable to the English psyche. Everyone who comes here for the first time can't figure it at all. It's huge, enormous, and low, and it just goes on and on forever. People describe it as not so much a town or a city, but a loose consortium of villages spread over seventy miles—and that's bang-on right. What's that line in the Dionne Warwick song? "L.A. is a great big freeway." The only way of getting anywhere is via some serious freeways that interconnect the different areas. It's quite nutty. To do anything here requires forty-five minutes in your car.

The place just didn't seem to have anything going for it. It seemed the furthest reach, it seemed impossible, it seemed ludicrous—it seemed like the place where the Eagles come from. Here I am in L.A. and I'm on the beach where Neil Young would walk barefoot with his acoustic guitar in his heyday.

Very quickly, it became "Why not?" What's wrong with that? At least these people don't seem to want to hate, kill, and despise each other. They're less combative with each other as musical entities. Maybe a bit of peace and love in the world, when it's actually truly meant, might not be a bad thing. After New York, it was like, "You know, what's the rush?" It was a mentality that I only gradually learned to absorb, but in the short term I realized there was a lot of fun things to do in a completely different way, in a different universe. This wasn't the big city, this wasn't full of nightclubs; everything involved travel and sunshine and getting up early—just finding other ways of exploring life.

People like Rod Stewart and Tom Jones apparently came here to disappear, to be a little fish in a bigger pond, which I kind of understand. It's definitely a "start all over again" vibe, getting away from the contaminations that you felt were somehow dragging you down from previous situations. It's exploratory. But Rodney did it with an enormous amount of money. I didn't, and I certainly never came here to be near the Hollywood acting scene. No connection, no interest.

I must stress, this was not escapism—it was to take risks, spread your wings, and go to pastures new, take on new challenges. Believe me, it's a dramatic challenge to uproot like that, and it's not done for comfort.

At the beginning, I went, "I hate the sun, urgh!" But I soon noted that I wasn't ill all the time, like I usually was, from head colds and sinus infections and ongoing spinal pains from the meningitis. The weather was actually better for me, smog included. It took time to adjust to the lifestyle: it really is all about getting up early and enjoying sunrise and sunset, and that being more thrilling than anything a squalid disco in Sheffield can offer me.

While we were holed up in Pasadena, Atkins started the ball rolling on getting a new band, by placing an ad in the Musicians Wanted section of *LA Weekly*. It said something like, "If you love PiL, and hate heavy metal," which kind of gave the game away, but was a complete lie, in the nicest possible way. In the past, I actually have said, "I hate heavy metal," but as you already know, sometimes I will say one thing to get a result, when I actually mean the opposite.

I was astounded when Flea turned up for the auditions at Perkins Palace, and he was fabulous. He tended to play bass in that "slap-bass" style, which wasn't great for what I wanted, but . . . "Yes! You'll do!" But then he said, "Well, I can't . . ." It turned out he was having problems with his own band, his fellow cohorts in the Red Hot Chili Peppers, who were really just starting out at the time. I think he just wanted to remind them that he could've moved on to pastures new, but fair play to the fella, he stuck with them, and the Chili Peppers are still on tour. I would've loved to have worked with him.

It was through those auditions that we found Mark Schulz (guitar, 20), Jebin Bruni (keyboards, 18), and Bret Helm (bass, a bit older!), and by October we'd got ourselves rehearsed and were able to tour the East and West Coasts.

Atkins and I, however, were not seeing eye to eye. I like Martin a lot. Every now and again we'd have a yelling session on the phone, and he always took it on the chin. He could be a bad bunny. I couldn't trust him, because he'd do anything for a fiver. But still I tried to be mates with him.

We fell apart because he pulled an issue the night before we went off to tour Australia and Japan over New Year 1985—the night before, he wanted more money and that was like, "That's not gonna happen, Martin. You've made your commitment." He didn't want to be an equal player in this anymore, he wanted extra, knowing I had no alternative at that late stage. But when it comes to them situations—and I've had it with quite a few different band members—they turn around and say, "Well, I never signed anything." That's a risky business when you try to work with someone like me. As soon as I think this is about greed—goodbye! We tried to make it up on the tour, but there was no bond there at that point. I practically isolated him from me.

Another thing was his misunderstanding of the idea of PiL as an umbrella covering other things as well as music. Every single person who's created a problem in PiL has always said they wanted to stop the music and do something else, but still draw the wages off PiL. And that something else was usually because they had their own band ideas, which is very disrespectful. Atkins was another one of those.

I'll never make that mistake again of allowing an openness with people in that particular way, because they then presume that the cash cabinet is theirs to plunder. I have to be able to keep enough cash in that chest of drawers for future operations, but I'm not a fool, and I'm not going to advance people on what I ultimately see as a deceit. It's where the line has been drawn every single time.

Before we set off on that Far East tour, Nora came out to visit. Almost by accident, she spotted this beautiful house up for sale in Venice Beach. It turned out that the bank had owned it—probably the previous owners had defaulted on their loan payments or whatever. Nora just grabbed it, for silly money. We hardly even knew what the house looked like—we just made this impulsive move. We walked in and we loved it. It's small, but look, we're not the kind of people that require twenty-two rooms. We don't run dinner parties, we're not like that. Three odd plates and two sets of knives and forks, we're happy.

We moved in, and cut all links with Larry White and Martin Atkins. From there on in, life here in L.A. was *fantastic*!

My youngest brother Martin is like a flag for people to rally around. When he was young, he used to be very loud and—*aaaww!*—a *real* troublemaker. As an adult, he's a very open, gregarious fella; he just seems to make friends wherever he is. He does all the things that I can't seem to get to grips with. He's open to contact. Maybe it's because of the position I'm in. I'm very wary when people want to talk to me. I'm very wary of what it is they're really looking for. Are they looking at Johnny Rotten, showbiz personality, or Johnny as a human?

Martin came to live with us in Venice Beach, because the place came with a little guesthouse. He became like our neighbor, sharing the place with our guitarist Mark Schulz, and from time to time Jebin Bruni the keyboard player and Bret Helm the bass player—so that became the L.A. band scene. In a weird way it was really very like *The Young Ones*. They were the nippers next door. I really loved Jebin, such a little character. He had huge pointy hair, black spiky hair—he reminded me of a hedgehog. I saw great potential in him on the keyboards—in fact he dragged in an accordion because I foolishly said my dad was an accordion player and he presumed I could play it. Wrong!

We'd always be saying, "Let's go down into the basement"—because this place has a basement. We'd put together the loose ends of a tiny little idea for a song and bash it around down there. We'd have to guess what the drums would be, until we procured some drum machines. And then of course you're limited by what the drum machines can do, because a proper drummer cannot be replaced. So these were approximations of songs. Great times. It was like just having fun, but great things came from that—great songs like "Fishing," "Round," and "Ease."

This was the beginnings of *Album*. By then Howard Thompson had signed me to Elektra in the U.S. and was under pressure to deliver my first record for them, not to mention keeping Virgin quiet back home, when I got a call from Bill Laswell offering himself as a producer.

I'd first hooked up with Laswell through that chap Roger Trilling, who'd also introduced me to the avant-garde label ECM. I met Roger on the club scene in New York. He'd overheard me saying, "I hate jazz! I can't get to grips with it at all!" He went, "Well, you would if you came to see my records . . ." That's a very tempting line to someone like me. There wasn't no fairy dust going on here, he meant the music, and I loved going round his apartment and just listening to all these different heavyweight jazz records. We formed a very good friendship, and that led to Laswell, who I think he was managing at the time.

It was Bill who was connected to Afrika Bambaataa, who'd wanted to put a record together with me a year or two earlier. That resulted in "World Destruction," under our joint alias Time Zone. This was 1984, early rap time. To me, it was almost like Jamaican toasting was what he was after from me, and I wasn't sure I was ready to go into that world at all. But Time Zone more or less kicked rap into the universe. It was a huge record in the clubs, opened people up, like, "Wow, that's interesting, new and different."

Afrika, what a sweetheart! His "nation," his whole peace idea—it was all glorious and wonderful. I've got to say, the words were all his—which annoyed the hell out of me!—but I came up with the chorus, "Tiiiime zooone!" I thought it needed a pivotal point in the song. In the silly cheap-arse video we did, I come across a bit bug-eyed but I'm trying, rightly or wrongly, to emphasize the thrill and excitement of putting the record together.

I learned a lot doing that track: there was a backing singer there, Bernard Fowler from the Peach Boys, who taught me how to hold a tone—technically, how to hold a note and not let it waver, for emphasis and power. I also did another song with Laswell with a band called the Golden Palominos, but that was their song, and I felt I was rushed into it a bit. But it was work, and it got me to think about myself again outside of the box. This was what I was crying out for.

So there it was—bingo! Bill's the bloke to work with for the new PiL album. Elektra were more than happy, so I called Laswell back and said, "Yeah, fine, ooooh, whoopee!" Bill had a world of respect wrapped around him, from the industry and from other musicians, and he seemed dangerously big-league to me. It wasn't an easy relationship. With someone as big as Roger Trilling looking after him, I was aware that Laswell's sensibility could become a contentious point in direction later.

Bill's vibe was, he had the beret and the beard, and a propensity towards black leather jackets. I don't mean short ones, like biker jackets, I mean like a suit jacket, which was very much part of the gangster look in New York—the Irish gangs, in particular, and some of the slicker Mafia mobs.

He was affiliated with that big-arse studio, The Power Station, where we started working in August '85. I brought my young band with me from L.A.—minus a drummer, of course—and it really didn't work. My little whippersnappers fell apart under the pressure. They couldn't cope with flying to New York, hotel rooms, rehearsal rooms, other bands walking in and out, as indeed happens in New York—everybody's involved in rehearsal studios, everybody is your friend—and they panicked.

By the time it came to actually laying down the backing track in the studio, they just couldn't get it right. It was very hard for them, for instance, to understand "Rise." I realized that they simply weren't at that level. It was driving me and Laswell crazy, so we just said, "Look, we have to replace them, it's not gonna work. We're gonna run out of money, taking far too long trying to repair it in the tapes." I panicked at that point, but Bill Laswell went, "Oh my God, it's getting late, let's start ringing people up and see if they'll work with you"—because neither of us thought anybody would want to.

Then we were presented with the dilemma of "Oh shit, we need to get some people in, and unfortunately they're all going to be 'names.' " My attitude was, instantaneously—and everybody involved agreed to this—that we *wouldn't* name names, nobody would be credited—it's not about ego, this would just be to the benefit of the record, so it would be judged on its own merits. Hence, on release, the generic approach in the artwork.

Thus it became a serious five-star catalogue of people—amongst them, Ginger Baker from Cream, and American's latest heavy-rock guitar whizzkid, Steve Vai. It absolutely shocked me that people of their status had respect for me. I didn't think I was respected in the world of music, so it was a real eye-opener for me that they actually liked what I was getting up to.

From there, it was like starting from scratch. The one thing it wasn't going to be was a jam session. The songs were already there, and these fellas were there to work on them. The only real guidance was the vocal track in my head, as I never got to lay one down with the young whippersnappers, so a lot of work was done a cappella, just singing alone, so Bill could get some vague idea of the songs. I found a tambourine, and I'd batter out a loose kind of chorus and verse, and we'd base a song on that.

God, was I happy to have those musos! It was fantastic! There again, everyone I'd recorded with before had been a novice. I'd never been in a room with this kind of frightener. It was an unbelievable pressure, and I understood how my young band couldn't cope with it, but there was no way I was gonna back down. I really had to sing this one. There weren't no fart-arsing about, blaming it on duff notes from the lead guitarist! I had to get it right, and I did.

Nobody could believe that I was working with Ginger Baker. A few years before, the *NME* had run a news story that we were working together—as an April Fool! "Hello, me derogatories, careful what you wish for, you just might get it!"

Ginger, I loved. What a nutter. People might've imagined that as two such strong personalities we wouldn't get along, but I know what he comes from. I know the working-class approach in his life, and I understand it instinctively. And that's where we hit it off, bang on the money. We'd be the first, if you put the pair of us in a room, to absolutely slag each other off, because that's our nature, because what we're doing is consistently challenging each other to do better, and not to rest on our laurels. I don't like achievers because they tend to hang on to what they've achieved. I like the struggle, and once you've reached that level, then there's the next struggle, on and on and on and on. That's the fun of it for me, and I can see that in people like Ginger.

Look what that fella did with drumming! That's from the bombed-out part of London, that one, right? And in the '70s he's off to Africa to live with Fela Kuti before anybody even knew what that place was offering! He was straight into it, because he loved his drumming, and he wanted to advance himself and challenge himself. Isn't that what we're all really doing this for?

Listen, he might say he's not playing fast but you can hardly keep scope of the movement and the instinctiveness that he's taught himself on his instrument. I'll never forget the visuals of making *Album*, just watching him in that studio, breaking drum skins, and bass drums falling apart, and cymbals being cracked and fucked. Can you imagine, walking into the room when Ginger's going off on one, and going, "Oi! What about my chorus?" That's how it was. He and I have *totally* different approaches to music—he's studious in his particular way, but I am in mine. And the two worked, we just gelled. He's a monster-raging-crazy-loony, but he gets it and he plays it, and he laid down the patterns that I so desperately required. The rigidity yet flexibility inside those beats allowed me to put the words into their proper perspective. At the same time I had the open spaces of Steve Vai, he's twiddling a thousand notes a second, but he's creating open spaces by flooding the area. It was a fantastic combination of events.

Steve Vai, again, was absolutely open. You don't get musical snobbery from these people, what you get is "Ah, what you're doing is possible, and different." They're interested and open-minded to helping you to shapeshift yourself properly, *without overeducating you*. Steve's name was a bit of a dirty word at the time—amazing, isn't it, what punk created? The very antithesis of what I was trying to explain to the universe!

There was another drummer on that album too: Tony Williams, who's now sadly passed away. What a sweet fella. He'd played with Miles Davis right through the 1960s. Again—wow, I would've presumed he was *waaay* out of my range, leagues above me in terms of experience and quality, but the longer I live, the more I learn, the better I get. It's not an age thing, it's an experience thing, and what you've learned from these amalgamations can only make you a better person ultimately.

There were others who dropped by, like Ryuichi Sakamoto, to tinkle some keyboards, but the biggest mindfuck was when this new fella showed up, all dressed up in his finest threads, like any good working-class scallywag on his Saturday night out. For years I've been saying it was Miles Davis, but I recently heard it may have been Ornette Coleman. It was hard to keep track, people were coming and going all the time. He played on one track, but we couldn't find space for him. The only words from the fella were, "*Eeeerrrr*, I play on my instrument the same things you're doing vocally." Me and him were hitting the same patterns, and that was an incredible compliment.

Putting together "Rise," we had Ginger going mad all on his own in a room, bashing the hell out of them skins—he broke everything! I've never known a drummer to play that hard and severe. Though, I'm not sure if he actually played on the final version of "Rise"—I think it might be Tony. And then you've got Steve Vai coming in there, and then you've got some bass rumbling in from this guy Jonas Hellborg, and then it's like, "Wow, you know, we need the folky vibe," and along came an Indian fella called Shankar, and he played Indian-type violin in there, and it just pieced it together very nicely. It brought the groove and the sway and the lilt to it, and the song took on almost what I would call a South African Zulu vibe—but also at the same time it's a song that would bode well on any Irish jukebox.

I'm so proud of that single more than anything really. It was such a confidence booster in the light of the slaggings I'd always had. How are you going to knock a record like that? Come on. Look at the work we put in, and look at the result. It's an anthem of freedom and, through all of it, I hit upon my ultimate one-liner: "Anger is an energy."

After Mark, Jebin, and Bret were off the case, we'd had to have really serious discussions about what the direction of this album would be. And we went for *hardcore*. Bill had a good background there with bands like Alcatrazz, in that heavy metal world, which is really how Steve Vai crept, and how we could push boundaries. It was a thrilling record to make; every single thing we did had to be tough, tough, tough. So that was my guiding light every time I walked up to that microphone. I would vibe myself up to be the most rigidly hardcore heavy metal singer in the world, but without any of the clichés of what that universe had to offer.

We went powerhouse on everything, full-on ferocious, and the way I delivered the vocals there was no doubt about it, this *wasn't* going to be soft ballads. We were all on the same planet, the same palette. That's how a record really, really works, it's when you're all there together in it.

I mean, hello! The "Love Song" era was over! What I'd done was, I'd indulged myself in power singing and learning to control the notes and deliver them in a particularly aggressive style that was not in any way an imitation of what I was doing in the Pistols. That's a completely different approach, and I didn't know I could do that until I did it. What a great experiment.

The songs were structured, but again, not at all in a Pistols-y way. Listen to the complications and the tonal changes of "FFF"—my farewell to all my fleeting collaborators of yesteryear! That song's up and down and around the roundabouts; it shape-shifts and maneuvers around the beats and the structures.

Because nobody was being credited we decided to take a generic approach to the artwork—Album, Cassette, Compact Disc, whatever. It was all inspired by the generic lines I'd seen in supermarkets, when I first came to America. Beans would be just "Beans," there didn't need to be anything else written on the tin. I really liked it. I thought it was a great approach to dealing with commerciality. Not branding. Tell it as it is, and that's it.

I also raised the money, out of my own bloody wallet, to put together a gift for the first fifty purchasers of the record, where they'd get a paint can with "Can" written on it, and a little PiL logo in the generic light blue and dark blue, and inside of that would be "Cup," "Pen," etc. It was a wonderful thing. Unfortunately, the only example I had of that tin myself got burnt in a fire years later—that still makes me sad.

On every level, *Album* is the all-time toughest record I've ever made—hard, in terms of, like a smack in your face. So it's kind of hilarious that, when it came out, it got a rave review by that culture vulture, Melvyn Bragg, on his show on British telly. Also, apparently Sting wrote a favorable review of it somewhere! What on earth's he taking time out to do that for?

There were some silly billies back in England who reviewed it, saying it was my effort to "engage with American culture." No, man alive, *no*! Lest we forget, boys and girls, heavy metal absolutely is an English thing—from Deep Purple to Zeppelin! Their music was the hardest of this universe, and the most influential, so there wasn't much I was gonna open my mind to from Americans on that score. You can't teach the British bulldog those dumb tricks. I'm not a heavy metal singer, yet I'm using that genre, turning it upside down on its head, and showing an advancement of the theory, shall we say.

It was a proper eff-off to a lot of listeners who'd made presumptions about me. I've got to say, I do enjoy those moments. I'm not deliberately out to antagonize an audience or spite them or anything like that, but if they adopt the attitude of "This isn't what we expected," then yippee, I'm gonna wallow in that, because you shouldn't sit back and expect anything at all. You can make the choice to like it or not like it, but if you're going to hate it because it doesn't sound like the previous album, you're not a John Lydon follower at all. You don't understand me. I don't follow myself so please—don't—follow—me.

Album went straight into the *Billboard* charts in America, and I thought Elektra would be over the moon about that. The guy I dealt with there was called Bob Krasnow, a quiet retiring individual full of business conceits. At his label offices, I'd think, "What on earth's all this ugliness in the corridor?" He'd growl back, "My wife's an art collector." So, modern art filled their offices, and unbeknownst to him the whole staff were going, "I know, it's so ugly, but it's his wife . . ." I suppose really it was investments, a tax write-off.

Bob Krasnow's house in New York looked like a brownstone from the outside, straight off the cover of Led Zeppelin's *Physical Graffiti*. Inside, it was mental, and uncomfortable. He'd totally stripped it out, taken out several floors, and put in an elevator going up two decades to wherever their bedrooms were. Everything was vast and expansive, and you felt a bit like you were in the Guggenheim, the art museum in New York, which I love because of its circular walkway. But no, his house didn't have the circular walkway, it just had freezing-cold modern art everywhere. Whatever the message in that stuff is, it's a secretive select language that they only share with each other. That's my problem with modern art: it cuts off communication to the rest of us.

Being that we hadn't furnished Elektra with a list of credits for the album, knowing that they'd print them on the sleeve against our wishes if we did, they were completely unaware of who was on it, or just how important a record this was. At the time, they were more concerned with backing their new signing, Metallica, to the hilt. When *Album* charted, they viewed it not as a welcome success, as you'd imagine, but as a threat to Metallica, their long-term prospects, and so they dropped me. They were horrified to find out later who they'd just jettisoned—not only Johnny Rotten, but Steve Vai, Ginger Baker, etc., etc. Duh, you've just sacked the heroes of music, you idiots!

If Elektra had kept us on, we might've actually toured with the band that put the record together. But the red carpet was pulled from under us, and that was now a financial impossibility. So, off I went to scratch around and find another PiL lineup. Start all over again . . . again.

WHO CENSORS THE CENSOR? #3

DON'T LET ME BE MISUNDERSTOOD

BBC Radio had a go at me for a while about Bob Geldof's Live Aid. I'd asked some questions: "Which army is he feeding? Is anyone aware that there's a Civil War going on in Ethiopia? What's this really all about?" That didn't go down well at all. Years afterwards, the questions were finally asked: "Why were the food trucks held up at the border? Where did the food ever go to? Was there any education system in line to teach these people how to farm properly? Rather than letting their goats eat everything and then wondering why there was nothing left."

I'm putting it in a very basic way, and obviously the problem is bigger and deeper and wider than that, but I thought these were valid questions. If I'm going to be asked about it, then that's what I will say. I wouldn't be hoodwinked into joining the whole Band Aid thing, because I wanted to know how accurate this line was. And Geldof's a mate of mine! Basically, if you didn't toe the line with Band Aid, then you were somehow a curmudgeon. That would not be the whole truth of it. Charity for charity's sake is not charity at all; it's pop stars showing off and feeling good about themselves. But if they bothered to dip into their own pockets, then they could raise more money than all the audiences in the world put together.

Band Aid was all smuggery and naked ambition and self-righteous patting-on-backs. It was unbearable. Not to solve any problems at all, but really self-aggrandizement. Ever since, charity has been sorely affected by pop people. They're dangerous to any real cause.

To be honest, I could've done with a Band Aid for myself. I came out with a statement saying, "I'm my own favorite charity and I'm the only cause I can think of worth donating to." I bloody well meant it. It's a hard enough struggle, but I'm not going to get on to the coattails of somebody else, just because that's what everyone wants to feel good about. It has to actually be genuine with me, and really *mean* something, really *do* something.

A lot of stuff I do is for orphanages. Those kids I feel really sorry for, and that's something I can do something for, and I try to do that undercover. I'll donate things and they'll raise money on eBay, but my name won't be directly related, therefore ego doesn't come into it. There's always people telling me, "If you allow us to use your name, it'll earn so much more." That's never gonna wash with me. I think it becomes damaging and egotistical, and that's a danger I don't want to happen. No one person should be bigger than the cause itself. If you're not prepared to help orphaned kids or starving kids or sick kids without a pop star's name on it, then you're an awful person.

Celebrity branding is nonsense, and it's dangerous nonsense too. Whenever anyone does something wrong, a week later there's a press blurb with the charities they're associated with getting a mention, and that's used as a cover. It's very profitable for pop stars to use all manner of charities for their own benefit, and I resent them doing that, because we should not be doing that.

I say "we," because I feel as guilty as the actual purveyors of the crime, because I should be saying more to tell them not to be doing that. But the more I say, the more I get stuffed and resented. There you go . . . that's the story of my epic journey, that's a great part of it, getting my head cut off because I'm the first one to stick it out. I ask for it, and by fuck I get it.

These days, in the modern world of Google, the letters that come in tend to be begging letters from fake charities. People telling me that their mother's dying of cancer. It's all just too much to take on. I can't champion every individual's case or cause célèbre. It's too much. I pick my own causes in that respect. I can't have my heartstrings tugged too much. There were a couple of them in recent years that turned out to be fake, and that just put a huge smear on the whole thing for me. It's of course so touching that somebody out there has no other option, but the responsibility that places on me is overwhelming. And are they for real?

At the same time, during all of this, there's a lot of death wrapped around me—my own band members dying, my own friends dying, my own family members dying. Some of them because of disease, some of them accidents or whatever, some self-inflicted. I'm not a saint, and I don't want to see myself cornering off some role as a procurer of good taste. Because I'm not, I can't do that, I don't have the energy for that. Hopefully I'll have a good effect on people's minds, but I'm not here to bolster their wallets or fill their coffers, because you read a lot of these fan letters, and the intent's selfish somehow, and they don't realize that it's not a rewarding thing to be doing to another human being, to demand that you sign their husband's birthday card just because he's a big punk and it would be great. On the one side, you look at that and it's a small thing, but where does that end? Then you become selective. Are you going to keep doing that? In which case that's a full-time occupation that don't pay well, or you just say, "No, it's got to stop, it's too energy-draining."

It's the same as signing things for people at shows. I genuinely love saying hello to fans. It can be great fun—we do meet some weird and wacky characters. We affectionately call them the "Lollipop Mob" after the PiL song "Lollipop Opera." I've literally spent hours after shows speaking to people, even bringing them backstage for a drink and a chat. But it was getting out of hand there for a while. People *expect* it of you every single night and when I don't I'm slated by them. They don't understand it's nothing personal—far from it: some nights I just need to get on the bus before I catch my death of cold and end up sabotaging the rest of the tour. Where do you draw the line with people?

There are also a lot of professional *dogs* out there who you know fine well will put it straight on eBay. They follow us around and ruin it for everyone else; it's an ongoing battle. And it's not just the professionals: we've actually seen people leave gigs early just to stand at the stage door. What, is my signature more important to you than the gig? I thought you were meant to be fans? Ludicrous. It takes the fun out of it all—it becomes a drain on you.

The energy goes into the songs—that's my commitment. That's my work, my effect on Planet Earth. I don't want to start charity-hopping. It's a repercussion of fame, or infamy. One line I actually did contribute to *The Great Rock 'n' Roll Swindle* was, "Infamy, they've all got it in for me!" That was Kenneth Williams! I've got such a great love of the *Carry On* thing. As a kid, those movies were hilarious. To a young mind at the time, they made a glorious fiasco of the rules and regulations of society. So, very early on when Malcolm was starting up the movie, long before it turned into *The Great Rock 'n' Swindle*, these are the things I'd be chucking in.

There's an awful lot you can learn from humor. Am I gonna learn from *War and Peace* or Norman Wisdom? Give me Norman, any old day. He's closer to my life's experience, and therefore relevant. And the in-depth angst of Russian intellectualism is very far removed from anything I've experienced. Although, I may yet get there. If you're dying of a terminal disease in a hospital ward, that's where Dostoyevsky can come in. It's so miserable, it can only cheer you up.

I did read *Crime and Punishment* when I was very young. I remember the TV drama, too, starring the English actor John Hurt. He looks a little bit like a rodent—what a great actor he is. He also played the famous homosexual Quentin Crisp in *The Naked Civil Servant.* I met him years later—I really liked him. He reminded me of Keith Levene, in the look, and the greasy-skin persona. But he didn't have the dead eyes. I found him to be an intelligent man and capable of really good conversation. I liked him. Smart bunny. His wife or girlfriend had died the year before, and so he was going through real sadness. It was very difficult for him to be in a public place. I felt his pain, and one way or another, it's the pain that goes into my songs.

Grieving in public is very difficult, and the worst aspect of it is when complete strangers come up and tell you they "feel for your grief." It just reminds you of the very thing that you're trying to lay to one side for a few brief moments. Eventually you realize that life can be shit, and it can only get worse, so you better be making it better. You have to deal with it. I do the best I can. Internally, inside the skull, when people reenact those moments, they mean well when they approach you, but they're reminding you of something that leaves you feeling very exposed and breakable emotionally, and isolated in a collection of strangers. That's a terrifying thing to endure. Happens a lot.

But let's face it, I've got a better lifestyle than any one of the alternatives that were open to me. There wasn't much option. The Antichrist is what I accidentally ended up as, but that's absolutely not what I set out for. I've always loved that song, "Don't Let Me Be Misunderstood." My intentions *are* good. The Animals cover, wasn't it? Eric Burdon.

I loved Nina Simone's version, too, and I loved her musically. I met her once at a Peter Tosh gig in L.A. and she was such a bitch to me. "Who's that white boy?" "*White boy?* I'm in my forties—who do you think you're talking to?" She was absolutely giving me the blank: "You don't belong here with black people." I suppose everybody has a bad day, but you've got to be so wary that your bad day can impose the same kind of reaction in others.

Oddly enough, here I am, a man who speaks openly from the heart, but I realize that you have to be guarded in them public situations. She's fantastic and I won't have a word said against her music; it was just "Don't honkify me! We all want a world where we're equal, and you're throwing that kind of garbage around. Wrong!"

Fair play, Peter Tosh, him being one of the original Wailers with Bob Marley and all that, wasn't having any of it. He tried to correct the situation and gave her a bit of a mouthful. Me and Peter were having a disagreement, anyway—it might've been to do with the Rolling Stones actually! Listen, we're all capable of getting up on our high horses, and sometimes those horses are a little *too* high. Peter was working with them and I was having my say on that.

I've always known, though, that Keith Richards loves his reggae. He's always been well rooted in it. Musically, that's not an uneducated fella. I've never met him. We probably wouldn't get on. But it's the same with, yes, Elton John—there's another man who knows his chops around other people's work and doesn't skip a beat on everybody's efforts. That always impresses me. It's a good mark, an indication—nothing to do with praising myself here!—of valuing stars, when I find out they're like librarians in their approach, that they want to know everything about anyone who works in the same field. That's how it should be.

So, good on Keith for liking reggae. I mean: in between yearly blood transfusions, why not?

10

HAPPY NOT DISAPPOINTED

"I never thought he had talent, I always thought Sid was the genius." That was the sad, sorry, silly indictment still coming in from Malcolm McLaren's camp. It was Vivienne Westwood's line, from way back in late 1975, when I wouldn't comply with her fashion dictates. Thanks a fucking bunch, bitch, you're flogging my fucking clothes ideas and you have the audacity to say that. The terrible thing is, a lot of people picked up on that, and wanted to believe it and still I'm ostracized from what we could call fashionable society because of attitudes that came down from Malcolm and Vivienne at that time. Tough tits, baby. I'm still here.

Eight long and hectic years had gone by since the end of the Sex Pistols. I felt like I'd achieved and proved so much in the meantime, but now finally my case against Malcolm was coming to court.

From the beginning, I was told by my lawyer, Brian Carr, that I had no chance on God's given earth of winning anything out of it and shouldn't go forward with proceedings. I insisted on talking to the barrister, because that's what you need to acquire when you get into these things in England. He said, "Oh, it's very risky. Are you prepared?" I just dived in.

Proceedings were concluded a few days before my *real* triumph—the release of "Rise" as a single. That song was changing people's perceptions of me towards the positive. We did a great video showing washing lines in some of the more tormented neighborhoods of London—stirring imagery from the kind of place I grew up in—which actually got us on MTV. We were on *Top of the Pops*, too, and everyone seemed to be connecting with the song—even the critics! Battling Malcolm just felt like ancient history.

The "trial" itself, on the other hand, lasted three days, and Malcolm advisedly settled out of court. The whole thing was like a damp squib, a wet fart. I didn't know what it was I really wanted. I didn't want any more resentment festering, and I didn't want to walk off owning it all. I wanted that sense of "share," and ultimately that's what we got: we four surviving members got to *share* everything, from the Sex Pistols name, to what was left in the band's bank account—a large chunk of which went towards an enormous outstanding tax bill.

Still, it set things up lovely: now we weren't going to be bankrupt forever and a day. All we had to do was work hard to stop things ever sliding back in that direction. But we took the name Pistols back off mismanagement and, ever since, not just me but the people I work with have been fighting to maintain a sense of integrity about what the Pistols really was.

People should understand: Malcolm was never out for the money. He was out for the accolades. That's really what the court case settled—it wasn't him what did everything, creatively.

Pretty much the last time I saw him was when the Clash were playing that residency at Bond's in New York, when we first moved out there in May '81. Bernie Rhodes was back managing the Clash, and he knew Malcolm was in town too, so he set up a dinner date for the three of us to try and talk us into making peace. I really couldn't take Malcolm serious. It was the most pointless evening ever. This was the ultimate indication of where

Malcolm was coming from: he says, "It's silly, Bernie, we're never going to like each other, why are you doing this?"—and we got up and left together. Outside, he turns to me and says, "Well, at least we didn't have to pay for it." To me, that was tuppence, not a big victory in life, but it was very much fundamentally Malcolm's lousy approach to people. Always trying to get one over on them, in that sniggly, ducking-and-diving way.

All the people who'd been around us at the beginning, all those who were "Friends of Malcolm"—like little Helen Wellington-Lloyd—had become "Not Friends of Malcolm." It had gone from the initial outbursts of "Oh, Malcolm's doing a great thing," to every single one of them realizing he was *not* doing a great thing. Malcolm was actually very destructive, to himself and everybody else, and tried to manipulate people's lifestyles when he wasn't really handling his own too well. A bit of a disaster. Poor sod.

Malcolm had a proclivity towards possession. The whole situation was very difficult for Paul and Steve, because their apartment in Bell Street in Marylebone was owned by Malcolm, and they were well aware that any decision on their part to change sides during the case could jeopardize their existence. They eventually switched ships once it began to look grim from Malcolm's side.

I didn't see it as winning anything, rather resolving things properly with Steve and Paul, because they eventually came to see that I had a correct and accurate point of view. Malcolm *had* sold us down the river contractually. Was there a big joint celebration, with the popping of champagne corks and group hugs? Oh, I don't think so! Not at all. They'd done so much damage to me with the bad-mouthing. All of that's still there in me. To this day, I'm aware there's a lot to be made up for here. But at the end of the day I've got no resentment or hatred towards Malcolm, so why on earth would I have any towards them?

At Brixton Academy, in May 1986, there was a mob of about 100 to 150 that absolutely came down to *destroy*. These morons considered me, wrongly, as The Enemy. They were spitting and chucking bottles; there was all sorts of crap whizzing past our ears! A bunch of them were constantly trying to get onstage or trying to climb on top of the stage-left speaker stacks. The bouncers didn't seem to be properly organized, or maybe they were just undermanned. Something just didn't seem right. It was a constant battle all night. All I could hear was "You're a cunt!" and "Sell out!" and "Boo, hiss!"—whatever. This was just shit they'd read in the papers—that I'd sold out by starting PiL, by making *Album* with some fine bloody musicians, and that I wasn't a punk anymore.

There was such a dark horrible vibe in and outside the building. I'd heard there had been some muggings before the show. Brixton in them days was a dangerous place. The locals saw the average concertgoer as easy pickings. Inside the venue the trouble seemed orchestrated. It was all bollocks, evil and nasty, but it shows how easily people's minds can be manipulated by the media telling them this stuff. Nothing had been easy since I'd finished *Album*—I'd had to find a live band very quickly, while the Pistols case was brewing. Now life was being made all the worse by these one-percenters, we'll call them, that were completely buying the media agenda and turning on their own. Quite frankly, I view them as government watchdogs.

It was such a shame, because we were offering the crowd something great, really taking PiL to the next level, and the vast majority were completely with us, responding to the positive energy that came from "Rise" and *Album*. It was just this certain clique—they were punky new-age squatter types—that were dead against us. A new breed who'd got it all wrong. They created an ugly situation. The security fellas on duty at Brixton that night were mostly from the local neighborhood. I wasn't overly impressed with them but some of them were genuinely all right. One of them who I happened to get talking to because he supported Arsenal said to me, "Look, John, they're spitting and chucking things at us too, what do you want us to do?" My answer was "Just *don't* crack any heads!"

It was a hard one to overcome. I had my work cut out. In those situations, you can easily end up with the responsibility of starting a riot. But, really—the hatred! The game was obvious: "Tear Johnny Rotten down! Who does he think he is!" To which, I'd respond: "You're doing the government's work, you fucks! That's fine by me, you ain't gonna stop me none!" But when it comes to my band being worried about really serious injury, well, I'm with them in that respect. Then you have to go about it a little differently.

It was so out of control, no one was listening. The only answer to it was to stop the set. The band went off, and I said, "We won't come back on until that stops. You know who you are—stop it! And if you know who they are, point them out, and we'll stop them." A big punch-up ensued, and a bunch were slung out.

At that time, PiL was too far ahead for them. What a pity. Still, we *were* good! It was very frustrating because we were getting into the discipline of playing for a real long time, and really giving you your money's worth, and trying to keep the ticket prices as low as possible. By doing that, we let in the dogs, the jealous, the absolute talentless lowest denominator. It's crabs in a barrel—that old expression I use. They just keep pulling you down to their level, because they don't have an answer. No empathy. It's a particularly British attitude to success—a hatred towards any of their own achieving anything at all.

It certainly wasn't easy for my new band, coming into this situation—the punk kickback. None of it was helpful to the agenda of getting started up again. Actually putting the new lineup together was straightforward, to my mind. It wasn't like running an advert in the newspaper this time, I was thinking in terms of sound, and kindred spirits, and some personalities who might actually get on. The people you meet backstage, the people in other bands, the ones you chat to, the ones you get on with—they stay in your mind.

So, after *Album*, I went back to London and found John McGeoch. He'd been the guitarist in Magazine, who I was a *big* fan of. And wasn't he astounding in Siouxsie and the Banshees? We fell into each other in a really great way; we were already very good friends. John sadly passed away in 2004, but hanging out with him was always hilarious. His humor was, "Where's the bar?" He'd start with a double martini, he couldn't care less if it was shaken or stirred, and onwards and upwards from there on in. He could be a difficult person—he was a Scot, indeed—and he had a lot of emotional wreckage going on inside him, but I don't mind working with difficult people when they've got the goods.

I will always think the world of John—just superb guitaring skills, and in different styles from what I was used to. I was used to more rhythmic styles and approaches. John was more of a note-y kind of guy, with the jazz chords every now and again. And it turned out he was a beautiful, beautiful person to work with.

Bruce Smith I'd known forever, really. I first met him at a gig with his first band, the Pop Group, and for some weird reason we Pistols thought they were threatening us. They thought we were threatening them, and they were terrified—but there was about seventeen of them! He went on to drum with the Slits. He'd been trained as a reggae drummer, and played on soul records, done jazz—quite a lot going on in there. He's a happy-go-lucky, amicable guy, but I soon found out he's a bit of brilliance in a rehearsal format. He's very rhythmically structured, and has this personality that can blend all these opposite forces together into something cohesive—which, frankly, is a quality I tend to lack from time to time.

Bruce brought in Allan Dias on bass, because they'd done the odd bit of session work together, and I trusted Bruce's faith in him. Allan became one of PiL's longest serving members—things don't always end up negatively. Allan was so easy to get on with, very good fun, and a ladies' man without a doubt. He has a certain quality of confidence where girls just fall all over him. A sexpot. This is PiL, we cater for everything, and everybody.

Lu Edmonds, on keyboards and guitars, came from another

strange meeting in London. I had forgotten completely that he was in the Damned, and I didn't recognize him as such—and I didn't hold it against him. There he was in his fisherman's cap doing his little roll-up cigarettes, looking just like a professional social worker. He's one of the most easygoing, greatest people to get along with. It's so strange, his brain and body are disconnected. His body is so uncoordinated with rhythm and yet he plays more superbly than I have ever heard from any human being ever. He loves ambience, sonorous rhythms, fractures, tonalities, chaos.

With those four in place, there was a bunch of people that are all incredibly different from each other, but I thought, "This could work, finally." No single one of us was dictating what the next vibe would be. It was a real sharing of talents, very generous, no dictatorship going on. It was a massive breath of fresh air for me, because up till then in retrospect, it had been like suffocation.

What these boys had to face on our first UK tour in May '86 was just terrible—very, very difficult, and not just in Brixton. The opening night in Hanley, some idiot chucked a snooker ball at me. In Edinburgh I was hit on the head with a lady's high-heeled shoe, a stiletto heel. Wow, did that knock the sparks in my brain. In fairness, Richard Jobson from the Skids came back afterwards and said he'd talked to the girl responsible, and she was really sorry. She was only trying to say, "Look, here's my shoe!"—she meant it well, she wasn't out to try and gouge my eyeball out.

In Vienna, McGeoch got a two-liter bottle of wine lobbed on his head. He ended up with something like forty stitches. We also had problems at another festival in Holland as well. The deal was, if it came to that, then we'd just have to go off. The band are standing there trying to play and they don't have a free arm to catch things. Myself, I got very good at catching stuff when it was thrown, and not skipping a beat. But it's not a game. You can get seriously hurt there. It only takes one or two at some of them large festivals, where the stage was low, and it could get very, very dangerous.

In Vienna, it was a support band responsible for throwing the bottle at McGeoch. They'd gone to the back of the bar and nicked these empty bottles and that's what they were slinging at us. It was like, who's stopping that? I'd go, "Police yourselves. Who's doing that?" Generally speaking, a crowd would point them out and off they'd skulk. You've got to do something, and make a stand.

At the time, I had no way of explaining this in the media. They really weren't prepared to listen and were more or less rallying the negativity. I was easy pickings, and I didn't feel that my record company was properly representing me with any sense of support, so that allowed a kind of journalistic freedom to unleash itself on me. At the same time, a negative review of, say, Madonna when she was starting up, would've been treated with very promptly by the record company by, for instance, threatening to pull advertising. I had none of that support, and therefore I was a free-range chicken, baby! Because my name was so up there and well known, wow, what a target. And not appreciated for that. Or for the music.

On that tour, we'd open the set with the Led Zeppelin song "Kashmir." I love that song, I really do. I don't mean the Puff Daddy version that came out some time after. I really wanted to sing it, but I never got round to singing it in rehearsals. I insisted the band rehearse it, and I insisted that we open the set with it, but every single time they're waiting for me to come on, and I'm standing at the side, and I never did it. I'd shit myself. I couldn't get it together—the very thing I'd set up and wanted so much—and it became laughable.

It was still a great opening song, a very lovely piece of music, and I got to like the idea of letting people hear it, unadorned by Yours Truly. "Hello, it's not the Johnny Rotten Show, check out that band!"

Deep down inside, I think I wanted to sing like Robert Plant. I love Robert Plant, a great fella. I've met him a couple of times. I've got nothing but good to say about him. He comes at you with no preconceived us-and-them attitude. He's very open-minded; he's everything good in music. I really, really like him. I mean, I don't like his hairdo, but so what?

Listen, he came down the Roxy in the early days of punk, when we weren't quite yet sorting ourselves out proper. The Roxy was punk's deep dark hole, the den of iniquity, but he had the balls to come down there—I think he came down with Lemmy from Motörhead—and it was great! I just made a beeline straight for them, and went, "Hello, great to see you!" Because it really was. What he was doing was giving us a pat on the back. Of course there were arseholes there, going, "*Ugh*, whatcha doing talking to *that*? He shouldn't be here . . ." "No, don't be telling me who should and shouldn't be here. Punk's open-minded! Abso-fucking-lutely!"

After all the violence on tour, we PiL-ites were feeling trapped with this idiotic element in the audience that was causing it. The success of *Album* in Britain, however, meant that we were very much back on Virgin's radar as a commercial prospect. The pop mainstream at the time was a terrible place, and we really didn't feel a connection with any of our peers. As in wider society, the entire period was about materialism. I had to squawk through it, and I was hated for everything I ever did.

I was trying to write about human emotions and political problems in the age of Reaganomics and yuppies. More than ever before, it was all "Yippee, let's do it for the money." Many of my alleged music cohorts in different bands were all the time on my case, going, "Why don't you just write a hit song?"—exactly that same old bollocks I heard from the first day I got into rehearsals. *No!* You write what you write, according to your experience and your humanity and your sense of understanding of the world, and if you try to step out of that zone, well, yeah, you might make the cash, but you're one lonely silly sod.

It struck me as deeply strange just how little music there was in the charts with any kind of relevance or political meaning. To me, someone like Boy George was the rare exception. All the people I like in music are the ones that have done something completely original, with a touch of genius, and I put Boy George in that bracket. He came up with something really great and challenging. At a time when punk had got staid and boring, out comes Culture Club. Fantastic. George would wear Indian menswear in a feminine way. The boy can sing, and he comes from the same background as me—the same hardcore rubbish. He's someone that stood up for himself, no matter what he got into, and he's intelligent, and therefore I like him. More respect, more power. He was the kind of guy there wasn't really enough of to make the '80s bearable.

The world I wanted was going back to early clubs like Louise's, where all manner of people would meet in the same environment and not cause a problem to each other, and not judge each other, and all be very different for their sexual agendas.

The '80s proved very negative in that respect for me, really just a bitter competition for who could make the most expensive video and show off the most. What a pity, such a shame. Because, as I said, I love Duran Duran. I love "Hungry Like The Wolf," but does it require hundreds of thousands of pounds' worth of promo video? All that was created there was a whole new monster of video directors, and they were arseholes to a man. The dictates that would come in from these people were just ludicrous beyond belief. The song wouldn't matter, the studio work, your lifestyle, your band, nothing. It would be "I have an idea. All you've got to do is pay for it!" The video was becoming more important than the music.

The biggest laugh in all of this era was—mullets. Again, it was "We're just having a laugh!" "That's all right, I don't care what your hairdo of choice is." But how vacuous it all was. Having said that, I was changing my own hair a lot at that time—though hardly according to fashion! I was the precursor to how many Beckham hairdos?

I'd started to stick bits of fur on the top of my head with superglue. I used fuse-wire to make certain fluffballs stand up high. They weren't dreadlocks, more like bunny tails. I had so much metal in my head, by Jesus Christ, you try getting through an airport with that. The machine that went "Bing!" would go "Bing-bing-bing!" every time. The "hotwire to my head" lyric in "Rise" was a hairdo reference, while also being a poignant reference to South African torture methods involving electrocution.

By default, I suppose what I was doing was protest music. I can't help that. It has to be done. The truth has to be out there. That was a very untruthful period. Quite amazing.

Politics has always been there in my writing. I can't help it. I just feel naturally inclined to help the disenfranchised, and I'll always feel that way myself. I know what those emotions are. I've never forgotten the endurance of my childhood, and so I have a great sense of empathy for people who suffer. To be ostracized from a society that should know better is not a great thing at all. So here I am. I will change society, I have changed society, I always will, and I'll also be the first one shot because of that. That's all right, I don't mind taking the first bullet, because there's enough people out there in the world of music, just music alone, who understand that, and that creates a great playground for the future. It's my obligation to stand up and tell it like it is.

When it came to making our next album, 1987's *Happy?*, I was very pissed off with the world, and I went very wordy. Rather than sing melody lines, I thought, "I'll shove as many words in here as I possibly can," and loved doing it.

After touring together, where we'd all got on so well, the rehearsals where we wrote the songs for the album together flowed amazingly. It really got me back to why I wrote songs in the first place. That's why we called it *Happy?*—it was almost hard to believe how "right" it was!—and it really surprised people, because it was a very confident step in another direction.

There was a great sense of working together, and loving the

work. I'd had enough of waiting around in New York for people to turn up. This was a room full of bright ideas which was heading towards a kind of pop sensationalism.

"Seattle" was an exceptional tune in that the band put it together without me. It normally doesn't happen that way but what happened was, I was stuck in New York with tinnitus, which meant I couldn't fly. The band went on to our destination, Seattle, and they had nothing to do for a week, so we agreed that they go in and record something, just a basic backing track. When I hooked up with them, they went, "Oh, this is just something we were messing around with," and I went, "What? It's bloody amazing!" I've still got the demo cassette they gave me. It just zinged, that song.

There's a lilt in there somewhere almost like an Irish folk song, so—bang, in I went, layering that sensibility into it. My favorite part of the song is the "palaces, barricades, threats meet promises" section which is dealing with the rioting that was going on all around the world in the mid-'80s, from Broadwater Farm, London, to as far away as India.

The part that goes, "Character is lost and found on unfamiliar playing ground," is very direct, but the whole thing is considering so many paradiddles of thought processes. A paradiddle is what a drummer practices. Every drummer I know, they're always in a corner going, "Paradiddle, paradiddle, paradiddle," tapping their knees. That's how intellectualism works too. And although we know that intellectualism is a big fraud and some of the biggest deceivers of mankind are intellectuals, it's also a very viable place to be. *Think!* And then when you think you've sussed it out, *think some more!* That's what that lyric is saying.

The songs on this album were really good, I think, and meaningful. "Angry" was a "look at yourself before you judge others" song. "Rules And Regulations" was on the "don't tell me what to do" theme—you know, lest ye forget, there's still a Rotten in here. "Hard Times," with its Charles Dickens reference, was an alarm-bell song about when national identity is corrupted into that siege mentality of us and them. Here I declare: we *are* them. All of us are them. We are us. All of us. And vive la différence!

"The Body" refers almost directly to a TV play by Ken Loach called *Cathy Come Home*, which I watched when I was very young and which really affected me. It's about unwanted pregnancies, and the sense of almost criminality put on an unwed mother, and what she would have to endure—the abandonment, lack of family support, and isolation. Terrible, terrible things. I was young but I felt it really severely. Years later at Gunter Grove, I went out and found somebody who had a reel-to-reel of it, and I played it back and just broke down and cried all over again. I felt so sad for Cathy, I just wanted to wrap my arms like wings around her. That's how I am, and I make no apologies for it. That's my basic approach to life.

The final track, "Fat Chance Hotel," was based on a true story in my life. Soon after the Pistols broke up in 1978, I was stuck out in L.A., and I met the manager of Gwen Dickey of the soul/disco band Rose Royce. The manager was English, and she had a child but no husband with her, and Gwen had nothing to do, so the three of us and the kid just rented an RV and drove to Mexico—me in full punk regalia, the tartan bondage suit, accompanied by a black gospel singer, for a want of a better word, and an English lady with a mixed-race daughter. We definitely turned a few heads.

Unfortunately, I had some digestive issues with the local cuisine—tacky tacos. I would also advise anyone going there: don't drink the water. So for a few days, I was stuck inside this dreary run-down hotel, and the song has some quite poignant lyrics about being bored in the brain, out here doing nothing at all—with a "splattery botty." To be able to find a way to write a song about having nothing going on except diarrhea, I was more than pleased when I heard the final tape. It certainly shouldn't turn you off a good holiday. It intrigues, with warning signs.

There's also a love of the desert in it. There's something about the silence of the desert, which isn't silent at all, it's the loudest silence you'll ever hear. So it's a very enjoyable song; you just close your eyes and drift in its space.

For me, the whole album was a very powerful coproduction with Gary Langan, from the Art of Noise, with well-balanced results. Gary was nutty, a bit of a genius, but with a laugh and a smile, and he always did things for the right reason.

At that time, Lu was obsessed with technology, and he wanted to somehow transfer modern electronic keyboards via the computer into the gamelan register. He wasted years and years on that, until he came to grips with "Well, you don't need to—just play the thing!" We maybe went a little far into the banks of keyboards: I remember us all being very, very in sync with each other.

It's a serious problem for me, all this technology. The people who've used it best would be Depeche Mode. "Your own personal Jesus!" Bloody 'ell mate, *they* got it! They were using the Casiotone effect and they wrapped a song around it, but they didn't let it dictate to the song. That's another tune I just absolutely love—I was so impressed with the bravery of attempting such a subject matter.

The front cover of *Happy?* was a nod to the German artist Friedensreich Hundertwasser, whose work I was truly impressed by. I don't know nothing about him, I don't want to. I just know that anytime that man put something together I was completely interested. He did architecture, too, and it was always interesting, for instance using gardens on the floors of skyscrapers, and altering the shape of a building and making it interesting to the eye for the observer—as on our sleeve.

How I got into Hundertwasser goes back to the job that Sid got me working at Heal's on Tottenham Court Road, when we had to clean up the vegetarian restaurant. Often we'd be bored and have an hour to spare—because you had to be there for a certain length of time, but you'd clean up that place in a minute flat—so then we'd wander round the store. I noticed the Hundertwasser books in the store's library section and *procured* some, shall we say.

His art to me is People Art, always about making creative and friendly environments that people live in. His paintings are always happy city scenarios where everything is brightly colored. Absolutely inspiring.

When we went out to tour the album, we used the multicolored-building idea onstage, with adventure playgrounds and runways, all in very bright, vivid, bold colors—greens, yellows, reds, oranges—and there we were, running around in it like happy children.

Through March 1988, we ended up on a U.S. arena tour, supporting INXS. Oddly enough, I loved their first album—honestly! I liked the emptiness in the production. I didn't realize they had a bass player till they played live. But there's something always exciting about Australian music; it's a really interesting place in the world. They're on the other side of everything and so their approach is different.

That tour got competitive. We were playing in 20,000-seater arenas, but only 5,000 were there when we were playing. It was very strange. Maybe that was the MTV apathy kicking in. In my early days, the audience would be in the auditorium right from the start and wanted to hear every single band and get their money's worth, but the MTV video world created a different environment, and the auditorium would be empty until the main band came on.

It all went horribly wrong in New Orleans when Michael Hutchence invited me after a gig to his "apartments" in the same hotel. What great rooms he had! Wow, I loved that. It had an upstairs, a downstairs, and a sound system. It turned out he just wanted to play me this version of "rave beat" that he'd been making. I just had no time for it. Because, hello, I like me rave, but I don't like it analyzed and interpreted and copied, and that's what it sounded like to me. And I had to say so. "Oh, I can do that," was the vibe, and we fell out. We never really spoke after that. And then he went and asphyxiated himself . . . Sometime later, I might add, nothing to do with me!

When people invite you over and want you to listen to their

latest record, it's never going to turn out well. It creates an ugly, uncomfortable environment, and I'm not ever going to be the kind of fella that's going to give you false accolades. My attitude is "You've made me feel uncomfortable. I might've liked this in a different time-space continuum, but that's a wrong move!" At the same time I understand that they're trying to share their sense of achievement. There's that going on, but at that moment that's not what you're feeling. You've just come offstage, you've done your gig, and you don't need to be impressed.

One good thing about Mr. Hutchence: he knew my voice was a bit raspy because I was overworked—and they toured with their own doctor! I got some really bloody good advice, and some medical hoo-hahs, which every now and again you need. Namely, a Vitamin B12 shot. There are many doctors that will say that's as useless as a placebo, but I don't find it to be so. I find it gives me the energy that I require up there onstage. It makes you very tired; you take it in the morning, up "le chuff." So, you've got a sore bottom, then you fall asleep for four or five hours and you feel very tired and you don't feel like you've got the energy, but the second you hit the mic—*bing*! It kicks in! Thanks for that one, Mr. Hutchence.

That summer, we played at a massive free festival in Tallinn in Estonia. I understood at the time there were 175,000 people in the crowd, but I've been told since it was more like 125,000. I trust two things here: other people's statistics, hee-haw, and I trust my emotions. I know what my emotions told me when I walked out on that stage. I lost my voice with the huge vastness and expanse. The sea of faces was endless, endless, just going on forever into the cloudy distance. And there were tanks each side of the stage, thank you very much, with the turrets pointed not at the riotous assemblers, who weren't rioting, but at *us*.

You got to bear in mind that this was just before the iron curtain lifted, when they were still a part of the Soviet Union, so it was a very tense period in Estonian history—and, I suppose, world history. And it was such a great audience. It was a free gig, so everybody had a reason to be there, and they'd traveled from miles around to be there. Yet we all knew the brutality of the Soviet regime could inflict horrors upon all of us at any moment. Fortunately, it didn't come to that; it became instead something incredibly special—and eventually it came to Estonia's independence.

It was amazing, terrific, but very odd. When we wanted to walk around the town, we couldn't do so without official escorts, and we'd see people being ushered away in the corners of the square. We weren't allowed to go up and talk to anyone. The secret police were not very secret.

We did actually get to meet some people, just before we got the ferry away from there. This is where Lu is such a great ambassador; he can break through any police cordon, just because they presume he's one of the vagrant locals, I suppose. These people had traveled not just tens of miles, but hundreds, from all the neighboring countries. When we left, I must've had 200 bloody vinyl records given to me. I love and treasure every one of them—not so much for the music, but for the actual thought and energy that went into giving me something that was shady and illegal, according to the authorities at the time. That's heartwarming stuff—fuel for my fires.

The reason I'd had tinnitus in New York at the time "Seattle" was written was because John McGeoch would love his amps to be so full-on, onstage. He was very into his heavy-metal amp stacks. It created real problems for all of us, it was just too overpowering. I wouldn't blame John for it, I'd blame us. We should've just yelled at him. We've since learned: Bruce and Lu are consistently onstage saying, "Turn it down." It gives you so much more freedom in the song, but we all come from that period—and heavy metal is always gonna be there, in our early-day psyches, that preconception that louder and louder is better and better. It's not. Lu Edmonds got it very, very bad. I recovered, but Lu didn't.

Lu was involved in the writing stage of our next album, *9*, but he literally had to pack in amplified music altogether for a while. He went all acoustic and traveled in the nether regions of Muslim and former Soviet territories—places like Kurdistan. He was gone from us for years, and it was a bitter loss.

With hindsight, we were beginning to fall into the treadmill of album-tour-album-tour. We initially started recording *9* in New York, with Bill Laswell. It wasn't at the record company's instigation, although they'd doubtless have loved an *Album Part Two* at that point. This was very much Bill volunteering his services. After a couple of days in the studio he said the band couldn't play and he hated all our songs. He said he'd written songs and I should sack everyone and use his people, and come out with a U2-type product. I told him to fuck off and we packed our bags and left. I was fully committed to the band.

The more I think about this, the more my memory grows about poor old Bill and what he had to endure with me. In his head, I was the lead singer he always knew I could be, but I wouldn't do it because I've got my own way. I've got my own learning curve. There is a point where I can take influence but I can't take teaching. It goes back to school really. Don't tell me what to do, tell me how to do it. That's how it works with me.

There were always personality quagmires with Bill, but my only serious problem with him was that, whenever I'd go over to his apartment, he'd be trying to show me his guns. I just find that to be all too ugly. These were early days for me in the American culture, and so I wasn't aware that when people are showing you guns, they're not threatening you, it's like they're showing you their art collection. For me, at that point, what he was presenting to me was very challenging.

Now I understand America very well, I think, and I do appreciate the fact that, yes, you should have the right to own a gun. Yes! To my mind, this lot could be a far more serious pack of killers, in proportion to the firearm potential they're wielding. In many ways I think Americans show an enormous amount of restraint, and of course, there definitely isn't anyone invading this country; they're too well armed for that. And what fun it is to go out into the hills, as I found out years later, to shoot things. It's not about wanting to kill people or animals, it's just the element of control. It gives you a great sense of achievement to be able to hit a melon at 50 yards, that you've got the power and the control to take aim correctly. It's skillful, and I like it.

People might be a little scared to hear about Mr. Rotten tooled up, but John ain't no killer. As I keep telling the world, I'm a pacifist until you stretch over the line and try to hurt or damage any of those I love. Then you've got a prob. Me you can slag off all day long. *Not a prob.*

So, it wasn't going to work out with Bill, and we ended up recording with Stephen Hague, who had been working with New Order.

There was also a guy involved at the production level called Eric "ET" Thorngren. Let's just say, at the time, that didn't seem like the better half of it. It always comes down to personality. Eric, nothing wrong with the fella, but I liked Steve Hague and his quiet, wispy personality, and his technical precision. It was a completely different approach, which is what producers can create. They're people you have to know how to pay attention to, and you have to know their weaknesses. Steve was a gentle kind of fella and his soft touch on top of our huge uproar made for an excellent result.

We were perhaps getting a bit too deeply involved with the new MIDI computer technology. The trouble with that is it takes away from the analog, the sense of live. But again, I think the songs were very strong and emotional.

"Happy," the song, is really an answer to the question that was posed on the previous album title, *Happy?* I felt like some deep well of integrity was being tapped by us around this whole period. "Happy" is reflective, it's a look back, it's a self-analysis—the upshot being, yes, I'm happy—God, it is possible!

One of my favorite songs I've written is "Disappointed." It's a spectacular, dramatic, and very forgiving song. If I could ever call a song a friend, "Disappointed" is one of them. As the lyrics say, it's truly "what friends are for." It's me talking to me on the lesson of em-pa-thy. It's one of those songs that somehow creeps into a PiL audience's psyche, and wow, do they "get it." It's overwhelming, sometimes, the tears and smiles and hugging out there. There's many reasons to cry—joy, usually, I hope, but there's sadness in the verses, the betrayals that lead to the forgiveness. You can see in people's faces, they truly know what you're saying because they've been through similar. You take on all the bad, you analyze it, and then you reach the better conclusion.

"Warrior," meanwhile, is a song of standing up, when you *have* to take a stand. I love Native American art. I truly loved exploring around the Arizona desert and seeing the Native American Indian paintings—they really seriously struck a chord inside me. It's very hard to explain but I finally found a way of expressing that feeling in "Warrior." If you talk about your Native American, you're talking of conflict, you're talking of treachery—you know, like, "Here's a turkey, now take my land and kill me." That's what the Pilgrims did—genocide, extermination—and there comes a time when you have to take a stand. Me, I'm a pacifist, but I can see that in the face of genocide and extermination, for all the values I believe in, I would have to stop that fate.

Native American art captures all that. It's a symbol of identity, of individuality. I understand fully Chief Sitting Bull's amazing line, "It is a good day to die"—that's powerful. No man is another man's farmyard animal. And so, in this respect, "I man a warrior." So our song is an appraisal of resistance. It's applicable to all similar situations.

It has lines like, "I'll take no quarter, this is my land, I'll never surrender." It's about defending yourself and what you truly believe in, rather than just leftie flag-waving. I back no political party, with very good reason. Never am I going to put all my eggs in the one basket of a politician, because the fucker will crush them. It's tough out there, baby, but that's the way the world runs.

Some of the "Warrior" remixes we had done were very big dance hits in the clubs. When we tour anywhere, Timbuktu or Japan, you'll often hear it playing in the background in shops and record stores.

"USLS1" is a PiL fer-de-lance, a deadly snake. It tells a story about the presidential plane, Air Force One, and a terrorist bomb on board, and the uselessness of that murder over a beautiful desert landscape, under a full moon. The pointlessness of it all and the sadness. Hear it, and you feel your mind exploring rather than just listening. Close your eyes and explore inside the textures. I thought it might be too up its own bottom to play live but in fact it really does motivate people. I watch their faces as we're performing it, they get well inside it and understand it, comprehend it. It's "Why are we murdering each other, for what?" For what? If you're going to murder anyone, then murder yourself first and let the rest of us off.

Every song had to have a point, a purpose, a direction, a meaning, and a humanity. We were utterly indifferent to the shenanigans of others, and it certainly wasn't hard to ignore what the rest of the bands were doing in that era. We might've done better commercially if we had—but, no thanks!

Everything I was writing at that time was very wordy, but there was so much running in my head that it absolutely needed to be. Very overcomplicated, even for me. Everything you do in music, if you really love it, is a clearance factory. You're trying to get these emotions across, and out, and by doing so you're stopping it building up inside. And at that point, the pressure was unreal to try and maintain an integrity here and keep things afloat.

Touring *9* was a happy experience, up to a point. Through the summer of 1989, we went around America with New Order and the Sugarcubes. It was fantastic fun, and great for the crowd getting three decent bands in one night, and all very friendly backstage—just a mass of people getting on with each other, and no pretension about who was headlining.

It became, oddly enough, by its openness, almost claustrophobic to me. It gave me precious little time to get ready for my bit—my moody moment where I have to find what it is I'm about to project onstage. I do need those moments of silence before I go on, otherwise I'm not grasping it, and I walk on there "au casual" and then suddenly it hurts because I'm not bang-on from second one, and you have to be.

New Order's singer, Bernard Sumner, was having problems emotionally and looked a bit the worse for wear. At one particular gig, they had to tape him to a luggage trolley, wheel him on, and prop him up in front of the microphone. When he came off, I went, "Bernie, you're now a trolley dolly." Nice fella, but never really got to know him well.

We got very close with the Sugarcubes; that was a band I loved and adored. I used to go to their gigs, long before we ever worked on the same stages together. I think I've got just about every Sugarcubes record ever made. I'm not so much of a Björk fan now. I find it borderline classical pretension; it's not interesting to me. Einar, their male singer, was a problem. Einar was a bit of an Einar, and he'd hit on me at every chance he got. I really liked him, but I couldn't bear the—"John! You must hear my new poem!" This was very difficult, backstage. Great fella, though—creative, bouncy, and sorely missing now in Björk's work. She's now left to wearing swans and making pretentious squeals and squeaks.

Also that year, I appeared at the "Hysteria 2" AIDS benefit at Sadler's Wells alongside the likes of Tina Turner and Dave Gilmour. I was invited to attend, I think, by Stephen Fry, of all people.

I was supposed to do a skit with Stephen, but I said, "There's no way I can learn lines," so when my time came, I went up and just made it up on the spot. I've no idea what I did, all I know is, when I came off, it was, "Wow! That was funny!" Everything there was great, fun, and wonderful—except that the snobbery backstage was appalling and really turned me off. My God, we're here to raise money for good causes, but what I was getting was the likes of John Cleese going, "Who is *that*?," pointing at me. What?! And a huge fuss was made for the arrival of Jerry Hall. Eh? That's just someone Mick Jagger bonked. Give it up! My God, was she tall and horsey-like. Very Texan. She walked in with an enormous entourage, and I was shoved into a corner, and not many of those alleged celebrities had anything good or nice to say to me.

I brought Nora and it was really hard on us; she felt the cold of it backstage so we left early. We felt it was just wrong, and I decided from that day on that I wouldn't get myself involved with these kinds of people, because they're not genuine—well, not all of them, maybe 80 to 90 percent. The few that are, like Stephen Fry, who's a crazy fuck and absolutely hilarious, are very busy, trying to keep the whole thing together, so they've got no time for you and it's incredibly unsociable backstage.

Through 1989, PiL became a proper, hard-touring band, but it went too far. We toured too much, *waaaaay* too much, to the point where we became distant from each other. Everybody ended up just getting up to whatever it is they wanted to do that particular night after the gig. Sometimes the touring doesn't actually pull you together, it pulls you apart. And poor old Johnny, being the old fart I always have been, I'm not one for going out after a gig, I don't have the energy to do that. Everything has been spent onstage and therefore you start not to hang out with each other and that widens the gap.

In amongst all that, certain band members were taking advances off the road manager. When you have to come in and say, "Look, we ain't got the money to be doing that anymore, you've had your share, and in fact you're already in debt even by the end of the tour when it all racks up—so stop it"—well, then you have a big problem.

I don't want to be the hardcore businessman, I don't do it for that, yet somebody has to. You're the lead singer, but at the same time you're the one that's at the helm here, and as much as you don't want to be bothered by financial dips, you have to be, and it can affect your work ultimately, and it can affect your relationships with band members. I'm not saying that was anything at all to do with little Allan Dias—he would have been the least of them, actually. Poor old Bruce Smith was the biggest problem—but we're still working together to this day. Bad bunny.

I know they say the captain of a ship should never be friendly with the crew—well, I'm consistently trying to challenge that perception.

The longer I worked with John McGeoch, the more problems he created for himself with alcohol. His nerves were terrible, much worse than even my own, and he became very absorbed in the idea of sticking rigidly to a schedule. Now: on tour, things change, buses don't arrive exactly on time, planes get delayed, but all of these moments would drive him crazy with his particular phobias, so he'd be starting to yell at the tour manager, "But the book says 8:30, and it's now 9:40, I can't work like this . . . I need a martini!"

I know we've all done it, but John got used to it, and then became unworkable. When Lu was still in the band, he'd have to take over parts that John would start to forget, or mistime. I had such a row with him on one gig, where he forgot a certain part in "Seattle," and he went, "Stop, let's start again," and we were already into the second chorus! I gave such a hateful glare. My message was clear, and my band knew this too—*you don't stop!* Once you've started a song, you do not stop it, not ever. People have paid money for this. Everybody makes mistakes, go with it. Wait for the beat to come round, there's your spot, and you're back in. That can actually be really good in a live performance, because it's adding flavor to the song.

After the *9* tour, Bruce Smith had his other things to get involved with, and I couldn't keep up a regular wage packet for him as a retainer, because there wasn't the money there for that.

So we had a pleasant parting of the ways, but I always kept him and Lu in mind.

By this point, our batteries were really low, all of us. We'd toured a lot, and the travel wears you down. You lose your sense of base, of home, and thereby, purpose. I'm a bit of a gypsy and a wanderer myself, but I do need to have the sanctuary of somewhere solid to go back to—to get a plug-in. I'd been "working in music," for want of a better phrase, for fifteen years now, and the smart thing to do is to take a break, as a human being.

Unfortunately, it's a very forgetful industry, where if you haven't heard of someone for a year, you presume they're dead. What kept us moving was "Don't Ask Me," a terrific single which Allan Dias wrote mostly on his own. He had a good chunk of lyrics there, which I felt needed to be bumped up just that extra notch but not too much. I was chuffed to pieces with it. You know, "What's it all about?/They scream and then they shout/Don't ask me, cause I don't know/No UFOs to save us/And do we really care?" It's a very long song vocal-wise. Hundreds of words but they're all poignant.

We did a terrific cover for it, emulating the Metal Box. It was in a small round little metal tin. We also did a great video, which was directed by Bob Dylan's son, Jesse. It started going up the charts, but somehow or other Virgin muffed the whole thing, by failing to press any more copies, because they weren't expecting it to be a hit.

Their big plan was to release a Greatest Hits compilation. It was an idea that came from Gemma Corfield at Virgin—a very crazy lady, and a friend of sorts, who'd always had something to say about my career, going back endless albums. Somebody did a Christmas prank on her, I think, and I'd seen it in the corner by her desk. It was a mocked-up poster that said, "Gemma Corfield's Greatest Tits," and there was a fake picture of her with her tits out. So obviously I wanted the album to be called *Greatest Tits*, because I thought that was exactly right. Right up to the last minute, that was going to be the approach, but then it was altered, not to my liking, and I was furious about it.

It seems a silly thing now, but it took away my willingness to go out and promote it. We didn't play at all, to coincide. What started us back up again was an offer to do a song for the soundtrack to a Keanu Reeves movie, *Point Break*. That absolutely puzzled all of us. God, what are we gonna do? What a challenge. Somehow or other we wrote this song, "Criminal," and it was actually used. My problem was, I sang too high. Maybe I was rusty, but I was developing these extraordinarily high notes that were unlistenable. Bloody hell, glass-shatteringly piercing, but that's the experiment I was running, juxtapositioned with this bassy woof/thuddy pattern underneath. It just about worked.

I couldn't really get any involvement from McGeoch in the studio. He was just into this layering and layering endless overdubs to form his familiar blancmange, and so there's a dissipated energy going on in that song, where it's not quite finding itself.

So, after that, when we got back together for the sessions in L.A. which led to the *That What Is Not* album, the idea was to leave everything at a bare minimum, and let the silence in between things actually fill your head with wonderful ideas. Less is best sometimes. You know, don't send 70,000 troops in with machine guns to break up a Boy Scout reunion.

I saw it not so much as a reaction, just the next level for PiL. I thought I'd been indulgent enough and learned enough in the area of technology and verbosity to know I didn't want to pursue it any deeper. That was fine for the element I was trying to express at that point. Communication is like this. You can get bored with the way you keep endlessly repeating yourself and so you end up thinking, "What if I change the sentence structure somewhat?"

I wrote the songs for the album and put the general music guidelines together at home in Los Angeles before we started recording. We did no rehearsals, because we were all living in different corners of the world.

We had a producer, Dave Jerden, that Virgin had recommended to us. I didn't like him, and he *hated* me. He got big because he'd worked with Jane's Addiction and Alice in Chains. What I didn't expect was him wanting a Slick Rick kind of poppiness. He was firmly planting his buttocks in the American pop of that time—like "Jessie's Girl" by Rick Springfield.

As soon as we started laying down the backing tracks, I had issues, and it was politely agreed via the record company that the best thing would be if Jerden would pay attention to what I was saying. I'd take a week off, and not be in the studio while they laid down the backing tracks—which was a great idea because that can be very boring. The week turned into something like two weeks, probably longer and, by the time I got back in, the direction of the songs had changed, the tempos were changed, patterns were altered—just not what I'd had in mind when I wrote them.

Jerden was trying to restrict me to "Yes, that's a great song idea, but I thought if we put it in this key . . ." I'd be like, "Well, did you ever consider that I can't sing in that register? And now I've only got two hours to do the vocals on three songs, and you've altered the range." This would be a serious fucking dilemma. Time and money was running out, and rather than just start again with someone else, which was my suggestion, with both arms handcuffed behind me, we had to make the best of it. So Jerden killed the life out of it, and that's why I called it *That What Is Not*. It feels restrained to me—the savage edge as delivered by ABBA.

Using name producers was always a problem. You try to avoid them, and produce yourselves, as we did in the early days, but then up comes an opportunity of trying it with someone else, and you think, "Why not? There's every chance this might actually work." You entertain the idea seriously and you go ahead with it, and then when you leave the studio that mélange of what I think is killing modern music creeps back in, and you're expected to adapt to that. It's like, "I didn't write this song for your version to dominate!"

I hated the rows revolving around the album, but I do still love a lot of the songs. I am still proud of it. There is some seriously underrated material on there. "Cruel" is the most perfectly tragic love song I've ever been a part of. I just adore it. Of course it's melodramatic, but that's the joy of it. It brings me back to those '60s songs that were all just so sad, stuff by the Bee Gees, at their best, or even Roy Orbison. It's a very sad song: there's the battle of the sexes in there, and all manner of deceit and treachery. But it sounds beautiful!

My other favorite is "Acid Drops." That was all about capitalism going out of control, and not serving the people. I felt like no one was dealing with it; it was all blandly ignored by every single other band out there, including those professional preachers called U2. Censorship's in there, too: "Who censors the censors, can I do that myself, make up my own mind, like anyone else?" That whole ethos was kind of negated in the album's very making, but then it wasn't, because it ended up being a most excellent song.

"Forget me, forget me not/Remember me like acid drops": I think I quite skillfully handle the venom in them lines as I'm delivering them. I was like, eyeballs firmly on Jerden—"Don't you yank my chain and don't you tug me strings, 'cause I ain't got none." At the end of the song I wanted the end refrain from "God Save The Queen"—"*nooooo fuuuuuture*"! That to me seemed absolutely appropriate. That's my satisfaction in delivering a thing properly. I hope he remembers "me like acid drops."

"Think Tank" was about the rewriting of history that was going on with all them idiot punk books, put out by people writing themselves in a bigger part in the story than they really had, and thereby altering the truth. Yes, I'm talking about Jon Savage's *England's Dreaming*. It was a terrible thing for someone who loves to study history—me—to see their own altered right under their nose.

Frankly, I look through that period at the turn of the '90s and think, "How the fuck did I survive that?" It was rigorous, everything came painfully. The punches were not pulled, and the backstabbing was nonstop. I was being painted as the all-time bad boy—but not in a good way. Written off as talentless, pointless, and actually not a big participant or player in my own universe, which is a fabulous lie. It does get to you; it makes you feel, "What is it they're really trying to say? Have they a point? Should I analyze myself?" And of course, being me, I do. Well, I'm glad to report that I came out of my own self-analysis rather favorably.

When *That What Is Not* came out in early '92, the media response, if there was any, was mostly negative. The vague idea seemed to be that I'd timed my run to coincide with grunge. Really, I'm not one for keeping my finger on the pulse of what's currently trendy. In fact, that just never bothers me at all. I'm absolutely oblivious to music working in things like trends. There's always a couple of bods out there with an original idea, and then there's 200,000 bands that want to copy that and then declare that those are indeed the new rules of music and everyone should sound like that. That's why trends do not interest me.

Unfortunately I've made myself into a bit of a trendsetter over the years and it's the one thing I hate the most about what I do—having to listen to the influence I've had. It's not at all rewarding. Imitation is *not* the greatest form of flattery. Somehow it indicates to me that the people doing it haven't clearly understood. If I'm trying to be preachy in any way at all, I'm telling you: find your own sound, find your own soul, find your own words, and your own way of seeing the world, and then share that with us. That's how we all live and learn.

In fairness, a lot of the bands in America absorbed the energy of punk, and loved its ethos. Of course, there were the Boo Monsters, the people forever stuck in the heavy-metal universe, but soon the punk influence began to shapeshift heavy metal into a much more broadminded thing, and so I've got a great deal of respect for a lot of the bands that did that. They stopped the heavy-metal crowd from hating punk and opened their minds to it. The English lot, like Def Leppard, definitely opened some doors there, and introduced the concept of unity. They're good fellas—they have come to our gigs.

A year or so later a guy called Tim Sommers—who I had been talking with at Atlantic Records about doing a solo album—actually tried to set me up for a meeting with Kurt Cobain. I was meant to meet him and Courtney and the baby at the La Brea Tar Pits. Well, I'd done the tourist shit, I'd lived in L.A. long enough, and I wasn't going to look at dinosaur bones one more time. In many ways, I'd have loved to have had a chat with Kurt—but about what exactly? It struck me as "Are they gleaning me for ideology, and then they're gonna dump me?" And my answer was, "Yeah, that's exactly what was gonna happen." So, bollocks—I canceled. Actually I think they canceled, which made me very happy. Mentally inside, I didn't want it. I don't like that kind of thing.

Going on TV and radio was even harder than getting a decent review. I did a beautiful black-and-white video for "Cruel," and *The Chart Show*, which was a program on British terrestrial TV with massive ratings, refused to air it, because it was shot in black-and-white. They said their policy was color only. You could go into the history and say, "Well, two weeks ago you played a black-and-white video, so it didn't seem to bother you then . . ." It certainly didn't stop U2 circa *Rattle And Hum*. It was a fabulous lie. You can go on like this, but you end up back to bad dentistry and pull all your teeth out with anxiety over it. The point was that it was a damn near relentless resilience against whatever I was doing.

I think it was just fear of what the content really means or is, without basically listening. Here I am now after all these years and there's still this suspicion, is this some kind of elaborate joke on my part? Am I having a wheeze on the wonderful world of music? Yes, I am. And what is wrong with that? Because in the meantime, I'm telling it like it really is. Like it *really* is, and I'm being honest and truthful with everything I do. Is that a problem? Is that so wrong?

And all the way through, from *Album* onwards—hello, we're PiL, we look damn fine, but because there was a social issue in our songs, we were ignored. So did the world erode the importance of Johnny Rotten, or did Johnny Rotten just carry on doing what he knows he does best, and therefore become too "out there"? I think it's six of one, and half a dozen of the other. I've got to be honest about that.

Everybody backed Sting and didn't back Johnny Rotten. But that's okay, you get what you deserve in the end.

There were some really good gigs, but I don't remember a lot of fun touring that album. Mike Joyce from Morrissey's band was in drumming with us for a short bit, and he and McGeoch were terrible with their Catholic-Protestant arguments. McGeoch would be, "I'm blue through and through," and Joyce would be going on about the wearing of the green! Jesus Christ, what is it you're going on about? I'd be going, "My favorite piece of clothing in the skinhead days was a green and blue mohair suit—remember them suits? Fine, excellent suits, tonics, which were double-shaded, so it reflected green sometimes, blue at others. My tonic was both blue and green! Think about it . . ."

In interviews, I carried on plenty about Virgin's lack of support. I never held back. I suppose I always had great support there on a personal level, but by that point there was definitely something missing at the top, which made it very difficult to be on the label.

The next thing I knew, Virgin was being sold to EMI. Of all people! Oh, for fuck-off sake! You wake up one morning and find you're back in the enemy's encampment, and Ken Berry's running EMI at the time—I mean, what?! At PiL's remaining gigs on the '92 tour we'd finish the show with "EMI." It was an absolute reminder of, "I was in this situation once before!"

We'd completed the eight-album deal PiL had signed up for in '78, and signed a new deal in 1987, but it soon became apparent EMI wanted rid of us. So I was in a kind of limbo. It's painful when you're in that position. It's like, what's this been all about?

So thank you, record label, and all my alleged friends there. I'd liked it at Virgin, but how many heads of department had changed on me through those years? After I moved to L.A., every two years there was a different chairman and different underlings, and I never knew a name or title for any of them. I began to treat them all as transient.

Then, of course, Virgin went into that robot answering thing, so I couldn't actually speak to anyone. The people I knew from the beginning like Simon Draper and Ken Berry—and they really were friends—they couldn't pick up the ball, because nobody wanted to accept the responsibility of working with me, because the blacklist had taken its toll. By the end, they were all going off to fresh pastures, and I had no real communications with the new people at the top.

I would never give up the ghost, but of course I felt defeated. But I knew: by sheer persistence, whatever it is I think I'm doing, I'm not gonna stop. Not at all.

11
JOHNNY CUCKOO

My life at that point didn't paint a pretty picture: my band PiL falling apart, the record company deals chiseling me, endless changes of management, exhausted from trying to keep the money together. I'm not one for self-pity, but some of it must have crept into me. I didn't see a way out for ever such a long time. The whole thing was just constantly on us. It was so hard for Nora to have to deal with this.

At the same time, I remained kind of creative. I kept writing. I always burn stuff I've written if it doesn't get recorded soon after the time of writing, but I've still got at the back of my head some of the songs from that time, and they're schizophrenic in the extreme. I've certainly never gone back and used them as ideas. But at some point when I feel clean enough that I can go back in and investigate what was going on in me there—*oooooh*, ha ha, that'll be intriguing!

All I know is, it was very pain-driven. I don't quite know what the pain was, other than *stre-e-e-e-esssss*. Stress, stress, stress. "Be the man, you've got to be the man in this situation, you've got to keep the band together, you've got to run the business, you've gotta do this, you've gotta do that—*aaaarrrrgh*!"

I managed to get a studio built at the house in Venice Beach with my brother Martin. But we got ourselves too bogged down with modern technology, to the point where we had no energy left to do anything with it. You can't, as a songwriter, be involved with anything other than the songs; you really can't. You're doing yourself a disservice. If you're getting clogged up with business, that's contaminating creativity.

You could spend the rest of your life trying to unravel this nonsense. Coming to America was no easy move. Having to deal with the accountants, lawyers, and managers, who were supposed to navigate me through how things work over here, was a living nightmare. Add in the record company not backing me, and the fact that I still exist at all in any stable form is astounding. The endless stream of managers—I mean no harm to any of them, they all tried their best. I trusted them that they were actually qualified to do the job, but they simply didn't have the tools.

There was a time back there when I was desperate for a good manager, and I actually thought about approaching Sharon Osbourne—would she look out for my career?—because I love Ozzy. I thought, "She's doing all right for him." And this is way before their TV program. People were saying good things about her, a tough, no-nonsense bird, but it never came to anything. A few years ago, I said something about Ozzy in an interview: I called him something like a "senile delinquent," which really upset his family. I didn't mean anything by it, I said the wrong thing. I have the most incredible love and respect for Ozzy.

For me, emotionally, the '80s and early '90s felt like an incredibly unrewarding period. I was going through managers like Smarties. It was very confusing. I needed help—just someone in the industry—but there was no one really prepared to step up to the plate.

I wasn't getting any benefits, I wasn't getting pats on the back, I was isolated. No matter what I would do, there would always be band members to go, "Oh, he's a bastard, he is." I came to thinking, a captain of a ship really can't afford to have friends on board. If there's not a difference there, then it won't work at that particular point in life. I've found out since that this ideology is unnecessary, because I now work with people I truly love and trust, and hopefully vice versa, but it's quite amazing how I had to go through all that, to get to *this*—where I am with today's PiL. To get to the real essence of what is me, how do you get people to pay attention? Are they really seriously only interested in the scandalmongering of a nineteeen-year-old? Or do they want to go on a journey of self-discovery? Well, no, they want the scandal. That's fine.

In the '80s the publicity machines took over. The frivolous hair-dos which we all suffered from in that period led to many, many things, and now it really is all about sensationalist headlines and no content. It's a curiosity to me, because I was accused of just having frivolous headlines and no content from the day I started, but I think the Pistols were *all content*. Every single song, and indeed every single word; that was a good springboard for me and I'm still that same fella. Obviously I've grown up, I've grown side-ways and frontwards, and I have a bigger vocabulary, and I can express myself far deeper and more meaningfully, but I'm still loyal to those basic truths in life.

By the early '90s, I was getting fed up with all the rubbish that was being put out there about the Pistols story, where my life and everything I stood for were being misinterpreted by an odd bunch of fellas that really should have known better—in particular, Jon Savage. I helped him with his book *England's Dreaming*, and when it was published amid loads of media fanfare in 1991, he'd cut out large amounts of my conversation, and just backed up his own philosophy. Well, who the fuck are you, Jon Savage? You were not a Sex Pistol.

His book was over-wordy: to understand it, you'd have to have a Latin-to-English dictionary. He used words that weren't pertinent to the scenario, and presented himself somewhat as an expert on what was going on. How could he? He was a complete outsider, not part of any inner circle, and in fact not much to do with it at all.

When he was putting the book together, I made him promise that it wouldn't be that kind of book, and I was horrified when it turned out to be everything that he promised me it wouldn't be. It was narrow-minded, insular, anti-women, misogynist, and hence had no real understanding of the driving force of punk, that it gave women the opportunity for the first time ever in the history of pop culture to stand on the stage, the equal of men. Up until then, women were just matching hairdos, a trilogy of singers, with nothing to do with the songwriting—they were just voiceovers. Bands like X-Ray Spex and the Slits absolutely went into it, like, "We're blokes too!" It was an amazing achievement, one that took forever and ever, and should be admired and respected—all those great, glorious punk women! Fantastic, amazing things they had to offer the world!

Savage's view of the world was Gang of Four, smug in-house student intellectualism. And all this coming out of the wordsmith genius of an ex-lawyer. The rumor was he gave up law to write music journalism. See? He could've been some use to all of us.

This kind of thing was going on and on left, right, and center, all around. I was furious with people deciding what I thought, and did, and how I was—that my opinion was somehow lesser than *England's Dreaming*, and how that was presumed to be *the* authority on punk. The agenda from all manner of bands of the day—not just the Clash!—was to outdo the Sex Pistols. It was an ugly, ugly world.

You'd have to be some kind of idiot if you're not going to analyze what your public persona is. That doesn't mean I have to read the sensational hate-lines on a daily basis. Far from it! But I keep my ear to the ground. When the truth is being swayed off into angles of hatred, for God knows whatever different reasons or purposes, there is a point where you have to stand up for yourself. And you pick the best moments to rally your troops.

So I thought that putting out my own true account of things in a book would be a good beginning towards counteracting it all. It was an issue I put off for quite a long time, until the offer came, and there was a lull with Public Image, so I actually had the time. Then I thought, "Yeah, I'll do this."

I knew it would be taking on a whole kettle of piranhas. It certainly put a full stop on a lot of the naysayers. I called it *Rotten*, with the subtitle *No Irish, No Blacks, No Dogs*, as I'd had to not be Rotten for so long because of the court case, and I'd been focusing on PiL—PiL—PiL, where I was Lydon all the way. It was like reclaiming that part of myself. It was a serious step in my character development, because, since putting that book out, I've become very much less fearful of who I am, and what I am. And even though people have tarnished and tried to steal that aspect of me, it *is* still me, it's an intrinsic part of me, and I've come to realize that as very relevant. I *am* the elephant in the room.

I gave over a lot of space to other people's voices. Despite all the allegations out there, I still made room for people like Steve Jones and Paul Cook. It proved the point: whatever you think we are, we're not. We will always still help each other out, because it's all about telling it like it is, and not just how others want it to be.

I also included some affidavits from the McLaren court case. I absolutely wanted them in there because it gave you a sense of the scenario—just read them and judge for yourself what you think was going on with all of that. There was Malcolm, declaring himself a competent manager—hmm, very novel. Then there was Vivienne's hateful nonsense—these were the adults in my life and they just weren't grasping reality at all. They had no qualms about trying to *fuck me out of my own life*. They came in hard with that ridiculous court case and left with their arses royally reamed.

Then there was a guy from Virgin going on about my "unrequited homosexual affection for Malcolm." Malcolm himself had put out stories—one in particular, in a German music paper—that I'd always loved him, but he had to say "no," and that was my problem. At the time, I thought, "He can have his little gay fantasy all he likes, I don't give two fucks." But when he hit back with the line that I'd always seen him as a father figure . . . that was too much, that's something my dad wouldn't like to be reading, and I don't think it should be said. It's wrong. I don't care what he says about me, he can lie all day long, but don't go into those areas. The spite in it and the childishness really ten-folded my contempt for him. I thought, "Jesus, listen, we were all really right about what an arsehole he is."

For me, it was a great book, maybe harsh, but then the situations I had to deal with in it were appalling.

One thing that needs correcting was something the ghostwriters rewrote without asking—they tried changing me coming from Finsbury Park to Camden Town! And their insane justification was that people won't know where Finsbury Park is, but everybody knows Camden Town because of the market! To try and alter where someone comes from is ludicrous!

After it was published in 1994, the most fabulous thing that happened was the improvement in the quality of the journalists that interviewed me. They were books-y people, so the conversations became ever so much more interesting. I felt, "God, I really like doing interviews now," and I've been that way ever since. Rather than feeling like I have to aggressively defend myself all the time, here was a whole catalogue of eminent whatevers actually giving me the time and respect to allow me to properly answer a really sensibly researched question. Proper lines of communication, and—the very thing I thought journalism was about—a respect for the facts.

In the following years, I tried very hard to get *No Irish* made into a movie. I got some kind of backing going there. I raised a million or two, but dealing with the scriptwriters was where it fell apart. It was all just people wanting to rewrite everything in a very "David Cassidy's life story on VH1" kind of way. It'd be, "Well, now we've got to have some romantic interest—your book don't say much about that, let's make some up." "No, we will not make some up. You'll tell it like it is, and that's that!" And no matter what it is, or whatever I do, I don't do kiss-and-tells. And certainly not just to sell books or cinema tickets for money. I don't hurt people like that, and quite frankly, anyone who has had any physical relationships with me must feel very hurt already. It wouldn't bode well for the reputation. They'd be contaminated.

My idea for the lead role was Justin Timberlake. I thought, "That would work." Because Justin at the time was out there, but he was coming to a rough end in his music. He was getting bored and starting to do acting, and he hadn't got any good roles yet. I thought this would be tailor-made for him, being so fluffy and nice and kind of simple in his approach to life. I thought this would be a real challenge for him to take on, the role of Johnny Rotten, and there was every chance he could sing the songs.

It started as a joke in interviews but then I got deadly earnest and serious about it. I might've been shooting at pie in the sky, but I thought, "If you don't aim high, you're not going to get results." It was the powers-that-be that wouldn't go there. They were talking like, "No, well, maybe we should use some unknowns." "What? What's the point of this, then?" They just wanted to keep the costs down, and thereby keep potential out of it. Who needs another B-minus movie.

It never got as far as meeting with Justin. I just shot it out into press statements to see which way the wind would blow. But even if they wanted to respond on his behalf, I knew that our lot, my backers, would've negated on it. But such a juicy backdrop of a plot!

I'd been quite outspoken about some of the more escapist dance music of the early '90s. There was plenty of rubbish going on. The floppy-hat teapot brigade were back; the lot in the loon pants who'd never lend you a tent peg at the original hippie festivals.

There was, however, another underground clubby scene, where the beats were harder and there was a load more urban grit going on in it—and that stuff I *loved.*

Leftfield, by chance, I knew through John Gray, because one of the duo, Neil Barnes, had been working at the play centers that I got kicked out of in the mid-'70s. He occasionally used to come around Gunter Grove with John. I was thrilled when he started up with Leftfield, and by the sounds they were coming up with together. I got to know Paul Daley, his partner in the group, and really liked him. It took quite a while to get us on the same page, from the first time they presented me with a cassette and said, "Maybe you could help us out with some words here." They were getting big in the British dance scene, and there were hardly any singers on the records at that point, in those circles. And so, in I came.

What I loved about that whole rave-y dance scene was that it was wide open, racially and culturally. There were no judgment calls on anything. It'd seemingly come out of nothing. It came out of itself, and created itself really well, and blended really well with punk. I was eager to see that develop, but certainly not move into it or copy it, which was one of my huge problems with getting into doing it. I kept telling them, "I don't want it to be perceived that I'm jumping on a bandwagon here." But the boys kept at me, "No, no, John, no one is ever gonna think that, you're the only person that can do this . . ."

As I say, it was a universe of music that was just about beats and things, but with no vocal direction on it, because it was almost impossible for singers to fit into the rigidity of the tempos. Well, we found a way. We found a very good way, and a very natural way. When I was ready, I rung 'em and I went down, and we finished it in a night. Fantastic. A great deal of worry went on on my part, I must say, because I didn't want to commit and make a twot of it. Fair play to the boys, it would be, "No, don't worry if it's rubbish, John, we're not gonna use it." Oddly enough, those words are never very comforting.

Lo and behold: "Burn, Hollywood, Burn," or, according to the official title, "Open Up," came into being. It was released in November '93—the same week that the hills of Hollywood were actually on fire! I had to deal with the press here in America going, "How dare you try to capitalize on a natural disaster?" Ouch! This is music journalists at work, supposedly in the know, but somehow oblivious to the fact of how a record is recorded a good while before it finally gets released. It was insane and twisted—ignorant fucks giving judgment.

The thing about that particular fire is, it came right down the hills. By then I had a house out in Malibu, on the coast, and the fire wasn't far away from the front gates. Nora and I thought, "What can we pack? What do we need?" We'd all been told by the police that we'd left it too late and we had to leave, so we thought, "Let's just get in the car and go." We drove to the other house and waited to see on the news what was happening—to see if we'd lost it all.

It's amazing, the vibe you get in those situations, it's almost like, "Well, if it's gone, so what, we haven't lost each other." That is such a rewarding, weird, strange emotion, where you realize that things like property and personal collections ultimately don't matter. If you're faced with a disastrous situation, it's the missus and the kids, isn't it? By no means throw your possessions out or sell them off or anything of the kind, because you'll regret that. But when you're confronted with what really matters most, that's a different situation. There's a puzzlement in there and I know I've got a song in there somewhere. Soon to be written.

On top of the fires, there'd been the Rodney King riots in L.A. the previous year, and all the folk around here in California weren't too enamored by touching on the theme of L.A. boiling over again. They weren't showing me any love for writing a song channeling a whole series of disastrous events. They weren't grasping it. I wouldn't say any of this is to do with psychic ability on my part, but I think I'm on the pulse of inevitability—consistently.

You'd have to be deaf, dumb, and blind not to see it happening. I think the problem is, most people in this world actually *are* deaf, dumb, and blind. Facing up to reality is not something many people do—certainly not your mainstream masses, and very few of the bands. In fact, whenever they do want to be realistic, they simply jump on the charity bandwagon, cause célèbre stylee.

The lyrics of "Open Up" are specifically about an aspiring actor trying to break into Hollywood movies, and they were actually autobiographical. As I said before, it was certainly not something I'd had my eye on when we moved to L.A., but things had been unexpectedly going in that direction, with the absence of PiL activity for the time being. I'd even been going to tryouts, learning scripts and turning up to be auditioned. I went to quite a few, and they were all, I've got to tell you, the most embarrassing and fundamentally soul-destroying things I've ever done. But I thought, "I'd really like to see if I can get on with this, because it's a good angle of work." It's an exploration of sorts.

But, no: massive humiliation in it. It's so hard to be rejected. I've got a great deal of sadness for big actors, knowing what they had to go through to get where they are. Not any of them had it easy. It's really hard to shapeshift yourself into the mindset of another personality and give up yourself so much, and give so much of yourself, and then face rejection for it. Wow!

One movie I tried out for was about a chauffeur and a butler kidnapping a toy dog from the wealthy owner they worked for, and the calamities that thereby unfolded. So it was mostly comedies like that. Another was a war movie. The scene I was asked to learn, I was dying—and, yup, that's exactly what happened. Oh, it was murder, I couldn't grasp it. I had the drive but I didn't really at the time have the depth in me to take it seriously, because still at the back of my head there was this little thing jumping around like a demon saying, "*Huurrgh*, acting's for arseholes." That was really a cover-up on my own behalf. It's not taking it seriously and I should have. I should've had more power over my own demons.

You know for sure that as a singer you'll be judged more harshly. I've no idea if I've got what it takes. I've gone through all the self-interrogation, which is more than enough for me to have to endure, but if it's there then I'm gonna take the challenge. I can't resist the challenge. It's harsh on me what I do to myself, but if there's something where I risk facing complete and utter humiliation and degradation, then I will waltz into it. I find that irresistible. That's what life is, a series of superb challenges, and in none of these aspects am I talking about selling my soul or copping out or denying my past, my present, or my future. What I'm talking about is there's a work thing here. A way of presenting whatever my message is in different ways.

I suppose that's what I was looking at, for the first time seriously in this period, since PiL was at a stalemate with Virgin. There was a thing flirting around for a while there with Channel Four in Britain about presenting their Friday-night youth-culture program, *The Word*. But in that case, why on earth would I give up what I do in music. That would've been the kiss of death. I would view being a TV presenter on a pop show as terminal cancer. What do you expect to get out of that when it ends? I'm well aware that there's many things I could take on, but obviously *Celebrity Squares* isn't going to be one of them.

Another interesting project that emerged that year, and brought in a welcome influx of cash, was a series of ads I did for Schlitz beer. They sent me heaps of crates and, surprise surprise, I liked it. There was no huge money for doing it, but it was good enough. I've noticed in life: something can come along that you least expect, and you've got to be smart enough to be able to take it on. It has to be the right thing, though.

The ads themselves were a junkified approach—a collage of events put together in a poster, based around my rubbish. There was a story going around that Bob Dylan had called the police, because the contents of his dustbin were being raided by this so-called counterculture scholar, A. J. Weberman. So there's the modern world for you. Thank you, hippie! Even our trash cans need padlocks. It might be articles we discard but it's still our own business. While I'm paying tax for dustbin men to collect the garbage, that's my business.

In amongst all these peripheral things, the notion of the Sex Pistols touring again slowly bubbled away. I agreed to it as an idea. I thought it was important that we as four people should not allow the band to have been broken up for good, for the wrong reasons—the subterfuge and the innuendos and the back-scene chatter and banter, being misled by a sarcastic management. After the court case, it was free of all that now, so there was room for maneuver, finally. For my mind, it was a matter of either putting an end to it, or taking it so far it would be impossible to continue. Either way was very interesting to me.

The first person I had a real serious phone conversation with was Steve Jones, and it was about Glen Matlock. It turned out that Glen was in L.A., and they'd met up. Steve was trying to convince me: "Oh no, Glen's not like that anymore, he's all right now. He doesn't drink like he used to, he's approachable. At least, I found him that way." So I was like, "Okay, where's this going?"

Steve convinced me to meet Glen with him in L.A. They came down to meet me in Venice Beach, and we had a walk around the marina. I pointed out the boats, and the things I liked in life, and, miracle of miracles, we got on!

I thought that if two people like Steve and Glen could come together on a project, then that boded well for the rest of us, because those two have never seen eye to eye. Not ever. There was never a time where Steve has not mocked Glen, or disliked him—it's always been that way. And the other way round. It wasn't all me: those two would create tensions that rippled throughout the work relationship with all of us. Paul would always side with Steve because they were so close. But if the band was going to get back together, I didn't want a different bass player. Since Sid was obviously no longer available, I myself wanted Glen.

So, it was a good meeting, and we were as usual very open with each other. There's not much subterfuge going on with us lot. The contempt is there right at the surface and can therefore be easily cleaned up. It's a much healthier way than burying it all and pretending we're all good friends. I'd rather have brilliant working relationships than friends, in these situations. Friends are important but not with this band.

We tried to ring Paul back in England, and left answerphone messages, then went out for food, and meanwhile he'd left messages in return. When I eventually got hold of him a day or two later, his approach was very friendly. I had long talks with him, and that pulled me back in, to the belief that there was something there to save and rescue.

The pressure was gone from the ugliness that we'd had to endure back in the early days. It became a valid thing to do. I thought, "I really like this, I don't want to let them down. We can rescue it, but not forever—just an affirmation of the quality that came out of that situation, and that the root core of all the problems of the band unfolding and imploding was down to the management, which was no longer involved."

And so it was supposed to be a "free I-self up" situation, as the Jamaicans would say. Still, I was horrified at the prospect, in all honesty. One big angle of the conversation was, I just did not want the word "reunion" to be in this. I said, "That's the first thing that the arse journalists are going to present to us; they're gonna call it a reunion." So I went through my dictionary of words to find every other possible vocabulary route around that. I didn't see it as a reunion. It was *re*-nothing. When it's on, it's on. When it's off, it's off. Reunions are deeply unhip, and in our case would've invited the accusation: "How dare you? Why don't you just go away, the world has had enough of your sort."

Then of course, all the young whippersnappers out there, being misled by the negatives, were in a world of resentment—"*Uuuurgh*, you old men! Isn't that all something that happened ever so long ago?"—without any of these sods paying attention to the fact that whatever freedoms they were now expressing, it was *us lot* that earned it for them. If you don't want to listen to us, just shut your mouth, go away, leave us alone.

Isn't ageism a terrible nonsense? It's back to poor old Pete Townshend again. He's such a good fella, and such a good writer—but that "Hope I die before I get old" thing has *haunted* him. And it's so misunderstood. I don't think Townshend meant it as an ageist thing at all, but the lesser thinkers out there hook on to that and then try to perpetuate the myth of "music is just for young dumb teenagers," because that way, of course, the magazine culture and the record companies can force-feed you whatever they like, using identikit impressionables. Anybody that shows any longevity is obviously going to be giving the game away, and that's not what the industry wants.

So, don't ask anybody to run away and hide and give it all up. Not when they've done any quality work—even if they just wrote one song that has crept into anybody's psyche, don't underestimate and don't negate. That's a talent, that's a human being at work—respect to them! I committed that crime myself when I was young: "What do you think of the Rolling Stones?" asked Janet Street-Porter, in a TV interview, back in the day. "I don't," was my answer. It was quite true, I didn't think of them at that point. I had my own thing to get on with—the Pistols. There was no hatred in it and I certainly wasn't asking them to stop, although it did occur to me that the idea of running up and down the stage at forty might be slightly absurd.

But I broke the forty barrier myself and I thought, "No, it's not at all, you're wrong there, John!" Never surrender! There's something terrible in British culture that teaches you this "Just give up, act your age" nonsense. Once I firmly came to grips and analyzed all of that stuff, I realized that's denying you your life! And *you must live your life to the fullest you possibly can*, until the very second your heart gives out. And don't let no one take that away from you.

It was fantastic. A hornet's nest, that's what we stirred up. No other band on earth would have to deal with negativity to the extent we had to. After the initial rush of euphoria, it became something of an endurance course. Of course, that bonded us together even more. In many ways, it was a great thing, you see—such a foul negative you can turn around on itself. There's not much self-pity in the Pistols. In that respect, it was a very healthy outfit to have come from.

At the press conference we gave at the 100 Club in March 1996, announcing our return, they were lining up to hammer us. It was a hostile atmosphere. In those situations you have to find a thought process that protects you on at least two sides. You put me down, so I cover myself left, right, and behind. "Oh come on, isn't it all about the money?" they whine. I don't want to run through my catalogue of answers, so I preempt them. "I'm fat, forty, and back. Deal with it. Next!"

It was a question-and-answer format, and I'm sharing a podium with Steve, Paul, and Glen, and it becomes apparent to me within thirty seconds that these three ain't got it. My own band, they couldn't answer questions properly, so I'd have to keep interrupting, and shifting the agenda. I felt the band weren't coming up to meet the mark. The questions were aggressive and negative and I just decided to body-slam them verbally. We had to make the point that the Sex Pistols was an ongoing force. Not that we were back again, but were still and always would be relevant. You know, game won.

"Ooh, but are you only doing it for the money?" these fools kept asking. For fuck's sake, this is an accusation from a pop industry that *only* does it for the money. The idea that it's audacious of us to expect to be paid for what we do! The most influential band in the world, at that point! Well actually, PiL had taken over from that, but you know what I mean? Asking us to explain ourselves, like we were in the dock at the Old Bailey, and these three wankers of band members actually going into waffling about it. Hideous. The cowardice of them was frightening.

By the time we got into rehearsals, we realized we really didn't like each other, all over again. Anita Camarata—who is Steve Jones' manager but also represents the other two on Sex Pistols matters only—was going, "Steve doesn't drink, so it's not wise to have alcohol around." I turn up with two crates. Those are mine. What's mine is mine. If I wanna drink, I drink, and if you don't want to and you're under AA, that's your business entirely. I absolutely declared where the lines were.

I was starting to re-familiarize myself with that old feeling of "What have they done now that I don't know about, but will find out soon?" Hey presto, there was an issue, Steve's manager had gone and got in a fashion advisor, to suggest what clothes we should wear. Hahaha, hahaha, hahaha! It went very well. I took it with great style. I thought, "Okay, this is not her fault, this fashion expert. There might be something in this for me," and lo and behold, on those racks, there was! It was mental stuff, and that's how I ended up wearing what was described as a Paddy's donkey jacket, a gray-and-white-check thing in nylon. There was a red jacket halter with a plastic front too—loved that—but that fell apart after two gigs.

So there were mad little pieces of clothing in there that I really fixated on and got into. What about the rest of them? "Uuuuurgh, where's the jeans?" They were right back at that basic problem that was always there right from the start with the Pistols: they didn't know how to present themselves. They didn't have self-confidence enough—apart from Glen, of course. Glen's *up there*, in a quiet way. Ladies' gear? Whoa, he'll have it. Glen's got a sense of style. It might not always be appropriate, but it's style, and it's him and therefore it *is* style. But as for the other pair? Basically, it was down to designer T-shirts and jeans. Ridiculous.

Steve, of course, was so conscious about his weight. What he'd done—oh, so foolish—was that he'd ordered a load of clothing from Vivienne. But he was shy about what his real waist measurement was, because Vivienne was a real cow for this. She'd be the first to tell you, "I don't make clothes for fat people!" So Steve lied about the waist and the chest and every other proportion, so everything that arrived, he couldn't get into. It must've cost him the earth. The only thing that fitted him was this stupid hat.

There you go, that's the fashion front dealt with. Do the public deserve to see what Steve looked like at the time—a mountain of butter wedged on two cowboy boots, with a Vivienne Westwood floppy cap? NO! I'm not a fascist dictator here, each to his own and all of that, but when I know that it's all down to a complete lack of consideration, and he's being almost cynical in his presentation, that's not gonna happen.

Once we hit the road, at one of the very early shows, this happened: you know when you get to your room and your luggage goes missing? Well, a case of Steve's went missing. He rang everyone up asking about it. It almost felt like he was accusing people, that's how it seemed to come across, making people feel really uncomfortable. It finally turned up and, lo and behold, he had just left it in the lobby! Was he expecting us to carry it for him? That vibe kind of set the tone for the whole tour.

My big stipulation for the tour was that Rambo would handle our security. He was an ex-army British paratrooper, an ABA boxer, and an extremely tactical and organized individual. I'd had enough of bouncers ruining gigs, and it was very foreseeable that occasionally morons would be in the audience. There's always that one percent of haters that you have to be aware of, and take precautions against, because they tend to hurt the innocent people.

I hadn't seen Rambo in years, but he turned up to see PiL at the Reading Festival in 1992, and he very clearly saw through what was going on: we'd worn ourselves out, and we were all very separate. The next time I saw him was about a year later, when I hired him at the signing launch of my *No Irish* book in New York, which he took charge of, and handled very well. I made a promise that the next time I went out on tour, he'd be on it.

Now we were doing a Pistols tour, I had to have my own man, someone I trusted, not a bunch of meat-mountains with walkie-talkies. Rambo wasn't there as a bully boy, but as a mate and an ally. He was brought in as my personal security, and was also the band's security while onstage. He would literally stand onstage throughout the set. Keeping an eagle eye that the venue security were not getting heavy-handed with the crowd. He was there to prevent trouble before it started. As I have documented already, there are a lot of people out there who want to hurt me. I've been attacked onstage many times, so I was taking no chances—we knew there would be arseholes out for trouble. Rambo would make sure runners did not get onstage and disrupt the set: stopping monitors, mic stands, and the likes getting knocked over and just letting the band get on with things. Over time his job progressed to other stuff like making sure the monitors weren't low, so we wouldn't have to interrupt the gig to get it fixed. I like to hear me! He'd check the time, song by song, so we didn't overrun and get the plugs pulled. And he'd clear the stage of unnecessary crew people; he likes to keep a clean stage. It immediately felt very different, and very good, having him there.

I think a lot of people think we only played Finsbury Park and that's it, but we did something like sixty or seventy shows covering the UK, Europe, North America, Australia, Japan, and South America. It was a massive commitment. We called the tour "Filthy Lucre," because that was one of the accusations from the newspapers when we were being paid to leave EMI and A&M, back in '77. My mate, Dave Jackson, who'd done our stage design in PiL, came up with the set idea of having all them sensationalist hateful headlines printed on a massive sheet of paper across the front of the stage, and us then bursting through it. It came at great expense, but there we were, smashing up the headlines—what part of that aren't you getting? These headlines are lies. We're not. And ultimately, if you don't even understand any of the infusions of the Sex Pistols in modern culture, please at least understand that we broke through the bullshit barrier. We survived Maxwell *and* Murdoch!

After all the media cynicism, the actual response at the first European gigs was just incredible. We were finally getting that respect. It was overwhelming and emotional. And the thrill of that and the audience absolutely being there right from second one, is that you don't have to become a parody of yourself. You're with it as much as the audience is. They give me as much as I give them.

The third show was a massive one for me—outdoors in Finsbury Park, actually in the park itself in front of something like 30,000 people. It was a very difficult gig to do, because my family were all there and close friends and everybody from the neighborhood, and they all want your attention. There were lots of pop stars and footballers whizzing about backstage, all wanting to talk to you.

The whole gig, however, turned out fantastic, one of the very best days. It was right in the middle of the Euro '96 football tournament, which was taking place in Britain, and a couple of the England team, Stuart Pearce and Gareth Southgate, came down. They were specifically warned not to come by the England manager, Terry Venables, but they came anyway, so of course I gave them my full support, and asked them to introduce us onstage, which they did. Some of the audience were not overexuberant in their appreciation of them, they were just puzzled as to what the connection was. "Hello, we're giving you proper England, you fools!"

All the preshow excitement *drained* my energy. By the time I got onstage I was exhausted, and I hadn't fully prepared. But we did good. My "fat, forty, and back" line seemed to go down well—a nod and a wink to that lovely old Tamla song, "Young, Gifted And Black," trying to bring a sense of humor to it. "Hello, I'm not here pretending I'm seventeen. Get it right! But this is what I've been doing, this is what I've done, and I'm proud of it and nobody else has the right to try and step into these positions that we created." That whole *Swindle* nonsense of Malcolm trying to get someone not only to emulate me, but steal my position in life, steal my own work—that was always an issue with me. I will always come back and get vengeance for that kind of activity, and rightly so. I am what I am, and you will get my bitterness if I see you stealing my hairdos, my lifestyle, my clothing, my lyrics, my music. I won't have much time for you, and I will say so. Other than that, I'm quite an easy fella to get on with.

It was such great fun being onstage, doing those songs again. Doing "EMI," after the label's recent halting of PiL's progress, I really, really hammered it in. I took no prisoners. I couldn't wait to get to that part of the gig—that was "Whhoooaaaarrrrghhh!" I'd be holding back on some of the more excessive screaming tones, and saving them for that. I made the delivery oh-so-vicious and attacking. Which is exactly the way I wrote it.

Them songs, still to this day I don't know how I managed to record the vocal lines all in one go. It was always hard to do live, and the longer I'm at it, the more astounded I am that I ever could get up to it like that in the first place.

A couple of weeks or so after Finsbury Park, we played a smaller gig indoors at the Shepherds Bush Empire, which prompted Alan McGee, then the boss of Creation Records, to pay for a full-page ad on the back of the *NME*, praising us to the heavens. That was all his doing; he felt really bloody angry that people were trying to slag us off or put us down, while ignoring our history. Very nice and neat of him, I thought. And nothing at all about a working relationship—that was never ever discussed with him, even though he was a big fan.

It's ironic, but it's delicious too, that, gosh, we've got a lot of enemies—and if I may say so myself, fully deserved. But that just makes us better, every one of us. What is it about us, or me in particular, that seems to annoy? Just telling it like it is. When it all comes down to it, that's my crime. I tell it like it is.

Once we properly got into the tour, there were bitter arguments. I'm the singer and I don't go on a tour bus unless there is an open window. I can't sit in there with air conditioner. It totally fucks up your musical instrument—your voice—and then you can't do the gigs. This is just a natural fact. The only instrument I have to fine-tune is my tonsils. I can't ask a roadie to repair them when I've been dehydrated for twelve hours on an overnight bus ride. You end up with all kind of medical problems from that, ripping your tonsils out due to nothing more than the lack of an open window.

Of course, this made me a spoiled brat. Certain members of the band found it annoying or precious, on my behalf. But these would be the same guitarists that insisted on employing someone to put the strings on their instruments. Well, an open window is the strings on my instrument.

It's the same with hotels: I have to have a room that has a window that opens, or preferably a balcony door, whether it be freezing or not. That's the kind of thing, for me, that keeps me going. Otherwise I start getting ill, I start getting run-down, and then strep throat comes in, then you need doctors and then it gets into Vitamin B12 shots. It becomes a downward spiral, and physically you become drained because of it. An open window isn't the perfect answer but it's nine-tenths the solution to the problem.

Then jealousy can creep in if the only open window is at the back of the bus—well, that'll be where I sleep then. And too bad if you wanna sit around there chatting. But I've got to tell you too, a lot of the time that's a problem in itself, because tour buses are very rattly things. When you have an open window, the sound of the wind rushing by—and the rattling and the banging—it's like being in an aircraft engine, and you're in a state of sleeplessness.

You can see how the rock-star private-jet phenomenon came about. You get there quick, you get it over with, and then you can stand next to your open window in a hotel room for eight hours. Rather than ten hours on a motorway, arguing.

We toured way too long. I realized very early on that I couldn't write a song for them anymore. I didn't feel it in that way, and every time a song idea came up, I was always thinking, "PiL!"

That would be something we could experiment with in a PiL way, not here, because this would be ten steps backwards, like a reenactment really, and I don't do those.

I enjoyed shapeshifting the old songs. I also enjoyed revisiting our anti-format of that time, and how we put those songs together. Fond memories—it would be there in the back of my mind how these things came together, as I was actually doing it live. Fantastic, really rich rewards in that. But then animosities started to creep in. Again. Sometimes, they deliberately wouldn't give me that cue for a verse or whatever, that musicians ordinarily do. Then you're standing there in front of thousands of people, wondering, "Where's the singer supposed to go here?" I caught them doing that more than a couple of times. I'd look round and go, "Where's me cue?" They'd all have their backs to me in an idiotic Status Quo impression, Steve and Glen doing this arse-waggling back-turn to me. Stuck in their little jam session bit. Ludicrous, lu-dick-erous.

Very early on, in Paris, I was in my hotel room, watching, of all things, George Formby doing "when I'm cleaning win–daaaahs." I heard this rustling under the door, and I thought, "What's that? Is that room service again?" because I do love my French onion soup every twenty minutes. That's one of the greatest things about the French. Wow! French onion soup—I'd die for it. But anyway, no, it was a cassette and a little note attached going, "Do you think you could put some words to this?" That was the general gist of it. What?!

They'd gone off, and laid down some basic tracks without involving me at all, and then presented this cassette, not to my face, but snuck it under the bloody hotel door. "Just put the words to it." I exploded on that one—to me, that's just, "What, after all these years, you don't think I'm good enough to be involved with the initial songwriting of these tunes? You didn't even invite me to that?" No. "Just put some words to it." To this day, I'm very bitter about that. Not hateful but just, I'm sad for them that they thought that would ever work with me, because they know me deep down, and they knew that would hurt very, very deeply. So unfriendly, and just not the thing to do, and a kind of "You're not one of us" attitude. Smug and pompous and, at the same time, going, "Go on, put the money-earner on top of this dross."

Rambo eventually talked me into listening to it, and it was awful. It was hideous, it was rubbish. Nonsense. Old lazy-arse strumming . . . bum-bum-bum-bum-bum-bah. The dullest thing. We never worked that way ever in the past and I didn't understand why it was being presented in that way to me now.

That's when I categorically knew: it's impossible for me to write songs with them ever again. From that moment on, it was compounded: "Who do you think you are, saddling up with Chris Thomas?" He had recorded Finsbury Park for our live album and was also recording the sound in Paris, you see. Again, their idea. I thought, why not? He knows the Pistols sound, but he actually didn't. But that's neither here nor there. They could have these little cliquey setups, and the bottom line was, however much was spent doing that without my attendance, I ain't paying for it.

Amazing, isn't it? Yet I still put up with them, but by God, that hurt. It really did, it hurt deeply. It was a very bad thing to do.

Some of the gigs were terrific, really excellent, but I kept feeling all the way through that I wasn't in the mood to write new songs for this, in THIS part of my life. That time had gone. More than happy to celebrate THAT part of my life, but not for it now to be a part of my present. There's nothing better for me than to get up onstage with them fellas—we wrote those songs together, so that's the proper presentation. Offstage, I spent a great deal of time *not* hanging out with any of the band at all. I was 24/7 with Rambo—quite frankly, it's the only way it really lasted as a tour.

Security passes were something I never felt the need to wear, particularly when coupled with Rambo. The way we were viewing it was "If our hairdos aren't our pass, well, we'll pass on the gig, thank you." Rambo and I got into so much wackiness with clippers and dye. No regular mohawk for me or John, it's not our way.

I always like a line down one side of my head, from the top to the back, but Rambo came up with some amazing wacky things on me—like the idea of "castle-ing," so the top of my head looked like the Tower of London, with checkerboard, black-and-white squares on top. All manner of squiggles and different designs—a gorgeous mass of matte colors. Rambo wanted 666 on the back of his head, and I eventually managed to do it well with some nose-hair clippers. So it all went up a notch from the local hairdresser.

At one gig in Italy there was this big gang of Gypsies with young kids, all dressed mad, punky, spiky, and some of the crowd—I think it may even have been some of the Gypsies—were throwing bottles at us. We had these young kids we'd allowed to sit onstage, because they were young, but there were a lot of beer bottles flying around. I was like, "Why are you trying to kill us?" Rambo moved quickly to get the kids cleared off the stage safely. I can't understand that viciousness. When the mob mentality takes over, it's hard to control. I managed to shut it up, and single out the leading elements—"Oi, you, you fat turd. You're the man, are you?"

Even more chaotic was the Axion Beach festival in Zeebrugge, Belgium. It was very exciting: my first time on Eurostar, and also Paul Cook's birthday, which I didn't know, so he bought some champagne and I bought a few bottles more—I wanted to, you know, *be friends*, quite genuinely, but unfortunately most of the champagne was left to me. I got a bit drunk, not mindlessly or violently, just a little tipsy. So after the two-bloody-hour drive to the gig from the hotel, we hung around—uuuuunnnggh!—went onstage, the gig's halfway through, I was standing out on a very long runway, right out in the middle of the crowd, a good 20 feet high and some venue bouncers started climbing up the side of the runway to try and get on the stage—as explicitly forbidden in our contract. These were really large blokes, and you knew they were doing this just to cause trouble. Their excuse was "fans running onstage," but Rambo was dealing with all that—there's no malice in it, we're not going to beat them up, just politely lead them back into the crowd, no harm done.

It very quickly turned into a brawl, where they presumed that Rambo shouldn't be on the stage—a major bad move on their part. I look around, mid-song, and there's these enormous great long muscular blokes attacking Rambo and then a whole pile of them charging at him, so I dived in with the microphone and did the best I could to get them the hell off our property. By that time one of these big lumps had already been knocked out. It was an invasion at that point. And in that respect we won. We cleared them off, the band carried on playing, and I said to the audience, "That's what happens when security take their job too seriously!"

The set went on, and I could see other bouncers still trying to get onstage and at that point apparently the band just stopped. The gig was over, it was going to turn into chaos and they left, but I was totally unaware and carried on singing . . . at least I thought it was singing, but in all honesty, I was croaking. Eventually Rambo came up and tapped me on the shoulder, and goes, "Er, John, the band have left the stage, they've all gone home." Oh! Gig over. The band had literally gone back to the hotel in the tour van and left us with no way out. So he and I stayed and watched Leftfield and had a good chat with Neneh Cherry and a few others backstage. When we finally got back to the hotel at about 5 a.m., the police were there, and wanted to interview me over the incident. We informed them of what our contract stipulated, and that was it, no case to answer.

Some of the gigs were spectacular. Others were ridiculous and weird, with cold indifference from the audience. Any time we played behind the old iron curtain was sensational, but you wouldn't be getting anything like that kind of joy and celebration of a gig in, say, Switzerland. It would be back to "What do you think you're doing, then, go on, I dare you, entertain me."

Still, I got to wear some fantastic outfits. I went onstage as Pinocchio in Japan. I turned up in a bright yellow/green skintight see-through top, with my nipples showing, and red suspenders, extremely short shorts, big curled-up-toe shoes like Aladdin, and an undersized red trilby, looking insane. The band went, "Oh my God, that's not punk!" "Yes, it is, if I say so!" Again, the audience go, "Oh my God." The Japanese are all dressed up, " 'punky' style," in what they think we'd currently be wearing. But clothes are to have fun with. Don't judge me by my clothes, judge me by the clothes I choose and why I choose them and know that's about something. It's audacity. The clothes ultimately shouldn't matter at all, but they *are* great fun. It's absolutely hilarious to me that clothes can distract people away from what's really going on. And that's how you sort the wheat from the chaff.

In Japan, I'd be throwing bananas into the crowd that I'd autographed, and the band took that very personally, that I was ridiculing the name of the Sex Pistols. Bloody hell, I was only having some fun. The crowd loved it. They literally went bananas. I was told some of them even tried to find a way of preserving their signed Johnny Rotten bananas!

Some of the best times I've ever had were on that Japanese tour with Rambo. We were there for four weeks—a long time in a small country. We played a load of gigs, including a couple at the Budokan in Tokyo, but we had plenty of days off. I'm not a keen walker, but Rambo had me out wandering the streets. We met this mob of Japanese skinheads. It looked like, "Oh, is this going to go wrong," but no, absolutely the friendliest chaps. Sometimes language isn't a barrier. A smile speaks volumes.

We traveled everywhere by bullet train, which was a pleasure every single time. You have a nice view, and you get there quick. Occasionally there'd be two or three days off, and you can get stifled out there. Our hotel in Tokyo had thousands of rooms in two enormous towers, and what felt like an underground city of shops and arcades down below.

Rambo came up with the idea of going to a traditional Japanese hotel, so we two gathered ourselves together and took the train to Kyoto, where we'd booked this little place run by middle-aged geishas. It basically was a granny's house, but with tatami mats. It was very otherworldly, very other-century. Nothing like couches or chairs anywhere—it was all kneeling on the floor at very low tables, and being plied with sake *relentlessly*.

Yet again I'm forced to go out by Johnny Rambo, probably a little the worse for wear, but sake is a very energetic kind of drinking. It gives you a creative buzz, shall we say. So out "walkies" we went in Kyoto, and came across a nightclub which, it transpired, was holding a punk night. "Right, let's go in 'ere," says Rambo. "I don't think that's a good idea, John, I'm currently touring with the Sex Pistols, we don't know where that could go." "That's all the more reason."

It was fantastic, from start to finish. When we first went in, they were actually showing Pistols videos, but it was very quiet, and people were very polite. In Japan, they don't just rush at you straight away and poke autograph books under your nose. They give you time, they wait for you to give them the signal—the signal being, I just looked over, and here they came like a herd of Japanese buffalo! Suddenly the DJ went bonkers and stuck a safety pin through his cheek, and everyone went crazy on the dance floor. I was signing people's bodies, signing their shirts, signing the bar.

None of it seemed show-offy or pop-starry, it all had a great element of fun and naturalness about it. It was one of those evenings where Rambo doesn't have to be on guard because there were no nasties in the house. There was no one wanting to stick a knife in your back or screw things up out of jealousy. The most perfect evening, the kind of thing you live for.

Back at the hotel, we were promptly ushered to our separate rooms, and shared our stories in the morning of how we were put to bed. We both had similarly horrible experiences—for me, I was practically stripped naked and pushed into the bath, which of course was too small for me, and then the same thing in the morning. They must stand outside with an earhorn, or have you on video, because the second you get up to go to the toilet, in they rush, roll up your bed, that's it, it's morning, it's breakfast time Japanese style!

When we played at the Budokan, there was an after-gig meal laid on by the promoter. The place he picked was a fugu restaurant. This we didn't know instantly: we were expecting a Japanese menu, but initially it seemed to be what I call "Catering by Motörhead"—cold soggy French fries and burgers. I was looking for sushi on the menu, but then I spotted fugu on there. Now, I was well aware of the dangers of that—poisonous pufferfish! I'd never had it in my life, but was soon persuading Mr. Rambo that's what we should be sampling instead of stale buns.

Now, it's not only that it's deadly and it can kill you, the taste of that stuff is horrid! It's even worse when it goes down, as it leaves an—urgh—inexplicably bad taste. Not harsh, just mildly muddy. Then it's like, "Rambo, let's have another!" We didn't realize they were murdering these things out of a big fish tank. So Rambo goes out there, and the one he picked was called Lucky—the one that had been there for years, the longest survivor on death row. They had to have glass screens around where they cut it up because the blood spurts from them so high, and it's deadly poisonous if it touches you or gets in your eyes. You better be trusting that chef!

So we had the second one and the effects started to come on. At first you feel very alert, then on our way back to the hotel, there was a slight numb tingle on your lips and tongue and the back of your throat. Then, at 7 a.m., you're wide awake, full of energy, almost "bouncing off the walls." Outside, in the street, we ran into a demonstration rally of antiforeigner sentiment. I'd never seen anything like that in Japan before. There was a surreal temple nearby, and some homeless people swept away under some bridges—very much like in Estonia where they swept the people away from you before the Wall came down. Japan has its shielded side too.

In the temple, there was the unreal vision of lines of Japanese schoolgirls in their matchy-matchy tartan dress outfits, all very small and polite, being led by teachers in this direction and that direction. All of these visions put together was overwhelming, and all under these wonderful ornamental trees. Then back to the hotel, try to have a nap before I do a gig. If I'm to tell you anything at all about being on tour with the Pistols, *those* were the finest moments. Not necessarily the gigs themselves.

"Go fuck the Queen!" and "Argentina, Argentina," chanted the seething mass before us in Buenos Aires. Always a lively anti-English sentiment in Argentina! I made sure I shouted down "el Presidente" and "down with the monarchy" in return. "Good to see you back, Argentina!" In the name of diplomacy, Rambo had put a sign outside his room saying, "Stay out of my room—Johnny Rambo, God Save The Falkland Islands."

This South American leg of "Filthy Lucre" was to be the final one, and it wasn't long before all the Argentinians were screaming along to "God Save The Queen"—a shared moment, from what could have been a whole pile of trouble. There was a huge football element in there, but we had a very good connection with each other, and kept it going with friendly insults and anything else you'd care to mention. Language barriers aside, and political barriers aside, dating back to that stupid, useless, pointless Falklands War, we found common ground there, the audience and us. It was a delicious experience, one I'll never forget.

The final night in Santiago, Chile, however, really took the biscuit. You could feel the tension in the air there. Looking down on this massive square, you could see the armed police lining up, and every hour they'd come out in their uniforms and guns and do a goose step, march around the square, blare trumpets and wave flags, and then go back in and firmly close the gate.

Our worry was, "Will anybody turn up? Is there any interest in us in Chile?" Well, there was a serious interest. It took a really long time even to get near the block the venue was on, and then even longer to get into the place, there was such a mass of people and so many of the maddest punks I've seen anywhere in my life. These were "full-ons," Chileans with mohawks, really into it, challenging the police, water cannons going off up and down the block. It felt like civil war was about to break out but, oddly enough, in a funster way. Like, I'm an observer in a scene of chaos that I'm partially responsible for creating!

There was a cold hour or two in the dressing room, but then—onstage, wow! The roar, the roar! These were committed fellas, and girls, and such a menagerie of all kinds. In one corner to my right, the ferocity, the heat of the yelling and screaming, was SUPERB! It was literally like being in a wind tunnel, but very hot and humid. Thank God, nobody had halitosis. To the right was a whole bunch of people who had decided to strip naked. So there were nudists and, up top, disco-dancing dolly birds, all in the big hair and the overdone mascara, and tank tops and very short skirties and high heels, and these full-on punks and mad football-y kinds, very young kids just screaming in tears of happiness—I LOVED IT!

And there I was dressed in this Dolce & Gabbana set of hot-pants and a black tight little plasticated waistcoat, the Aladdin slippers . . . and my hair looking like an orange and blue cockatoo. Rambo called me Johnny Cuckoo.

People want Johnny Rotten in punk regalia, but what they've got to know is, *it's all punk regalia, if I'M wearing it.* So I bowled out in this outfit that Miley Cyrus would be ashamed of wearing. I felt *really fucking hard* in that outfit, I was delivering the songs in a venomous, detailed way, wearing that. Love me, not what I'm wearing. Geddit?

There were police on the left side with riot shields and truncheons, but the main problem was the local police-type security—I'm not exactly sure who or what they were—they were attacking the crowd with their mini-truncheons. Rambo had to clear the stage of this hooligan security before we came on. But this crowd weren't having none of it. They would not show any back-down, they were really admirable. Many times over I went to tell the police to stop it and finally they backed off. Rambo certainly had his work cut out with the crowd; they were constantly trying to get onstage all night.

We'd heard some of the fans who couldn't get in had tried making a hole in the ceiling and attempted to rappel down into the crowd, absolutely brilliant! Some of them may even have made it. I really wouldn't have put it past them. During the gig, I'm sure I saw little pieces of plaster tinkling down.

It was from start to finish an insane, mental, MENTAL gig—one of the best gigs I've ever done in my life. The songs just felt right. I was at my toughest best with this crowd—proper bloody Johnny Rotten stuff going on there! The band played bloody great, too, it was good ol' rootsy stuff coming out of us.

Near the end of the set, Steve had a problem with his guitar and just walked off. Never said a word to any of us. Just stopped playing and walked off. Oooooo, what's that all about? He just left us out there with thousands of screaming fans but, you know what, it didn't matter anyway. The crowd just kept singing, so I got all a capella on it with them. We had a great sing-along. It was looking like Steve wasn't going to come back on, but then I think he realized he wasn't being missed and reappeared.

Then we go off, as you do, and you wait for the encore, because you need your breath and your cigarette. So we asked Steve why he went off. "Uuunnh, I cut me finger." Unbelievable! Then up stepped Rambo, and showed him his own leg. The monitors were metal-framed and he had run into one while getting fans off the stage and torn the skin to the bone between his knee and his ankle—he lifted this huge flap of skin to expose the leg bone.

Fair play to Steve, he went, "Oh, my God, okay!" Rambo got a roll of gaffer tape off Frankie the tour manager, and gaffer-taped his skin flap back down, and back on we went. And the encore was much more insane than anything we'd done prior. It was truly, truly an amazing experience.

The camaraderie I was feeling with the band and the audience was terrific—the whole point of doing this thing in the first place. It happened a few times on that tour. Let me tell you, there were many times when it *didn't*, and you'd feel like the shutters had been put down on you. Just making you feel like an outsider in your own thing. But Chile was fantastic, and it's just a shame that it was the last gig for some while.

The next morning, Steve, Paul, and Glen were flying separately to me and Rambo, and they didn't even say goodbye when they left. This left a real sour tone on me and that's where it is with us. It shouldn't be like that—especially after a gig like that, my God! Must you really all rush off to bed early and then leave without saying goodbye? Apparently, they must.

Rambo and I duly traveled as far as customs in Florida together, then I had a connecting flight to L.A., and he had a connecting flight to Memphis. Because of his leg wound they had reluctantly put him in a wheelchair! I even got to wheel him a little, he hated it, but it meant he got whisked right through. When I finally got through, I ran into him. There he was, walking around the airport without the wheelchair, and we had a real laugh, and a proper goodbye.

And that's a moment in time I should have been sharing with my band, but they don't give you those opportunities. What can I say? It leaves a blemish.

WHO CENSORS THE CENSOR? #4

DO YOU WANT MY BODY?

I've always had bad teeth, from my early youth onwards. The dentist's was the very last place any of us in my family would go. It was where my mum and dad had all their teeth removed. They were given money by the state towards getting a set of dentures fitted, which they were told would solve all of their problems for the rest of their life.

This policy, which was obviously all about saving the government from paying for proper dental care, created nothing but trouble for the patients who took them up on it.

Come nine or ten on an ordinary night, they'd take their teeth out after dinner and soak 'em in this vile liquid, Steradent. Otherwise dental hygiene was unrequired. And it wasn't just my mum and dad, it was my aunts, uncles, and everybody I knew.

Once the teeth had been extracted, however, the gums would recede and the dentures would require all manner of sticky-back plastic, shall we say, to keep them in, because the gums had dissolved to nothing. Every time they laughed, their teeth would fall out. It was an even bigger problem when Mum and Dad would throw a party at our house. They'd all lose their teeth from dancing, from all the jumping up and down. My job was not only playing the records, but finding the teeth and working out whose was whose.

So that was how it was presented to me: I needn't bother brushing because when I grow up I'll have a fresh set ready at the dentist and I could lose them on the dancefloor like everyone else. So I would naturally avoid the dentist. Also, because of the pain. Dentists were very brutal back then. Yes, it was free on the National Health Service, but the cost in trauma was incalculable.

When I was about thirteen, I had a very bad experience. At school I had this toothache, so bad that I was screaming with the pain, and school actually booked an appointment with my local dentist. She was Polish and insane, and she had a "Brünnhilde SS" kind of vibe, with the hair pulled back tightly and a bun in the back done in a braid. Short, chubby, very blonde, very Germanic in her approach, and very, very volatile. She absolutely wouldn't listen to you squeal in pain. She had no time at all for any of us children. She scared the living daylights out of everyone.

Anyway, she immediately decided she had to pull this tooth, but when she ripped it out, she broke a blood vessel. She gave me a cotton swab to hold on the wound, but it just kept bleeding. The dentist's was on the corner of Holloway Road and Seven Sisters, and I caught the bus to go home, but I actually passed out on the bus. They stopped the bus, and the conductor took me home—basically dragged me, all limp like a dead body. I'd lost a lot of blood, I was covered in it. Luckily, my dad was back from work, and he rushed me straight to the hospital, where they stitched up that side of my jaw.

From that day forth, on the upper left side, I had a huge gap between my teeth where she'd taken quite a lot of the gum with her. I could make dolphin noises, I discovered, by sucking air in through the gap. The budgie loved it. The hamster never responded too well, but the cats and dogs loved it too. So it became like a party trick of mine. I actually used it on the Sex Pistols song "Submission"—the "pffffmmmmwwwwp-p-p-p" noise in the bridge section is all from that. I've since had it replaced, so I can't do it any longer.

Understandably, that Brünnhilde experience left a really negative impression, and strengthened my aversion to oral hygiene. At the dentist's, you always got either that dreadful gas mask, which would make you feel like you were being gassed to death, or the injections, or just the sheer violence of pulling teeth, which seemed to be their main ambition. After a while, fillings were more the fashion. They'd drill holes in every tooth, and fill 'em full of mercury . . . and then they'd pull the tooth anyway! Just more pain on pain.

So, I always had bad teeth. The concept of brushing them never occurred to me, and I can't blame Mum and Dad for that. There were toothbrushes in the house, but I'd only ever seen Dad use one on his work boots. I suffered a lot of ill health because of it, and I was naive not to be aware of that. It took me forever to catch on that my teeth were one of the things making me feel so ill all the time.

By the time I was joining the Pistols, the second I smiled, it was like, "Oh my God, look at those teeth on him." It was Steve Jones who went "Uuuuh-uuurrgh, you're rotten! Look at you, your teeth are rotten!" The front two had this green mold on them. It wasn't just like I'd eaten some spinach or something. If it was, that stuff remained stuck there for a real long time. Between the gum and the tooth, there was a green line on the front two, like slime, and on every other one, there was that horrible yellow stuff that I never understood—plaque.

At the time I thought it was exceptional—a good thing. Nope. It wasn't. There I was, you know—"Why don't no one wanna give me a kiss?"

So I was known as Rotten, and the nickname stuck—for life! I know it's a bizarre thing for me of all people to say, but, really—*take care of your teeth!* In this one respect, don't do *anything* like what I did! Through all those years of ignorant behavior, I was slowly but surely poisoning and killing myself.

Much later, once I was in California, I spent what ended up being a small fortune getting them fixed. You don't even want to know the cost. I had to have a whole series of operations, because so much was going wrong up there. Eventually I started to listen to what the professionals were telling me, but it was mostly pain that guided me. It reached a point where I was really seriously ill. It became so painful that I'd rather deal with the pain of the dentist, than put up with the pain I had every single day.

I was poisoning myself with constant abscesses, and had near-permanent headaches. To remedy the situation, I've had everything—crowns, you name it. In 2012 I had titanium screw-ins, because there was so much damage going on up in the bone region, it all had to be rebuilt and replaced. I had to have all the bones realigned. I pretty much had my jaw realigned—major, major stuff.

I put it off for so long because I thought it might affect the way I sing, or the way my voice sounds. I took the risk, because I thought I'd rather live, and live without pain.

I'm really pleased I did it. I couldn't imagine having anything like this done on the National Health in England.

It should be possible. All I know is, I paid through the nose, so to speak—that's where most of the injections came through, from the inside of the mouth, upwards!—and it was a great deal of agony.

It's different now. I don't get ill as much, I don't get run down as much, and I notice the difference physically—very, very seriously. A lot of the perpetual illnesses just stopped overnight. I have a lot more stamina. A lot of that hunched-over-the-mic early Johnny Rotten posture was like, that boy was dying. And taking the long, painful, slow way about it.

You're probably thinking, "Yeah, and everyone in California has a pretty smile." Well, I haven't got one of them. That I would not allow. The replacements and screw-ins are all the same gray color as the rest of them were, and there's no sense of matchy-matchy about it. As Rambo said to me, "Your teeth are like Lego bricks." He's full of the truisms of life.

I'm still not too used to the toothbrush, though. I know I should be, but I have to remind myself. The only times I really brush my teeth are when Nora catches me before I go to bed, or on tour when one of the band will go, "Jo-o-ohhn!" So it's not like there's suddenly mouthwash on our rider. I generally always use brandy instead, particularly onstage, to clear my pipes for singing, but that's a fallacy too—apparently, it's not helping in many respects.

Since I finished our last tour, and have been off the brandy bottle, I've lost a lot of weight. It's a bugger, because that's the only comfort and joy when you come offstage. So there I am going on about my mouthwash, but I was really just washing out my innards!

I know I let myself go for a bit there and I really spread out like a balloon on tour. I got so happy and content, I just ate everything in sight. And of course, most fatally of all, I gargled brandy onstage and then finished the bottle when I got off, and that put the calories on. So I've stopped all that, and the weight's just dropped off me and I feel physically better for it.

I'm not the kind of person to do any kind of physical exercise. One, I get bored, and two, it's bad for my heart. I've yet to discuss it with my doctor, but I'm sure they'd be in full agreement if I paid them well enough. That's how it works. Gone are the good old days when you could get doctors to pump you full of narcotics at the drop of a hat. Now in this health-conscious universe we live in, it's a nightmare. More so than ever here—it's really strict. Or maybe I just know all the right people—depending on your point of view.

I wouldn't go that way anyway. That's like, "*Urgh*, authorized inebriation? No!" I hate that idea. It takes the fun right out of it. As long as there's that innuendo of contamination and naughtiness wrapped around it, it wards off lesser mortals who wouldn't be capable of handling such chemical devices. Of course it attracts others, but those people are like iron to magnets, aren't they? If it wasn't drugs, it would be something else. It would be politics or religion, which is far worse.

I did once have a personal trainer. Bless him, the poor thing, he didn't get on with me at all. I'm afraid that I have no motivation for developing muscles. *Pffff*, three of them tummy-tuck things and I go, "What's the fucking point?" I'm here to enjoy my life and not get trapped into somebody else's impression of what my physical shape should be.

At the same time, there's a lot to be said for the notion that if you maintain yourself somewhat healthily physically, your brain will work better. Because of the climate and the sheer dampness of everything in London, you eat in a comfort-food way, don't you? You just stuff anything warm in you—missus! It's the old joke, whenever I go back there: the English always look gray and sickly.

But praises be to Allah, Jesus, or any one of them fellas—here I can open the door and walk out and pick a lemon off a tree and have a delicious drink in three and a half minutes. I love that. And I loved that about Jamaica too. There's fruit just growing by the side of the road. Wonderful.

I used to have a bit of a thing about Pot Noodles, but I've gone way off them. I actually stopped it on tour, because I'd have one when I'd come off. Now, I can't bear the chemicals in them—I can literally taste them. How on earth did I ever manage to eat that amount of salt and sugar . . . and God knows what else, what powdered substance.

At home I cook, but not anything you would call a set menu. It's really just whatever's available. If it takes longer than ten minutes, don't bother. No need, because if you take longer than that, whatever you're cooking, you're killing. Unless it's a turnip! I love turnips—I really do. And rutabaga—rutabaga is my favorite, even though I always used to call them turnips. They take forever. You cannot possibly eat one unless it's been boiling for forty minutes.

I've always had a firm love of brussels sprouts, particularly having been called one by my father, which I now realize was somewhat of a compliment. But I love the taste and the texture, all of it, and that's without salt, butter, or any condiments. I always used to watch my brothers scream at the prospect of them. You know, "What's wrong with them? Yum-yummies!" Mini-cabbages, I used to call them.

I never forgot the taste of blood from that ordeal with

the Polish dentist, yet I'm absolutely one for a good rare steak. I eat meat very rarely now, and it's not because of any health agenda or stupid diet cause, it's that I find I don't need to. Every now and then my system will go, "Oo, we'll have a bit," but after that I won't need to for a while. I quite naturally now follow my instincts and that leads me to maximum vegetables, which I am absolutely in love with.

At the moment, I'm experimenting with my moles. I got an iPad last year, and I've fallen in love with it. I was looking up on moles. I found a home remedy to file them down so you take off the top edge. So of course I did that—I gave one of mine a good sanding, with sandpaper—and it bled like I'd opened a vein. Now every day I'm putting apple cider vinegar on it, and it seems to be working. Where I just had a mole before, I've now got this huge pusy septic thing. I'm beginning to worry that it might turn into a cancer. Like, "Oh fuck, what have I started here?" But I'm going to carry it on to the bitter end. How stupid, huh?

But don't you dare go praying for me! I'm not having that. What an awful thing to do. Let's hope I'm around when the book comes out. Left to my own devices, there's every possibility I won't be. For God's sake, there was no mention that it would bleed like that. It really is quite literally like an open vein. It's horrible. I've had it all my life, but I just open-mindedly one night thought, "Oooh, why not get rid of it? Really, why not?"

I've always been quite proud of it. "Does my mole annoy you? Good!" It used to annoy Steve Jones *very much*. He'd always go, "*Urgh*, that's horrible, that is, you want to get rid of that!" "No, I don't—not anymore!" I don't get rid of things just to please other people. I'm very self-determined about my moles in particular, and now I've determined that they have to go. There's only room for one face, and it's not gonna have a cauliflower on it. Or a bloody stump. That's what it's actually started to look like, a broccoli stump.

But yes, I remain Rotten to the last—a reluctant toothbrusher. I know it's a stupid move but there it goes. I won't have a toothbrush on my tombstone. In fact I won't have a tombstone, I'm very spooked by that whole body-in-a-casket burial thing. I'm thinking cremation, or selling my body to scientific research. Have a bang of this number, baby!

Apparently, I don't have anything like liver damage. I've had full medicals and in that way I'm remarkably healthy. You would presume otherwise, but no. I'm physically very, very fit. It can only be the singing. It keeps the lungs inflated.

12

YOU CAN LOOK TO THE FUTURE WHEN YOU'RE CONFIDENT

I did the Pistols tour for many reasons, but one of the most important to me was a promise from Virgin. I'd just completed my solo album, *Psycho's Path*, and the promise was that if I did the Pistols tour and withheld the release of the record until afterwards, so that they could fully back the Pistols momentum—in return, they would help launch *Psycho's Path* in a much bigger way. To me, that sounded like a good deal. Stupidly, I believed them.

I struck the deal with a fella I really liked, who was in charge of my file at their L.A. office. I didn't see the two projects as being equal and opposite in the way that the record company obviously did, but I thought, "Okay, that'll work out. I'll put my solo album back a year." Of course, when the Pistols tour ended, I found out the guy had moved on to another job, without passing my file on to anybody else, or informing anyone of the agreement we'd had. Nobody at Virgin seemed to acknowledge me, and, in fact, their whole staff seemed to have changed. And they didn't do such a good job backing the Pistols either.

That left me between a rock and a hard place. There was no one holding that album or taking care of it, so when it finally came out in July 1997, it was a minimal release with virtually no press coverage. They all but buried it. Insane. But at the same time, they also destroyed the Pistols live album of the "Filthy Lucre" tour through lack of promotion. Most people never knew it even existed.

Psycho's Path really came about from building a studio at home in L.A. Having my brother Martin living in the guesthouse was extremely useful, because he's very technically minded, and he practically built the whole thing himself. Back in Gunter Grove days, he was very into radio. He'd bring old radios around, stick nails into them, form little pirate stations, and start all manner of chitchat with other like-minded people. Very early on he was into computers. Anything technical, anything with a serial number—he's got it *down*.

I can't remember my own phone number, but he can sit there and blather away the serial numbers of every computer model that's ever been released since the beginning. He can't really read or write to any great skillful level, but by God can he understand a manual—that's a skill process all to itself. If I stare at these serial numbers of things—20 megahertz, blah blah blah—my brain just freezes. He's very good for those intricate, complicated electrical details.

And he loves all the little boxes that go "Blink!" So we just built up a collection of odds and sods—sound effects and a mixing board, a 24-track thing—pony, by most people's standards, what you call cheap and cheerful, garage-stylee, but that's the way I like it.

By the time we got to make *Psycho's Path*, we'd made the whole thing digital, because I was always thinking about getting into film soundtracks, which is obviously extremely hi-tech. The idea of putting music to scenes has always been very interesting to me. I wasn't just geographically well placed for it, I was also mentally well placed for it, because I view things somewhat with an artistic eye, like an artistic photographer, and I love to paint. Whatever it is I do musically, it is a tapestry of sorts, rather than just the words alone. Words alone can be fairly empty and open to misinterpretation, as we know with the Koran and the Bible. And, indeed, the *Sun* newspaper.

I also used that very fine studio we'd built for some quite odd and amazing things. We made an advert there where I sang a version of "Route 66" for a soda-pop drink called Mountain Dew. Because we were up-to-date with the equipment, we could do it digitally, which was what was required. It's great fun, working on things like that. I did voiceovers for all manner of cartoons. I had a syndicated daily radio show in the States called *Rotten Day* where I took a sideways look at the history of music. I also had a four-hour online talk radio show in the early days of the internet. In fact, I did many weird things that you wouldn't think I'd be into at all, but anything I think will offer some kind of creative "new" to me, I'm bang-on into.

Martin had started a family. He'd met a young girl out here called Renée, of Mexican descent. She's got Japanese in her family too, so, very international. Once they started having kids, to be honest, it became a bit wearing—too many souls clonking around the house all the time. I'm not being mean here—I loved the kids, I loved them running around—I'm just saying it can be overwhelming when you are trying to work.

We'd also have record company people coming down, constantly begging me to do this Pistols tour, and that's deeply frustrating when you're so involved in a project. From that point of view, *Psycho's Path* became very difficult to make. It was an album fraught with problems, but somehow we got it together.

Martin actually played little bits and pieces on the album—we all did. The producer, Mark Saunders, played little bits. One day, he brought in a very strange guitar. He was a difficult fella to fully comprehend or understand. The draw for me was he'd worked with Neneh Cherry, and Neneh and me, we're mates. We know each other from way back when, from even before Bruce Smith, who was married to her briefly in the early '80s. When Neneh first came over to England, she stayed with Nora at Nora's place and hung around with the Slits. She kinda grew up around us—she was only thirteen or fourteen—and so I always viewed Neneh as family. And so: you've worked with Neneh, this must be good.

During recording, I discovered that a friend of mine knew Todd Rundgren. Todd rang me up and asked if he could come over. That was a great evening. We just sat there drinking, and I really liked him. I love Todd Rundgren's records. I was basking in his glory, it was terrific to be in his company, and we swapped musical faves. I played him all different kinds of music, showed him what we were up to and, wow, hooo, haha—it was amazing, he gave me that pat on the back, like, well done, keep at it, this is different, this is what we want. Todd Rundgren has always strived to break the mold.

I never thought about it like, "We should work together!" That's never the way it is because you're always wary that it might take over the situation. I already had a gameplan in mind for the album, so it wasn't even raised as a question or an issue. There were all kinds of possibilities of working with lots of different people. Steve Vai rang up, saying, "Look, if you need me, I'm more than willing," but I wanted it absolutely stripped down to the minimum. There's only really three people in there—the producer, me, and my brother Martin—and that's how we welded it together.

Frequently, we made music out of nonmusical situations. The drums for "Sun"—that's cardboard boxes in the front room—cardboard boxes! I loved that sound! Or throwing an accordion down the stairs like a Slinky. You know, those coiled springs that can crawl down the stairs? That was the accordion Jebin Bruni from the mid-'80s PiL band gave me, used in a terrific way, I thought, and then stretched out technically—a completely different approach to music.

"Grave Ride," that's completely nonmusical in every way, shape, and form. There was no set melody in it and deliberately that way. Let the ears pick up what they want, but nothing specifically played, like a rigid drum kit, or "the guitar part." I was never saying, "Here's the part, Steve Vai, where you play *that* bit." Nothing was like that.

There were a lot of loops: we used a Kurzweil synthesizer to make loopy patterns, let them run randomly, and you'd sit there for hours listening to these loops, until they clashed in the right place. "Stop! We use that section!" Very "amateur hour," but that's the most fun you could ever get putting a record together. THE MOST FUN! Absolutely ignore the disciplines of musicality, and you'll get the better results.

The only problem was that most people didn't know that the album was even released. It could've done something, that record. I loved "Sun." To my mind, selfish though I am, it's one of those festival anthems you hear over the PA when the roadies are setting up the next band, like T. Rex's "Life's A Gas," or Dave Bowie's "Memory Of A Free Festival." That exhilarating happy vibe like when you're going to have a party and the sun is going down.

An escape from Alcatraz, is what the whole album was. The lyrics are odds 'n' sods of curiosities. The song "Psychopath" is based on John Wayne Gacy, the serial killer—the famous one, the clown. How many hundreds he must've murdered. In my darker moments I've thought, "But for some kind of inner sensibility, I could quite easily be that way. I could go and kill people, aimlessly and pointlessly, and take some kind of gratification."

I'm analyzing myself here and seeing that it is possible to be a serial killer, as indeed it is possible for any human being to be exactly the very thing that you think you hate and despise in someone else. What you're really doing when you're over-judgmental about those things, is you're taking it out on yourself because you know your inner possibilities. We all are capable of the most ultimate evil. And because we are also capable of analyzing that, that is exactly why we're better.

Once the album was finally coming out, I put a band together to go out and promote it. We were due to start the American dates in August followed by Japan in September—where *Psycho's Path* had actually been released a few months earlier than everywhere else. I had Rambo back on board for security and he was becoming more and more involved in the day-to-day running of my work. I'd rehearsed the band for a good three weeks, and then, a few nights before we were due to fly to New Orleans for the first date, the drummer, Robert Williams, pulled the same stroke as Martin Atkins back in 1984—"I want more money or I'm not doing it." This time, rather than try to iron it out, I just told him, "Go home, goodbye—I'm not going to be blackmailed twice in the same position."

Then, for whatever reason, there was an alleged flurry of fisticuffs. I would call it more like a girly slap-fest myself, but he ran to the police screaming "criminal battery" charges against me, which they immediately threw out. He then decided to take out a civil suit. Apparently at this point the producers of *Judge Judy*, a courtroom TV program on CBS in America, somehow got wind of it and approached us. I was cynical about it at first, as it was Mr. Williams who was pushing for it, of course—you get money for being on it, and you get publicity and these are all the things he was angling for. But what many people in other countries don't realize is, *Judge Judy* counts the same as a proper court case, it's contractually arbitrated: what she says goes. The whole thing just seemed bizarre but it was just too much of a wheeze to pass up. And what am I looking at in the press? "It's a fake court case, it's John Rotten just trying to get some publicity for himself." Well, no.

This guy, Robert Williams, was someone who'd been recommended to me by my brother Martin. He'd played with Captain Beefheart, which obviously was fascinating to me. Then stuff like this happens; you go to the TV show, and you think, "What the hell is happening with the world?" The whole thing backfired very badly on him. It was rather stupid of him to brag about being a black-belt judo and karate expert, and then declare that fat old

Johnny Rotten—the world's most renowned pacifist, with the philosophy of Gandhi—beat him up in the car park!

My brother Martin let me down because he wouldn't even appear, and my manager at the time, Eric Gardner, and the tour manager let me down because they hummed and hawed the whole way through—they were both very scared of the TV cameras, and nobody would make any commitment or tell it as it really was. Those two lemons could have dragged me down. Anyway, Judge Judy Sheindlin saw it for what it was. This Kung-Fu Kitty was a very, very greedy person jeopardizing everybody else's work—an American and Japanese tour—and he lost and I won. It was a false accusation. I say to the world, "Believe what you will, but the truth will out."

As for promoting *Psycho's Path*, all this nonsense totally fucked it in the head. While I was waiting for the "trial" to happen, the tour went ahead. Luckily, I'd known a drummer called Otis Hays, who was hanging around the studio, an African-American fella, and he quickly got on it. "I can do all that stuff," he said, and he could! In Japan, however, Otis behaved like an insane person. He was young and very talented, but went mental around anything female and Japanese. When you're being taken out to dinner by the record company—don't do that! And he did it. I can't really blame him. It was all definitely a new experience for him, but he did a good job for us.

Eventually, I had to knock the whole thing on the head. It was difficult, because Martin had asked to play live keyboards, and he wasn't up to that, not by any stretch. Martin was great to work with in the studio, but not to play live—he couldn't cope with the pressure, having to remember the sequences and the patterns. The flaws don't bother me, or mistakes, but acting inadequate does. You must never show fear onstage—that's pointless. Never back down—never, *never*! No matter what mistakes are being made, go with it, keep going, find yourselves, and don't any one of you suddenly decide, "I don't know what's happening, so I'll end the song." Wrong move.

• • •

Finally, I really did think, enough was enough. I thought of just retiring from music forever. That's happened quite a few times. When it gets down to the malarkey of false accusations and career-hoppers and users and abusers, they really do leave their mark on me, unfortunately. They take away my joy and love of my fellow human beings. They smear and tar everybody else with that same brush, by that activity. I know it's wrong of me to view it that way but I take it very deep. I don't understand them.

I certainly don't consider the release of any record I've made as a mistake. I've always been proud of what I've done. I've tried to make the best of knowing I was stuck in these bloody record deals. All I kept hearing were the words "recoupment, recoupment, recoupment," but I always kept my independence, never allowed them to dictate the content of any record, and I suffered accordingly.

At this point in 1997–8, the power brokerage at the top at Virgin Records had shifted, and in came a particular duo who were from that '70s flower generation, but were businessmen to the hilt while still wearing the Grateful Dead tie-dyes. I remember one particular absurdity: one of them had a Victorian toilet imported from England to the L.A. offices, one of the ones with floral designs on, very elegant, in the shape of a seashell. Hundreds of dollars for a loo to pee in? Oh, come on! I liked them fellas, oddly enough, although the Grateful Dead tinge was in there. A lot of them Grateful Dead followers are really all rather selfish lawyers and accountants, of the mind frame, "I've got mine, fuck you." And it's very frustrating when you're struggling to keep your career together and all of that, and expensive items like that are being shifted about into office spaces. That's a waste of money.

I don't like flamboyance, I don't like people with overelaborate flashy cars and jewelry, because I think deep down inside they're a bit of a cunt. It's all about an audience they're looking for. Hello. I don't even drive but one of the first cars I ever bought was for Nora, and it's a Volvo. We love it, and we still drive it. It's falling to bits, but it's damn fast and it's relentless—a T5R turbo, yellow, with a go-fast wing on the back. I watched it in all the rallies around that time, '93/'94/'95/'96, when Volvo were really doing well in world racing. I just loved the shape of it, it looked like a pram! No sleek curved lines, this thing was right angles, boxy, built round a steel cage for safety. Fantastic. Two major accidents that car's had, and neither of them our fault—the last one Nora got hit by a concrete mixer truck. That would probably kill you in most other cars. All we needed to do was replace some paneling. Wonderful. These are the kind of things in life that impress me, the practicality of it.

If I want to drive a Ferrari—and I love the idea of those cars too—well, I've got *Real Racing 3* on my iPad. Somebody's got to buy 'em, I suppose. Myself, I can't drive at all. I know! I keep putting it off. My dad set me up for a driving test once, but I turned up for the first lesson like a complete arsehole, drank eight very large Heinekens before I got in, and drove straight under a lorry on Gunter Grove. So that was my driving experience. God, was that instructor furious! It was quite a bill I racked up on that one! That was a lesson, a serious bad lesson.

The shame of it was that it didn't end there; he thought that I might've still been capable of making it round the corner, and indeed I did, and drove down towards the gas works in Chelsea, and there was a huge empty acreage back out there in them days. He was trying to teach me how to maneuver, and it was hopeless. I was so *not interested*. It all seemed really pointless. The hand and leg coordination wasn't in me at that time. There's a great many problems I have with learning in these kinds of situations. I'm really slow. I get there in the end if it's at my own pace, but I can't be pushed into it, it doesn't work. I have an auto-resist button, I don't know what triggers it; it's one of the few things I'm learning to try and control—you know, an auto-rebel.

Back in L.A., without a music career to attend to, what I learned to do instead was boat driving. Nora and I really learned that through necessity. Living on the ocean, the draw was overwhelming. Initially we rented big yachts and filled them with friends, and went off for a couple of days to the islands, to Catalina, just as a party of friends would. Instead of going to a nightclub, we'd do it on a boat with bunks. And that was great fun, and Nora and me really got into it. Loved it.

Aaah, I've always loved the sea. It all started with rowing out to catch herring with my granddad—my mother's father in Ireland. He always had rowboats in the yard, all rotting and God knows whatever. He hardly spoke to me, barely even sentences, but we'd row out into the Irish Sea—actually no, it was the bloody Atlantic, because they were in Cork—and throw out some breadcrumbs and wait for the herring to come. You'd throw out a huge net and there'd be your dinner.

The sea clears your head. If you look at the ocean, when there's nothing but ocean all around you, it just wipes your mind of all the rubbish. When I was very young, when the TV stations would close down, I'd just stare at the screen, the static, even though my mum and dad would belt me across the back of my head going, "Ye'll mayke yerself go bloind"—but that's what the ocean feels like to me. That wonderful static. Brainwashing, in the most delightful positive way.

We eventually bought a boat called *Fantasia*. It's small, 32 foot, and goes really, really *fast*! And we love it, we go deep, deep, deep out at sea, and the whole joy of me using the GPS, and knowing where we are, is *fantastisch für meine Pussy-Frau*. Nora's incredibly brave with all of this, she's utterly fearless. I'm the one trying to give her my tuppence worth of knowledge about "That's a big wave we're heading into, I think if we went in the other direction it might help!" She plows on through, but the bounce on the back side of that wave—oh, that's earth-shattering and painful. It will hop out the water and—*bang!*—that's ten and a half tons slapping the water—it feels like it's gonna crack in half. For a boat, that's not heavy.

We had an instructor when we first bought the boat, but he was eighty-eight years old—Captain Something-or-the-Other. He was as dithery as they come, and we found out halfway out at sea—in very rough waters, in 12-foot swells—that he wasn't too used to motorboats. He was a canvas man, himself. Yes, so it was sink or swim. In fact he ground the gears once: he was standing up on the helm leaning out pointing where we should go, and he accidentally knocked the gears out and the boat ground to a halt and a *huge* wash of water came over the back. We could've drowned at that very point—the boat would have just bubbled down to the bottom. From that you actually learn. You learn: that ain't going to happen again.

As soon as we moved to the coast, I loved L.A. I didn't like it up in Pasadena because it's desert—baking hot, no breeze, and carbon monoxide. It's very much like Beijing, but obviously not as bad. As soon as we moved down to the ocean, that was it. I realized—seriously for the first time in my life—how much I loved the sun. It's a beautiful thing to wake up at five, six in the morning—sunrise! It does wonders for me. I love being alert in the daytime, and well tired and exhausted and ready for sleep at around 10 p.m. That's the way I like it. I'm really not one for staying up and watching the late-night chat shows. I don't see any joy in them things, I see them as formats and incredibly dull.

L.A. became the perfect place, because if you look at it on paper, you'd think, "Oh my God, there's no reason to be in that environment at all. It could never work, it's the last bastion of hippiedom and mellowed-outness and sensible 'new-age food.' " And I found it great, refreshing—the idea that you don't need to stay up all night! It's equally if not more entertaining to get up very early in the morning. For instance, that's when all the best CNN reporting goes on, before they censor it. That's an absolute truth, to this day. If you catch the CNN news reporting early in the morning, it's far more open and detailed than by the time it reaches the afternoon, because the censors have come in, clipping and editing so there's less information in it. That's equally as rewarding to me as noshing it up in a nightclub.

Mainly, living here is all about the ocean. I just love the sound of the sea, and being near it, and also being wary of its terrifying power. Ocean-bound changed everything. We got into boats, from those short little cruise trips to Catalina. There's nothing more glorious than, when everything just seems to be grinding you down, getting in your damn boat, going out, losing sight of land, and working your GPS all the way back to shore.

It's what's needed, because the pressures of what my life has become can be overwhelming sometimes. Soul-shattering. Physical exhaustion is one thing, but mental stress and exhaustion, that's something else. You just need to be reminded that you're here to live. Not to work, to *live*. Work is a pleasant intrusion—keep it that way.

It's kind of like the weather's too nice to be walking around pretending to be angry all the time. Let's face it, aggressive clothing really is for colder climates. It's really hard to go round being angry in flip-flops and beach shorts. Though I hate bloody flip-flops! And then: speedboats and sunshine—you show me a working-class kid who would say no.

It was here in California that I got into my nature side, the love of the wildlife. If you just sit still and calm down and stop having to rush around the whole time, you'll find that the bunny rabbits will come up to you! And that's quite nice, because you don't have to kill them, you know, because you don't resent them.

Then there's all my little nature jaunts. I mean, I still love TV watching. I'm not out there admiring the dandelions all day long! You should never overdo any one stretch. Don't make your life a prison, and don't get locked up in a routine, or else it can get like, "Oh, I'm off for my nature walk again!" Again and again, it can become a drudgery, when it should be an excitement.

I learned to ski, because Nora could ski already, and she said, "Oh, you should learn it." For years I put it off, then finally one weekend we just drove up to Squaw Valley in Nevada, and I loved it. I learned to fall down a mountain at various different speeds. There's no sense of losing your dignity just because you fall over a lot. In fact, you're making people laugh, and what's wrong with that? I love it, because everybody, I don't care who they are—they could be an expert all their life—they're gonna end up on their arse at some point. We've been going for years now, and there's no sense of improvement, and indeed no care for it either.

With all these outdoorsy things, when you just leave the pressures of living in a town behind you, and all those daily business dealings that seem so overwhelmingly important, you become less anxiety-ridden, and you find the answers. In fact, you find a great deal of clarity just being an isolated nobody out there in the wilderness. It does work. I can well understand, for instance, not the weather aspect of it, but why people move to, say, the craziness of Alaska and live hundreds of miles away from anybody. I *can* understand that. I know what it is that makes them feel so content. But at the same time, too much of that would drive me crazy. Nora and I? We're three days of isolation together, not three weeks. We quickly get that out of our system and then come back to the drudge, which doesn't seem like a drudge anymore, it seems exhilarating and exciting.

I can't seem to get a suntan to save my life, though. I'll just burn. Two days, I'll think, "Aw, that's looking great," then it'll start to peel off, and then I have to deal with freckles. I'm just naturally too pale for life. I suppose that may be why they liked me in Japan originally, just because I was so dead white. That's their vision of beauty, isn't it? Death white, with the blood drained out of you. I don't suppose amphetamine stopped the pallor any. My only problem with amphetamines is, I never got around to doing enough of them. That's been a terrible waste in my life. It leaves a sense of longing. He said laughingly. There'll be some fool who'll read that and take it literally. That's the world we live in. Humorless fucks. Quite frankly I wish they would take it literally, it'd give me endless hours of entertainment. Much better than a line of speed itself.

But, oh God, I still keep getting these colds. My next-door neighbor has just cut the grass, which has added to my agony. I get up very early and I'm straight on the bloody over-the-counter medicines. Because here in L.A., if the wind blows, there goes pollen season right up my nostrils. There are times when I've got rips around my eyes, because they're so teary and itchy, I just have to get my fingernails in there. I literally want to tear my eyes out.

I tried all the usual nasal sprays, but what I use these days is mostly saline solution, salt water, just up your nostrils. That or a highly chlorinated swimming pool, that seems to work, it just burns it out. Dying my hair actually is a good way out of it, because you can't help but breathe in the fumes of the bleach and that seems to be useful. So my nostrils are a lighter shade of pale on the inside! Prescription stuff never works, and it just makes me really down and tired and lazy, and I hate that because I'm naturally run-down and lazy anyway.

After *Psycho's Path* it seemed pointless to try and be making records because there was nowhere for them to go, but I put out quite a lot of music incognito. It was mostly dance-y things, instrumentals, and I deliberately kept my name off them. I wouldn't put my voice down: I didn't want anyone thinking it was like, "Hope you like my new direction." I went through a period there where I thought my name was poison. But I love a good dance night—you know, going to a proper dance club. It was nice to hear some of my grooves were there rocking the dancefloor, and nobody knew it was me—there was a very nice enjoyment in that. Making people happy, but they wouldn't have been so happy if they knew my name was attached to it. Now, I might be wrong here, but that's how I was feeling at the time.

I don't want to name names of tracks or anything. There were no major hits, but plenty of dancefloor hits. I liked the anonymity.

In fact, that whole rave culture—the anonymity was what I loved most about it. It's a shame it turned into superstar DJs, but them early-ish days were fantastic fun. And I could understand fully people saying it had a punk DIY ethos about it, that anyone can do it, and I loved it for that. And indeed anyone *did* do it. Hello, I can be anyone.

The freedom in that was fantastic, so why go back and stir that pot? I thought I made a mistake when I did that record with the Golden Palominos, years before. I didn't want anyone to know I was on that—it was a version of some old California-scene record. I was really annoyed when I got back to London, and I ran into Mick Jones of the Clash, who is always good company, and he went, "I've heard your latest release." I was furious! I didn't want it to be known. It's an attitude I have which I think is actually healthy. He'd bought it because he knew I was on it and that was kind of defeating the point. I was experimenting, but it was just the name that was the pulling power.

It's a dilemma of sorts: should I namedrop when I release a record and remind people of who it is, or what I am, and what I've done? I don't know. You'll always get the record company wanting to do that—and always the manager! For me, I was quite happy when I set PiL up proper in California, and it was presumed that we were just a brand-new California band—for quite a long time! Nobody out here had any knowledge of us having anything to do with the Pistols at all. It was utterly fantastic. But when that cat got out of the bag, then I started doing "Anarchy" live, and I got bored with it, because I thought the audiences were coming in for all the wrong reasons. It's amazing that I've achieved two entirely separate audiences in my life for two major bands that I've been in, that for quite a while there—particularly in the States and certain parts of Europe—people didn't realize were fronted by the same singer. So I was both Johnny Rotten and Johnny Lydon.

These were the things I was dealing with. How I could "get away with it." Just that sentence alone, there has been an awful lot of that about me in the music press—thinking that this is all some elaborate hoax or joke on my part. Well, the joke's not on you, if that be the case, the joke would be on me. I don't see what I do as a joke at all. I see it, for me, for my own personal point of view in it, as insightful, not only into how I work and operate as a human being, but how you all do too, by your reactions. By you, I mean the broad expanse called the human race.

Of all the reasons that I stopped making music, the most important was the arrival of Nora's grandchildren, Pablo and Pedro. In 2000 they suddenly came to live with us indefinitely. Their mother, Ariane, better known as Ari Up from the Slits, had been bringing them up in Kingston, Jamaica, and had more or less just let them run free. They were very wild, and they needed help and support. They couldn't really read or write, or even swim, at fourteen and a half, when they came to us. They had no comprehension of speech or formulating proper sentences.

When Ari would go on tour they'd be dragged about with her. If she had a new boyfriend in a different country they'd have to go and live there. They were in a permanent state of confusion about where they belonged. In Kingston they'd be surrounded by a changing lineup of women and boyfriends—a very confusing position for boys trying to grow up.

They'd stayed with us before, many a time, and we were well accustomed to each other. Nora and I certainly hadn't been planning on being substitute parents, but the bottom line is, the twins really needed help. You can't have any kids in the house and not be paying attention to them. These two were especially needy. So you dedicate your life towards their life.

The whole scenario came right out of the blue. Ari had moved them all to New York and a situation exploded. She had a huge row with them over money that had gone missing in a New York apartment. I think a great deal of that was to do with the boyfriends and her own clique, and her having no sense of a regular dinnertime or food or anything for them.

Ari rang Nora and said, "I can't cope with them, I'm throwing them out of the house!" And Nora said, "What do you want me to do?" We talked, and while Ari was still on the phone I said, "Send them here immediately, we'll take care of them. I'm not having those young human beings abandoned because of your lack of care." It was a big row with Ari. It made life very difficult between me and Ari for a long time. But it always was difficult between her and Nora.

My heart just broke for all three of them—the twins and Ari. The twins needed a family unit at this point, and God only knows where Ari would've sent them if not for Nora and me. They would have been cast off to one of her many distant friends. Even they were apparently all telling her that her lifestyle was incompatible with raising children. It's very hard for a young mother to be a pop star—I use the term loosely—and pursue a career and, at the same time, particularly in Jamaica, 'fess up to the fact that you've got fourteen-year-old twins at home. You're not twenty-one, regardless of what you tell the press. She would deny lots of things, and it caused great pain to the twins, as kids growing up.

It was very hard for them when Ari's out there living her Rastafarian dream, being a woman warrior, "save the dolphins" and all of that, but showing a complete neglect for her own children. She just didn't have the time for them—as a single parent with a libido. That's very difficult, that aspect of life. "Meet your new dad."—"No, I won't!"

For so many reasons, being a single mum was too much for her. She found Pablo and Pedro to be uncontrollable. For instance, she'd put their hair in dreadlocks. "Mummy, I can't live with this hairdo, I'm getting bullied at school." Ari could be quite dictatorial in that way, and very unforgiving and rigid.

It's very difficult with children; you've *got* to let them find themselves. You can't be inflicting dreadlocks on a fifteen-year-old. It's just not going to work, unless they're fully regimed in that particular religious dictate, which of course they're completely not. They're so non-religion! It's one of the things I'm extremely proud of in them.

I'd be saying, "Come on, Ari, Rastafarianism is a religion that doesn't accept equality for women!" All this time, Ari was living off Nora. All this "I'm free!" Oh yeah? Somebody's always paying for that kind of freedom, and unfortunately that person was Nora, this constant financial drain on her.

The twins' names were some ridiculousness that Ari came up with. She didn't want a name from the Bible, so she found Pedro and Pablo, which is indeed Spanish for Peter and Paul. *Aaaargh!* Hippie-minded liberal parents can cause such problems to their kids. It's been very difficult for them, because they're not actually Mexican or Hispanic, and living in Los Angeles, that's a real problem, because it's automatically presumed that they can speak Spanish. *Nooo!* They'd come out with Jamaican patois as an answer, and it didn't go down well. But Ari wasn't to know that problem would arise.

We quickly got them into school, and it was very difficult for them to catch up, because they were basically illiterate, really way behind. Ari had this attitude that education was "Babylon system." That's all well and fine, but you've got to be educated enough to know that. In many ways, taking away a child's opportunity for education is absolutely *corrupting* them into "Babylon system," making them unemployable. And, in fact, antisocial.

It was very hard for them when they went to school here. They still had dreadlocks at that time, because Mummy insisted on it. It was hell on earth for them, with Jamaican accents in a Los Angeles school, with a high rate of Mexican immigrant kids . . . Confusion.

Nora and I differed severely with Ari on the dreadlocks. One of the first things the twins wanted to do when they came to live with us was cut them off. They were very upset with what they had to endure at school with that, because it labeled them, and it made them stand out for something they couldn't quite justify. It wasn't their own belief system. It was also extremely uncomfortable in the heat to carry a bag of hair dragging down the backside of your head. I think all religious aspects in any human being should be self-determined, and so we gave them that freedom. We gave them permission to snip 'em off! It was the happiest day of their lives, not having to be dressed up like Victorian dollies, according to Ari's whims.

Of course, that drove Ariane nuts-crazy-bonkers. She was furious. I'd be the big bad bunny in all of this but, you know, you've got to free yourself up from the dictates of parents at a certain point. You know, it's their life. At sixteen they had the right to determine these things, and self-image is very important. It's where you start declaring yourself as a human being. Of course they're going to make foolish mistakes, and look like daft twats, but that's the privilege of youth and you can't take that away. You can't treat your children like they belong in a monastery. They're not property, they're actually human beings of self-determination, if they're lucky enough to get good guidance.

Poor things, they went through hell there, but they gave us hell in return. It was very hard for them to adjust to the kind of solidarity that Nora and me were offering them. It left them feeling very confined, when they were used to running in the world, left to their own devices. The reminder came from us: "Well, we're not paying for that, so that ain't gonna happen." What they needed was boundaries, as do all kids.

We soon moved out to Malibu for a better school, just to help them catch up. Up until that point in their lives, Pablo and Pedro had mostly been in free schools, and Montessoris, which run along the hippie guidelines of "Oh, you know, they'll work at their own pace, one day they'll just want to learn . . ." That—does—not—happen. It doesn't teach a sense of independence and drive—quite the opposite.

So, ugh, I did my best. I spent quite a while trying to do Maths with them. Me, of all people! I also did sentence structure, which is more my thing. Then, as that progressed with them in school, they started to learn things that I didn't know, and that helped them a lot. "Wha'? Ya don' know dat? Bu' I know dat!" I'd be like,

"God, you're so much cleverer than me!" Then the self-esteem would creep in. Mainly, it was just about teaching them a respect for others.

One time when Nora was away, I had to go to a parent-teacher association meeting. Haha, the argument I had with the English teacher, who told me sentence structure didn't matter! "What? You're telling that to a *songwriter*?"

So the twins have been with us all through the 2000s, and they're still wrapped around us one way or another. They've all gone different ways, and they're all now different people and not yet fully realized, I don't think. They were still in a complete state of ruin for many a year, to my mind, and it's a problem that they're still to this day trying to work out.

Also what happened was that Ari's younger boy, Wilton, came to live with us. He also doesn't understand what the guidelines are, because there's never been any. "That's what *I* want!" "Well, what about the rest of us? Why should we suffer because of what *you* want?" We've been trying to introduce that sense of empathy with other people.

You have to know that I was never being spiteful or resentful or jealous towards Ari. She may have been an inspiring stage persona for many people, but it's what goes on when you get back home that really matters. It's hard to get people to understand that, but such is the simplistic nonsense of pop music. It's a beast of our own making, all of us that are involved. It's our own fault that we don't show a more open side, but it's very difficult because you're judged every time you walk on a stage, you're judged in every interview, in every public representation of yourself. You need time out, and that only comes when you get home and, unless you do a Kardashian on it and have cameras following you about morning, noon, and night, you're never really going to understand what the full experience is. Indeed, then the cameras become your reality and that is in turn a non-reality. Again, Kardashians.

Ari and I always had a deep respect for each other, always.

She said many things over the years, but who cares? It was very important that I went to see her in the hospital in 2010, the day before the cancer finally got her and she died. What a fantastic, amazing reunion it was—we sang together. The staff were very generous with us, because we've both got big mouths, so the songs we were singing together were loud! "Four Enclosed Walls" was the major feature, with the "*Aaallaahhh!*" I was trying to get to grips with where the notes were, but Ari was a very good singer and a very good musician. She got it bang-on, even then.

As a life pursuit for the two of us, parenting for Ari's kids was almost depressingly educational: you have to realize that they come first. That's very draining on the creative selfishness of being artistic. Because there is an element of selfishness in writing songs, the luxury of saying whatever you please. You have to get a little more in tune with the domesticity situations, and understand that you might not be quite bang on the money in that department. When I'm ranting and raving, there are other aspects to consider, and consider them I have.

If anything, it means I'm much more thoughtful now when I put a song together. I love hurting institutions but I don't want to actually hurt people. Yes, now there is some forethought, some perspective. *Whoooaaarrr!* How do you say, "I'm a very nice person," without dying of laughter? Quite frankly, if I wait for others to say it, it's never going to be heard. There'll always be that element out there imagining I'm just the world's nastiest bastard. And why not? I suppose there must be that element to me, somewhat, and thank God for it, because it's him what did it, Officer. Not me—blame God. That's a double-barreled gun, you see, that whole thing of "God made me, I'm an act of nature." "Thank you! Then blame God, it's him what did it . . ."

The house was certainly cozy with the kids around. I loved that time, because it gave me that sense of family. I love kids around. I'm one of those people—not always, but mostly—who doesn't mind the screaming and the shouting and the squealing and the yelling, so long as it's happy sounds I'm hearing.

However, the situation with Martin's family living in the guesthouse in Venice Beach did get unbearable after a while. Sometimes, trying to live with family members is a waste of time. Because you all end up at each other's throats and the over-familiarity can lead to all manner of problems. There's no hate in this, it's just that when Martin developed his family around us, and then the twins arrived—the whole thing just piled up into a zoo-like atmosphere. You try concentrating in that. Having a recording studio built into the room between the kitchen and the front room is like trying to make a record on a subway station.

So things fell apart in that way and, at the same time, the necessity to make records in my own living room became less and less interesting. It becomes too much of an infringement on your personal life. You have no escape from it. There's moments when you've run out of ideas and you need the pressure lightened. A moment of reprieve which you don't get when the equipment's all wrapped around you. There's nowhere to turn, so it's not a good idea, home recording. Not unless you've got a shed at the end of the garden—something not attached to the place you actually live in. Maybe get an allotment. Because it turns you off the very thing that got you involved in it in the first place—the love. The love becomes cumbersome, and that's not right.

Martin's got two children, and they're both now working here in Los Angeles. Martin himself just does odd jobs here and there. Very early on, he got American citizenship. He decided he never wanted to go back to London. There was nothing there for him. You know, "No future." Quite literally, just no jobs. Nothing. No interest. There's no way out of the trap. The tedium of council flat existence—it's an incredibly unfair universe, Britain. You're led into a life of crime, accidentally, against your wishes. It's the only way to make the money.

So Martin's seen no reason to go back, and never did. He's happy here, and he turned out to be a really marvelous father. It's wonderful to watch. As for the eldest of my three brothers, Jimmy: as madcap as he is, he's a wonderful dad too, and so is Bobby, the middle one. These days, Jimmy's painting and decorating, any old job that a working-class lad can get up to. Bobby's moved to Northern Ireland. He married a Northern Irish girl—out there, he repairs burglar alarms, he did a bit of plumbing, he's a bit of an electrician—a very technically minded and quiet person, if prone to the killer one-liner. Wit does run in the Lydons.

Like Martin, Jimmy and Bobby both raised kids really well. Smart kids, not one of them is a duffer. I'm really proud that my family endures and brings forth very good people. That bodes well for humanity in the future.

Me and Nora, we couldn't have children after problems when we were young. Nora had a very difficult childbirth with Ari and, to cut a long story short, it hasn't been possible since. Looking back, our lives wouldn't have allowed it, anyway. I was constantly touring in them days with PiL, just constantly too active to give the proper attention to what a child would need, to be in this world. It's 100 percent, a kid.

Rambo never saw himself as management material. I always pushed for him to step up to that role. He started out just wanting to do security, but eventually, round about the year 2000, he moved up. As situations developed, the opportunities were there, and I'd keep pushing him because I know he's got what it takes. Anyone who can organize a coachload of football scoundrels can do anything. To keep that well in order, that's a skill in itself.

In the past, before he came on board, situations would arise where I'd just go, "I can't cope!" and just sit around for years. I got very tired with everything being done in a businesslike way. I find that if it's done by the book, it doesn't work. This kind of industry is not the place for set rules. Problems can shapeshift on you instantly and you have to be able to react to that and not just be pedantic. That's not slagging anybody off, just me myself—I can't work with a strict businessman, it slows the process down an enormous amount. It becomes very tedious.

Rambo's a stickler; it's one of the greatest things of working with him. I can tend to fob it off with an ill-phrased expression. With me, the paddy can lurk in there and I'll hum over the details. That can lead to confusion. Sometimes, what I see as relevant is actually not. When you work with a bloke like Rambo, it's got to be 100 percent. His stamina and commitment are irrepressible. He holds himself fully accountable, and that's an amazing concept.

Doing my security, he was the best. He taught me a lot. Like, you don't want to leave yourself open 360 degrees, so you narrow it down. When I was on Skinner and Baddiel's *Fantasy Football League* on British TV in 1998, it all got messy. I weren't gonna toe the line with those pompous stuck-up alleged comedians, mocking old England players, turning football into the farce that it's become. In the press the next day they even started accusing me of attacking a producer.

What really went on was that I wouldn't play by their rules—they didn't like me outsmarting the presenters. When I went off during the break they locked the doors and wouldn't let me back on the show, so they brought in their security which involved all these lumps trying to come and smack my head. Rambo quickly took charge of the situation—in the corridor there was a right turn, so we went in the corner, and they couldn't get at us, all these professional lumps. We stood our ground and held firm. There was only one way to us and that was full-frontal into us, and they couldn't do it, they couldn't risk it—*wouldn't* risk it. It's football-crowd skills, but you can apply that mentally. That's a physical reality that you have to deal with, and that's a very useful way of lowering the odds of getting hurt. We slowly marched out the building with our heads held high—with a "good night and a good evening" to the security.

Rambo's a difficult fella for you to get around and understand, but I'll tell you what, once you do, you will love him forever. He's a fucking diamond, a man as good as his word. There ain't nothing wrong going on in him, and he's the most hardcore fuck there is. He knows me not as Johnny Rotten, he knows me as John the human being. A small amount of people would just have him down as a thug ambassador, but he's not, he's much, much more than that. He's John, the Rambo.

Around the time that Rambo took over, I was getting more into the world of film and TV—far more than I ever did, ironically, when we were talking so much about it in early PiL days. I got this offer to do my own program from VH1, which was like an MTV for older viewers. Given the chance, I obviously jumped straight in.

Immediately it came down to scripting. The stuff that was put in front of me—ha! I kept some of it because it was so silly, but I look back at it and think, "Thank God I avoided that!" So basically I approached the head office, in a big board meeting, with the idea that there'd be no scripts. Scripts are not my thing, they're too limiting. I don't mind a rough idea, but as for writing dialogue and expecting me to *verbiage* that back at a camera? That's never going to happen. It doesn't work like that with me.

The leeway they gave me was this: okay, do what you want, but we're not paying so much for it. "Okay, that's fine with me"—I can make an excellent, low-budget work. I loved the challenge. I also liked the people I had lined up to work on it with me: a guy called Rob Barnett was very good fun, and a fella, Jay Blumenfield, who's gone on to become quite a big producer and director. Together, we just had fun with it.

The most difficult part was putting together the presentation reel. That was a pretentious lack of fun, limited to this studio-type warehouse place which had a big multipurpose screen at the back. I was expected to walk across in front of different images projected on it, waffling lyrically about why this was a must-see program, with no actual programs already made. It was like advancing the theory before you even have the theory. I understand that for people forking out money, you're not gonna be stupidly risk-taking willy-nilly, so I'm not moaning about it. I kind of like budget restraints, I can work very well in that, but I cannot work under content restraint. That I will never accept.

Again, it came down to problems about my dialogue. "Here is a list of words you cannot use!" It was all the usual ones: "fuck, damn, cunt, bastard." "Twat" is a big word not to use on American TV. To me it's a casual expression: "Oh, shut up, you silly twat!" Here in the States it means a vagina. I suppose it does in Britain too, but not to the same degree of "I've never been so mortally offended." The level of angst the Americans place on that word is *ridiculous*.

In the first proper episode, I sat in a tank and exploded a blow-up doll covered in priceless Sex Pistols memorabilia. That was one of the best things. I've always wanted to be in a tank. I used a wonderful Alan Stivell theme tune for the show—I love Alan Stivell's music; it's Breton Celtic, very old traditional folky stuff, but with electric guitars—used in a bad way, sometimes, but mostly very interesting. It can be quite cinematic. And it blended well with a bloody great big tank, let me tell you. Plowing up a field with all manner of armored vehicles was just a dream come true. We picked out targets, and of course it would be blasts from the past, like some of the no-no's of the Sex Pistols—like Sid's alleged suicide note. Why not throw bombs at all that? If anyone can, I can.

I really would have loved a studio audience, but what it turned into was filmed escapades—just going out and doing interesting things, and trying to find people worthy of a conversation. I could talk to the Devil, but once the camera's on, I find I'm looking at this person and I'm not the slightest bit interested in anything they have to say! I don't want to hear the same old answer that I've already researched in my notes. I like the element of surprise and unfortunately with cameras, people become incredibly unnatural. I know this happens to me from time to time. They make you dive into a box of phobias.

On the other hand, I had the opportunity to turn the coin, so rather than be interviewed by what I thought were halfwits, as had been my experience for twenty-five years, I'd actually indeed be doing the interview. It was role reversal, and I found that I didn't like it at first. It's something where you have to dig deep inside yourself to find the motivation. Now I'm perfectly fine, but at the time it was all too much pressure on me. It seems silly to now think, what the hell's difficult about sitting down and asking a couple of questions out of anybody? Well, at the time, quite a lot, because all my worries were "My God, look at that septic spot, I'll be laughed at." And, you know, "Are my ears too big for a camera close-up?"

There was definitely an outbreak of spots when we went up to the Sundance Film Festival. I had a serious food allergy at the time, and it only got worse freezing up there in a fabulous ski resort in Utah. We were up there because the Sex Pistols documentary, *The Filth And The Fury,* was getting its premiere at Sundance.

A lot of work had gone into that movie. We dug out some amazing old footage, and we Pistols all appeared as blacked-out silhouettes on camera, which I think was just down to the low-level lighting that was available. No, I jest. The idea was to do it like a *Crimewatch* kind of program, where informers are blacked out to protect their identity. It was the same kind of thing as the *Never Mind The Bollocks* cover looking like a blackmail note—that teasing with criminality, but hopefully never actually being criminal—except musically! The thing is, we're seen in all the video clips, and everybody in the world knows what we look like, so why give you more of the same? Try and put a fun twist on it, as opposed to Malcolm in a bloody rubber face mask.

I think I was very open in the parts where I was interviewed. The tears for Sid—I *meant* that. The death of anybody bothers me, particularly friends that I felt really close to. There's no point in trying to fake a thing like that. I'm like that, I cry like a baby at funerals, I cry like a baby at the death of anybody. It really gets to me. I feel a terrible sad loss even for complete strangers.

I didn't think it should come as a shock. People try to chisel you into a cartoon image of yourself. The narrow insular selfish little git that people wanted to believe Johnny Rotten was. Mr. Annoying Man. My songs were echoes of rebellion and empathy for people, and certainly not as the work of some sneery, selfish little turd.

Anyway, so through *The Filth And The Fury* we basically blagged our way into Sundance, and documented it for *Rotten TV*, all wrapped around the most wonderful ski holiday I've ever had. At the hotel we booked into, I could ski right from the front door down the mountain and take the cable car back up. I'd do that three or four times every morning, then off to do a little bit of work.

The film was received extremely well at the premiere, although I really didn't think it would be. I was introduced to the stage by Danny DeVito, who was involved somehow with the money and investment—wowzers! There's a smart bunny! I think he's absolutely up there as a comic actor. What a giggle to be sharing the same stage as fellas like that. I'm starstruck just as much as anyone else. I've never viewed myself as being up in their leagues—although I'm far from humble!

He did a really funny interview for *Rotten TV*, but the only conversation we had that mattered to me, even to this day, is that he loves those old Dinky toys—little cast-iron replica cars. I'm so on the same plane with that. We talked Corgi, Dinky, railway sets by Hornby, all of these model toys.

While we were running around up there, I interviewed Aidan Quinn and James Woods, who were a lot of fun, then there was Cyndi Lauper and Christopher Walken—I didn't have much good to say to either of them two. I also managed to miss one actor. He's very big and popular, he'd even won an Oscar in that period—Kevin Spacey—but I didn't know who he was. I since found out, and I really love his films! They say he has a photographic memory, he can remember dialogue on one read—well, let's hope he don't remember me. When we went back looking at the footage, the VH1 lot said, "Look who you just ignored, John!," so we snuck in some text at the top of the picture, saying, "Oops, you missed one, John!"

VH1 asked me to do a TV advert for the series, and so around comes this crew to the house out in Malibu, and they turn up with a script, and I say, "That's all very interesting, here's what I want to do." I led them out to the backyard, where I had a lemon tree. I had an orange in one hand and a lemon in the other, and just held them up to the camera and went, "Oranges or lemons?"—and then I stuck my lemon into the lens. "Lemons!" That's it, that's the advert for *Rotten TV*.

For the third episode I went into the American electoral system very heavily. I traveled to the Democratic and Republican Party conventions and interviewed the likes of Newt Gingrich, the former Speaker of the House of Representatives, and Jesse Ventura, the then governor of Minnesota and a retired professional wrestler and Navy frogman! Only in America, folks. Ha ha. I loved that side of it; I loved talking to these people on a one-to-one basis. I found Newt Gingrich well-read, clever, devious, but bear this in mind: politicians have to be devious—it is their nature. You don't want a simpleton halfwit running anything at all.

At VH1's own awards, I met Paul McCartney. He was always Glen Matlock's hero, of course, so for that reason alone I'd developed a sort of instant dismissal of the fella. I met him at one of the after-parties and he was great, he was really, really open. He came up and went, "Hello, John!" I absolutely loved him for that. I'm always open to people that have an open heart, and I could see it in him. He's got really honest eyes—childlike eyes, really. That reminds me of myself, because I think that's the driving force in a lot of people in music, that we're incapable of corruption. It's what keeps us going on. It's a glorious thing to see that complete innocence in people. How long has he been in music? And he's completely innocent.

There was a situation there previously in London, around the time of the *Happy?* album in '87, when McCartney and his wife came face-to-face with me and Nora. We were going to my brother Jimmy's; it was a Saturday afternoon, and our cab got stuck in traffic outside Harrods. Up came Paul McCartney and Linda, running up and banging on the cab door, going, "John! John! It's me! Paul!" How embarrassed I felt at that time! I couldn't deal with it. I'm going to the cabdriver, "Quick, swing into a back road!" And the cabbie goes, "Bleedin' hell, I've seen it all now."

So, there was a bloke that was always chased by people, and now he's chasing me! He meant so well, and I did so wrong, and handled the situation so badly. I don't know what was in my head at that point. I should've just wound the window down, or opened the door. No—instinctively, I just fled. Nora's going, "What are you doing?" So, meeting him in New York years later at the VH1 party, there was a lot to say sorry for. I took the opportunity, and he said, "Oh, don't be silly."

In the end, only three episodes of *Rotten TV* aired, before VH1 pulled the plug. It's funny, because while it was running, they did a research poll to find out what kind of people would like my show, and oddly enough it wasn't young alleged rebels, it was housewives that were *really thrilled*, and liked the up-in-your-face-ness of it. So there's a lot to be said for bored housewives, they're not all for Des O'Connor. So there you go, the mums love me. I'm housewives' choice.

Afterwards there came all manner of shenanigans because MTV were moving in on it. They used to practically run VH1, according to the information at the time, and they wanted to use the format of *Rotten TV*, keeping the name, but replacing me with Beck, which of course I raised a huge stink about. I don't know if he personally was up for that cup, but that's what MTV were thinking. I wouldn't allow it. It's got Johnny Rotten printed all the way through it—how are you going to ignore that? So it got into a legal wrangle there for quite a bit, then it faded away, so I thought, "Move on!"

I found that making a TV show was a hell of a lot more complicated than you could ever imagine. The pitfalls and the calamities of the hierarchy in the offices is quite a serious minefield to wade through. TV people? They're a nightmare, they're so indecisive, and at the same time full of determined advice. And if you ever followed any of it, they'd be the first to give you the boot. They're all very good at taking praise when a thing works, but *phwoar*, they won't take the risk, they'll put your head on the chopping block and then wait to see the results. Very fake business. Full of lovely-dove, "hi darlings" and all of that. Backstabbing is the order of the day. You'd think I'd be well used to it. I'm not.

Many TV shows have since garnered that approach and turned out rather magnificent things, tapping into that element of surprise, which chat shows will never have—things like *The Daily Show* with Jon Stewart and *The Colbert Report*. That's the best TV in America at the moment, those two shows. I'm not saying that they copied us, but they're heading in that direction. It may be an example of great minds think alike; at least I hope that's what it is. If indeed I have a great mind . . .

What do our kind ever achieve? What do we get? Nobody's out there offering us an easy lift, so any one of us from the manor, if they do good, stand up for them. Although . . . a lot of bad things will happen in the manor. You *will* get the proverbial crabs in the barrel. If one crab gets to the top, the rest will try to pull it back down. That's the rampant hatred and jealousy of success that you get in Britain. Any American arsehole can tour in the U.S., by contrast, and they're a genius.

When I was offered an "Inspiration" award by *Q* magazine in 2001, however, it wasn't really that I personally wanted the award. It was just a nice excuse for a party, and I brought my dad along for the ceremony, along with quite a few of my mates. The cry went up from Reggie, "Johnny betta win one"—and Johnny did. I arrived in an old-fashioned "rag-and-bone" horse and cart, with every kind of old iron and junk you could imagine in the back, like broken bicycles and toilet bowls. What a fantastic day! We totally took over the place and made a massive splash. I gave the actual little award to Dad, and he put it on the front windscreen of his lorry. He was so proud of it. It breaks my heart to think of it. "Moi son won dat awoorrd!"

The Sex Pistols was still dribbling along. In July 2002, we wanted to play for the Queen's Golden Jubilee, but all that was available in London was Crystal Palace National Sports Centre—a run-down running track from some old Olympics in God knows when. Maybe it wasn't even an old Olympic site, just a running track. Still, we flooded that place with all of the alleged villains and hardcore Sex Pistols fans that Britain had to offer, and no trouble was had. We were friends amongst each other. Yet at the same time, all day, up on the hills there's the riot squad and their shields just waiting for something to go wrong. But—nothing—went—wrong. Hello! Don't need ya. We police ourselves.

In the official program that day, we mocked up an advert answering rumors out there that I was now making my living as a real estate agent. The idea came from a conversation with Scotty Murphy, who was working with us at the time, and was later conscripted into running our websites along with Rambo. Then all of us chipped in, playing up to it. I thought, "Well, let's really give it to them then." If you look at the pictures of the alleged properties for sale, one of them is a run-down caravan, the other's a couple of sticks of wood in an empty field at the edge of a cliff—absolute impossibles. Some of the nasties out there actually took it at face value, and believed that these places were up for sale. If I could sell them properties, I'd be a bloody successful real estate man!

HUGS AND KISSES, BABY! #3

NORA, MY "HAIR-ESS"

It was anger that brought my memory back, after meningitis. Anger at the nurses and the doctors and the way they'd talk tough with me in the hospital, and how they'd told everyone to be like that with me, because that would make me fight and get my brain to click back—rather than making me comfortable, so I'd settle into a happy nonexistence for the rest of my life. So anger became a seriously important energy for me.

In my songs, I'm trying to duplicate verbally the pain or the joy or the emotion that's going on inside my head. Whenever I achieve that—and this is no word of a lie—it takes me right back to when I first woke up in hospital and could not speak. I thought I was speaking, but I was just mouthing words. Whatever came out was a jumble, a juxtaposition of noises, squeaks, and bubbles. I'd forgotten language. That's the pain I was in at that moment, and that's how I want to be when I represent my songs live—right back at that moment of anxiety. Everything in me cycles around that horrible feeling. And, of course, the shyness too. I mean, bloody hell, they'd have to pull the sheets off me; I'd keep still for hours hoping no one would notice me, because I felt I didn't belong here, and I didn't recognize anything. "I don't know why I'm here, I don't know who I am, I don't belong here, I also can't speak any language they seem to understand. Why not?"

All this is really the core of my makeup, and it's something Nora understands absolutely intrinsically with me. She *knows*, and this is why we are so tight. I know her pains from childhood, she knows mine. You can't explain this stuff even to your best friends. And it's even harder to explain it to you now. I wouldn't be here without Nora and the support system she's given me emotionally.

The age difference between us matters not. We'll be together forever, regardless. Our hearts and souls are in the right place. You know when you've found your soul partner in life, you just know. We've had some of the worst arguments mankind has ever known. But we're the kind of personalities where, after all the insults have been fired, we can laugh at it because it's preposterous. When you really love someone, you can practice hate in an enjoyable kind of way!

You've got to bear in mind: me and Nora, we don't like the same thing musically. Sometimes we do. But I could never live with a mirror image, or a reflection of myself, and neither could Nora. She's actually quite athletic, and I'm not, and somehow that helps.

After we moved to Los Angeles together, we kept the house on Gunter Grove for a while, but eventually we decided it just had to go. The endless people that would parade on the front doorstep became overwhelming and very difficult to handle, so we moved out completely to free ourselves up. Instead, we bought a place in Fulham, putting our monies together. I put in what I had from selling Gunter, and Nora put what she had in, and from that day on we became a complete couple and have done everything like that ever since. What's mine is hers, what's hers is ours.

We get bored in L.A. from time to time, so then we'll go back to Blighty, and try a little bit of that just to lighten the load. This situation out here can become a bit sterile, which of course it always does with kids involved, because your life's all about them at that point. When we more or less adopted the twins, everything changed—it had to be all about them. Unfortunately their attitude towards us was, "Yes, we agree, it *is* all about us!" That's how you get tied down.

Nora and I are not ones for the "happy family babby thingy." Kids are great, so long as they're not your own. I might've had a different attitude if we did have children, but there's definite reasons we don't. So we're now firmly like, "*We* are each other's babbies, and that's it." Although, I've got to say, we both *love* children around us. And I'm world famous for throwing the best, best parties for children. I love it.

Every time I'll dress up, any excuse, even in the neighborhood, right up to this current moment in life. Here I am at fifty-eight years young, and I'll buy a box of Lego and fill the front yard with kids. And I'll be there rolling in the dirt with them or in the grass, I'll be playing—I love it. I think like a kid in these respects. I don't act the adult.

I'll dress up as anything, but never anything that will frighten a child. Never a ghost, say. I don't want to introduce the concept of fear. I want them to realize it's comedy. This whole world of people teaching their children to be frightened of the dark is very much the opposite of me. Embrace the dark—it's the best place to sleep in. There are no ghosts, there's just human beings—watch out for *them*.

Nora and I are both of the same opinion that we could never have pets, because that commitment means you can't travel, you can't just load up and shift when you really need to recharge your batteries in a different scenario. We're gyppos, but without the limitations of a dumb caravan.

It's not an inactive lifestyle we have. It's one of never getting too used to your surroundings, because then they get uninteresting and become almost prison-like. It's nothing to do with money—it really isn't, because Nora's very good at getting cheap flights. That's one of the most wonderful aspects of Nor': she won't have me spend money foolishly. "Why pay that for that, when you can get it for this?" Absolute discipline, but total mutual respect.

Other times when things get too much, we just go out in our boat till you can't see land, and then play with the GPS and hopefully find a safe harbor. The engines'll stop, and we'll look at each other—who put the gasoline in last?

You've got to bear in mind that in amongst all of this happiness, there's family members dying on us—not only mine, but Nora's—and having to deal with all the pain of that.

Nora's father ran a newspaper in Germany after the war called *Der Tagesspiegel*. There's a difficult thing for many people who're born into a wealthy scenario, family-wise, to get their head around: it doesn't necessarily mean any of that cash is coming your way anytime soon, if ever. In fact, usually, the wealthier the parent, particularly the father, the more dictatorial and mean they'll be. If you don't toe the line, then life ruination be upon you. It's a serious oppression you have to escape from, and so that's what Nora did. Utterly amazing to just kiss all that goodbye and say, "Fuck it, I'll get my own life." And she did, and eventually came to England.

Her father put Nora and the rest of the family—her mother and sister—through hell. He was a very argumentative, abrasive *führerbunker* of a fella. To my mind, he was someone that really didn't learn the lessons of World War Two, because he thought he could run a newspaper in the same way as that lot ran the country prior to 1945. His crowd were the big money people of Germany. He was quite politically tied in to people who in my opinion would be called corrupt.

Her father hated me. He'd read the tabloid rubbish, and being a press man himself he should've known better than to believe it. We had no contact, never spoke to him, never made any attempt, and just left it that way. If anything, that whole situation would've caused a problem or a rift between me and Nora, but our bond was so tight by that point, that all of these things were just silly acts of indifference to us.

Rumor has it that Nora inherited unimaginable money. Not true, it's barely imaginable. She is routinely described in the media as "an heiress," which absolutely has her in fits. She just finds this hilarious. We've got a standing joke that they've spelled the word wrong—it's "hair-ess." H-A-I-R. Because she spends so long combing it.

However, but for Nora not being able to pack a suitcase in time, we could have been on Pan Am Flight 103 that got blown out of the sky at Lockerbie on 21 December 1988. An hour before we were supposed to leave for the airport, we were nowhere near finished packing, so we canceled it and booked it for the next day and just went back to bed, because we'd been up all night trying to sort out suitcases. We decided that day not to answer the bloody phone either—just too tired from being up all night worrying about packing. By the time we did get round to answering the phone and checked out the message machine, it was just full of—oh my God!—family and friends presuming we were on that flight.

What a shock that was, knowing that we were minutes away from a wrong decision. We'd have been blown to smithereens, and for what? What point or purpose is the destruction of another human being? My view of terrorism is quite cold: if they're going to go to that extreme, don't be locking them up, give them the death they so wished upon others. And don't take your time about it either, push them to the front of the queue. Such savagery and poison, it's inexplicable.

The first person I spoke to was my brother Martin from America, who was meant to be meeting us at the other end. He was like, "Oh my God, thank God!" "*What* did you wake me up for?" That's when I went into the answer messages. So then I woke my dad up, I rang him back and he was extremely grateful, because that's a terrible thing, and then all my brothers, and then all my friends. Quite literally, the next two days was spent ringing up and apologizing for not being blown to smithereens, and not telling anyone we'd changed flights. The lesson I got from it is, tell everyone what you're doing all the time, and don't jump on a plane or not jump on a plane without informing everyone of your movements, because it's damn well irresponsible to do that, and the pain you can put people through. It's best they know a thing accurately, rather than let the imagination run wild.

When we found all this out, we were so nervous about it that we changed airlines. There was no way we were going Pan Am, not anywhere, not ever again. We'd flown Pan Am a lot in them days. Then of course all the rumors coming out in the press didn't help: that Pan Am was secretly ferrying around American spies and the CIA and assassination squads for the American government. The whole thing was just "Oh God!" This is a world not of our making, but unfortunately it's one that all of us have to live in. The spiteful, precious political views of a few. I'm very, very wary of extremist political or religious agendas; they are the world's most stupid and dangerous people.

Ultimately, you're at the mercy of life's luck, and sooner or later your luck runs out, whether that be a killer disease or a car-crash victim situation, or whatever. Or maybe you just run out of steam and croak it. But in the meantime enjoy it to the full. All the problems Nora and I go through, they're just problems—nothing can take the sun away.

We're the closest that I can imagine any two people ever being. It's beyond words. It's one of those situations that's very difficult to describe. I've tried in songs, like on "Grave Ride" from *Psycho's Path*: I used the backdrop of that horrible Bosnian war to explain my bond, and how the situation of a war, a calamity, separating us would be so earth-shatteringly destroying to me. I don't know if it's quite the greatest thing ever written, but it's a song that makes me cry, and I don't like to attempt to do it live because I know it would really hurt.

The idea of losing Nora is unbearable. And we're coming of the age now that we have to consider death, because all my peers are dying around me left, right, and center! I look at all of them, and I think, none of them have done anything like what I've got up to, and they're all kicking the bucket rather sharpish. I must have sustainability, and that has to be because of the positive influence of having Nora in my life. She is such a positive person.

Our worry is, how we are going to manage to die together, because if one goes before the other, it's going to be absolutely murder on the survivor. But the way we look at it, in terms of statistics, women have a greater longevity, so we should die at exactly the same time. That would be just perfect.

13
NATURE DISCOVERS ME

It was Rambo, my now full-time manager, who conned me into being a contestant on *I'm a Celebrity . . . Get Me Out of Here!* They'd actually tried to get me on to this British TV reality show before, but I'd backed away and didn't even want to consider it—in fact, I never even bothered to watch a single episode. Warning bells went off in my head—it's just fading celebrities wanting to get on TV for any old reason, like game-show panelists.

After numerous rejections I finally said, "Oh, all right!" I had no idea what I'd committed to, other than family and friends wittering on about a bunch of C-listers being imprisoned in the Australian jungle. Rambo, the git, kept saying, "Nah, it'll be good, John, it's something different, it will make a change from touring." So I went into it wide-eyed, and dumb as a plank of wood.

Of course, there was instant uproar: "What a sellout, he just wants to be famous!" "No! I'm already *infamous*! I'm doing this for *me* and my love of nature, you daft a'peths!" But thank God for people like Johnny Rambo because, after a huge amount of arguments, he ultimately reminds you that you've got to challenge yourself.

When me and Rambo arrived in Australia to do the show in

January 2004, there'd been a fake story put out in the press that I'd caused a scene at the airport in L.A. and hadn't boarded. All made up, absolute nonsense. Gold Coast airport was swamped with paparazzi and when we eventually made our way to the hotel, of course they'd given up the rooms that they'd reserved for us, so I wasn't staying at the same hotel as the rest of the cast—the Versace, on the Gold Coast. Fine, anywhere will do. Plastic Roman and Grecian statues really aren't my taste. It took them all day and way into that evening to find Rambo and me a pair of rooms, so what was poor old Johnny to do? The bars were open.

The next morning, there was a meeting, and the day after we went straight into it. That first meeting was ridiculous, because everybody was embarrassed to be in everybody else's company, and at the same time we were being fitted for our jungle clothing. I can tell you: not one of us would admit our real waist size. It was all done out in the open! Very hard. There was a great deal of whispering with the wardrobe woman. You'd bark out, "Oh, I'm 34 waist," then whisper, "Really, I'm 38!" And of course her being Australian, she'd yell, "Whassat? *38?*"

They told me I could only bring one luxury item into the camp, so I decided on a jar of Vaseline. I knew the ants could bite the living daylights out of you, if you let them, so the idea was to rub the legs of my bunk with Vaseline so they couldn't crawl up to do so. I knew everyone would take it wrong, and raise an eyebrow, but it was such a good tip they passed it on to the other celebrities, and a couple of them followed suit.

The whole thing immediately felt like a setup. It stunk of agenda. They'd put me in with people like BBC TV's former royal correspondent, Jennie Bond, and Lord Brocket—the living embodiment of the upper-class black sheep. I didn't know at the time that this is the bloke that hid all them Ferraris in a lake. What a dastardly cad!

I had no idea what to expect other than I'd feel like the odd one out, which is my normal state of affairs anyway. I got serious after-burns from their zip-slide on the way in. I thought, "Oh bollocks, is it too late to leave, because this looks like daft shit!" It was just a collection of people all moaning about their sorry lot out there. The whining and the whingeing and the weeping and the wailing and the gnashing of teeth. "My God, you mugs, haven't you ever roughed it? Look, the whole thing's a calamity, a farce, but the money goes to charity. What the hell are you moaning about?"

I didn't *not* get on with anybody. It was a bunch of socially inept people that somehow managed to like each other until competition raised its ugly head. I liked the girl Kerry Katona very much, I loved her energy, but I've got nothing to say about the so-called glamor model, Jordan, one way or the other—just nothing. As I said at the time, "It doesn't contribute." She wouldn't lift a finger: "*Uuuuugh*, there's no *wat-aaaah*." "Well, boil some!" "The fire's gone out . . ." "Well, you're next to it, sort it out!" She can't actually connect the dots. With her, there's nothing there, so there's nothing to bother about.

I liked the guy she started seeing in there, Peter Andre, the pop singer. Although it seems a ridiculous persona he has there, I think it's fairly genuine. He really is ridiculous! And happy—he brings no harm or hurt to people.

Their music-making was a horrible sham. For me, the moment of absolute terror was around the campfire, when somehow or other somebody pulled out an acoustic guitar. Oh no, camp songs! That's the very last thing on God's earth I want, so I just walked off into the wilderness in the dark. I could still hear it from way off out: Jordan rehearsing her new hit single, with help from Andre. I thought, what a setup that is. Their perception of what they think music is, and to think I was going to share in this moment with them—impossible! It all felt very contrived and unnatural. How indeed did a bloody acoustic guitar turn up? Cue Camera Two, introduce the prop!

By that point it began to dawn on me that, although it's nice and wild out here and all of that, and the animals are real—and the lizards and the insects, they're all *seriously* real—the situation is *not*, and I was losing interest in it. It just seemed like foolish escapades.

So how do I entertain myself in a situation like that, when there's allegedly nothing to do? I go and get firewood, I go and get water, I keep the fire running, I boil the water, I keep the water trough full. For me, that's great, it's an activity. I view myself as indolent, and that's actually the thing that makes me get up and do things. "Bad Johnny! This relaxation will kill you, get up and do something!" I do have conversations with myself, and in a situation like that I found that I had the most thrilling ones.

The actual camp itself was fairly dreary and overcast because it was surrounded by high trees, so you had no idea what time of day it was. That was, for me, very frustrating. I loved wandering away and just knowing that every beast around could bite and cause me a serious problem. I liked it—I didn't think I'd be able to do that. Fantastic—the wildlife, the snakes, and knowing these are all killer things, but just letting them swish by. They'd look up at you and let you know, "Don't fuck with me." And indeed I didn't, and so had a great affinity there.

Having the cameras on you all day and all night, on the other hand, was a very good lesson in how to let go of that false perception you have of yourself, and not feel the need to protect yourself, and just *be yourself.* You have no choice. We were on camera twenty-four hours a day—get used to it! What a great training camp it was.

I learned not to be camera shy. It was so damn hot there, I said there was no way I could walk around wearing all that clothing. I wanted to jump in that bloody pond. I didn't care what was in it—crocodiles, swamp rats, anything, I needed to cool down. I knew cameras were on me, and of course you're thinking, "Oh my God, they're going to laugh at my waistline here, and my fried-egg breasts!" You have to give that up and just be yourself. It's a useful tool to learn: all those body-conscious perceptions are fake. Other people don't view us that way, if we're happy. What you gather from the face is how you judge another person. If you can spot weakness in the face then everything crumbles for that person. And it doesn't matter what your body's condition is. If you're happy being yourself, that's how you will be seen.

You can get realism, or you can go to the gymnasium like Peter Andre. But where does that get him? Unnecessary muscles in all the wrong places. I've always seen the laughability of that side of how celebrities want to present themselves. I definitely do not want muscles attached to my person, because they *will* turn to fat. I'd rather my fat be seriously earned. I binge-drink, and I love it. I also love the hangover because it reminds me not to do that again for a couple of months. Look, I'm Johnny, I ain't no weightlifting gorilla.

Being comfortable in your own skin is the ultimate reward. I love that kind of person, and I love being in that kind of company, when people are the exact opposite of body-conscious, where all the gifts of being alive are all coming from the brain—that's, for me, a fantastic human at work. Crafting body definition—that's so ridiculous. Unless you're an athlete by trade, what on earth do you want to do that for?

I got lumbered with the seriously ridiculous challenge of trying to grab ostrich eggs out of an ostrich pen. It seems absurd, but that's actually rather a dangerous thing to do. These birds are incredibly birdbrained, their head is so tiny, and those beautiful eyelashes on them. I've got to tell you too, their feathery backsides are just *so* comfortable, but then you've got them bloody turkey legs of theirs. Wow, what a creature! However, they kick, they'll tear you apart, and their beaks are like serrated carving knives. My back was covered in bruises and cuts. The power of the ostrich.

There were problems raised about that after. The camp doctor said it was a dangerous thing they did there with me, because it was feeding time. When you picked up the fake halves of ostrich eggs, underneath was the birdseed. They're just thinking, "Why are you trying to steal our food?"

I obviously hoped, and indeed believed, that I was going to

be voted out. Up until that point, there was no idea of audience feedback—you had no idea how you were being perceived. When the viewers kept on voting me in, I called them "fucking cunts" on live TV. This time, compared to Bill Grundy, there were very few complaints. I think the Great British public actually understood, those were words of "Oh no! More of *this*—thanks!"

I liked these people a lot and could not accept the loss of any one of them, but secretly what was really driving me crazy in there was that the producers wouldn't tell me that Nora had arrived safely in Australia. She flew in separately a few days after me—the twins were still in our care, and sometimes you can't just "up bags and off." Beforehand, the TV company had promised they'd let me know she arrived safely, but they wouldn't even give me a hint.

It began to really grate on my mind. I said, "Is everything all right?" "We can't respond to that." "Oh—now what's that about?" Just a word of comfort would've been pleasant because I was well aware everybody else was getting little treats on the side.

Knowing about Nora's welfare would've been a great treat to me. I kept going to a little hut you went to and talked into a screen. Sometimes they would respond through a speaker, sometimes not. I just kept asking, "Why won't you tell me? You know we almost died at Lockerbie, it's really important that you let me know she arrived safely!" It went on for days and days.

That's why I decided to walk when I did. They were just trying to turn me into that horrific personality that they so wanted out of me. "Well, no, go fuck yourselves, bye-bye. There's nothing to win here." I never viewed it as a personal challenge to be the last one left. I was very sorry to see people go right from the start. I for one suggested that we did not accept anyone to be removed from the camp. I continuously kept this thing running that I wanted to beat the system. What could they do if we refused to leave? They couldn't starve us—we were starving already. One for all and all for one . . . but that's not the way celebrities work. Is life a party or a parting of our ways?

So I told them, "I'm out . . . now!" and down came an escort or two, and led me up to another camp at the top. Nora, of course, was fine. The following day they took me back to the camp to be interviewed. The presenters, Ant and Dec, were going, "Oooooh, why did you leave, Johnny? You should've stayed, you could've won it!" They really are sweet-natured fellas.

And then up came a TV screen, a prerecorded video with my dad on it asking me why I left. I thought "Bloody hell, Dad, do you really need to know?" Bloody obvious. I spoke to him much later and found out the reason he'd been so upset: "Whoi did yer leave, Johnny? Oi had money bettin' on yer!"

Of course, Rambo was disappointed that I'd walked, and Nora too, although she knew I'd be heartbroken not knowing she was safe. It turned out to be a great thing to have done, because I learned a lot about myself. I'm a survivor.

Whatever waist size I went in at, I came out a damn sight thinner. Did the pounds shed off me! It really thinned me down, in such a short time. I'd reached that thin-enough waist to wear designer clothing, and I hadn't had that since I was eighteen. I thought, "Wow, I've got to stay this size forever!" Of course, I didn't. The first thing you do when you get out is stuff your face.

I'm a Celebrity itself was positive, it was just the surrounding media nonsense I had a problem with. One of those British papers had followed Nora around before she left L.A. for Australia. We'd moved the twins into an apartment of their own at that time, and one time when Nora went over to see them some journalist had followed her there and wrote an article saying she was meeting her young black lover. "Hey, that's my grandson!" So that was the nonsense I had to deal with when I emerged.

I know that going into TV World, you're somewhat asking for it, and let's get real—Johnny Rotten is going to get it, *dans le chuff*, seven ways to Sunday, from these rags. They will always be looking for the nasty in me and that's just the way it's always going to be. But when it gets into my family and my friends and my personal life, that's then a wicked line that I think they have no right to cross. I don't view myself as innocent as far as a target for the press goes, because I've more than asked for it, haven't I? I don't mind my own head on the chopping block, just don't execute my family. Can you imagine what Pedro and Nora made of that, when they had to read that shit.

I don't think the public take too much of it too seriously, really. But there is the one so-called friend who'll ring up and go, "Oh, that's terrible, Nora's got a black lover," stirring it up. The problem is you don't have any real comeback or revenge (there's a word!) on these journalists. Ultimately, their driving force is jealousy—very evil. That whole thing of spying on people—that to me is unforgivable. To just try and trivialize and squander away another human being's life because you resent their popularity—it's very suspect. The lesson you learn is, if you keep away and you're shy and you don't get involved in that world, and you don't invite it by doing *Hello!* magazine covers, the badness of it wears off.

I don't read my own press, and I haven't been able to look at any of the footage from *I'm a Celebrity* either. When I got back to the hotel they had a showreel waiting for me. "No, John, you did good, you must look at this." "No! I don't look at myself, I *am* myself, and quite frankly I've had enough of me. I've got a whole lot more of a life to live, and I'm not going to get bogged down in admiring myself on camera."

I can't see any positives in keeping abreast of that stuff. It'll make you conceited and contrived, for one thing. Or it can really upset you and bring you down. There's enough going on in the real world to contend with, without that vacuous condition occupying a large part of your resources.

In the light of what's been revealed in the Leveson Inquiry into press misbehavior in Britain, I know damn well that my phone was tapped during this period. Everybody was tapped. If you are going to tap my phone all day long, you aren't going to get any dirt. I don't even use it as a chat line. I don't like the phone. I like one-on-ones. We don't do dirty dealings; there's no subterfuge in how we operate. So, fine—come one, come all! I should publicly broadcast our business line. You'll be bored by the purity of it all. The best they could hope for is a discussion on the room allocation for our Eastern European tour. Imagine them sniffling into that.

One thing I learned long ago about touring arrangements is to always have Rambo with me, and in hotels for us to have interconnecting rooms. We'd learned about all that sex-scandal nonsense in hotel rooms. As a result, I *never* answer the door, because that's how they catch you, and try and pin something on you. As a by-product, I can't order room service because John'll be aware of it, so that keeps me thin. Everything is for a point and a purpose.

There was one hotel, probably in Hungary, where we drove up and the driver said, "Uh-oh, that's the press waiting to catch someone." Outside, there was a girl in a mac, and it was obvious, even looking out through dark windows, that she had nothing on underneath. There was a cameraman next to her, so she was going to run over, and it would be a sex scandal shot. That's the kind of stuff that you can't have happening. So much of it is fake. There are certain hotels in London where you know damn well that's the kind of thing that may happen. We used to book in this one place all the time. It must've driven them mad. No joy from us.

Whatever you do in life, you know somebody is paying attention. So pay attention, because you're being paid attention to. A lot of them alleged celebrities involved in the scandalmongering were probably quite thrilled. They're really looking to profiteer from that angle, because it guarantees a media profile.

Sometime after I came out of the jungle I had a meeting in London, where the producers of *I'm a Celebrity* said, "Everybody's got a TV show from it—what can we do for you, John?" My reply was, "Oh no . . . please, nothing!" Their suggestion was along the lines of that I do one of those question-and-answer sessions, where a big celebrity stands up there and you discuss your life, and a celebrity audience fires questions at you. They wanted me to do *An Audience With Johnny Rotten*. So what did they think the content would be? "Well, if you can focus it on *I'm a Celebrity . . .*" "Goodbye!" It was the most disjointed evening dinner I think I've ever attended. I brought Nora and Rambo, and we just sat there and listened to this uncaring, unsympathetic, pointless vision of what they think I do. I don't run with the rest of that crowd, and I don't want to be perceived as needing celebrity more than reality, because, indeed, I don't.

Doing the show also gave me the opportunity to raise money for charity, and what a mad one we picked. Me and Rambo had both watched this nature documentary about a chimp sanctuary in Sierra Leone. In the show they followed this little albino chimp that had been rescued from the wild—it's tough for albinos out there, apparently—but at the end of the show the poor thing died. It broke our hearts so we signed up for money to go to the sanctuary to make sure the other chimps were looked after.

There was no great trading of phone numbers between us contestants. I've thought about it over the years, maybe I should try and ring up for a laugh—"Oh, wasn't that a giggle!"—but then I'm thinking, "No, it really wasn't!" Beyond two minutes, you've got another half hour on the phone going, "Pfff, you know, you're not really important!" And I mean that from both sides, we're not important to each other, so why try and make it something it isn't?

A few months after my appearance on this ratings-topping TV program, an alleged half sister of mine crept out of the woodwork. Was the timing a coincidence? I think not—it was glaringly obvious. Her story was that my mother had had her out of wedlock and sent her off for adoption, before my mum and dad got married.

From the off, she was too pushy. She approached my dad first, which I found odd, then she barged into my Uncle Jim's funeral in Ireland. I couldn't attend that one myself because I was on tour, but apparently she got involved with my Auntie Pauline there. So she'd approached Dad and Auntie Pauline, but neither of them chose to pass the info on to me. I only got wind of it when it hit the newspapers—she went to the media with her "story" before I'd even had a chance to talk to her, which really didn't feel right to me at all.

When you try to do this stuff through TV networks and the press, really, how can it be anything else but a money hunt? You know, at the very least, "Why don't you just hang back, and stop trying to push your way into our family?"

Around that period, *The Richard and Judy Show* approached us. I had been on the show a few months previous promoting something or other—I actually appeared on that show a couple of times—me and Rambo always got on really well with Richard and Judy, and the producers had been very friendly with us too, but this time they said, "We hate to say this to you, but we talked to this woman yesterday, blah, blah, blah . . ." So they wanted me to respond to this on TV, a situation I thought she should never have gone to TV about, because if I wasn't Johnny Bloody Rotten, it wouldn't be in the TV framework—so obviously your angle is a little crooked here, *babby*. And she'd made no approach to my other three brothers, by the way. She don't want to know about them, because they ain't got cash in the bank. I felt like, "Is this someone else claiming an inheritance from me if I kick the bucket? Sorry, love, I didn't grow up with you, I don't know you, I don't care. My family is my family, and my money stays with us."

Normally, I might've sat back and considered her situation, because I'm very empathic with adopted children, and I know the pain it can bring to not know who or where you came from. In our case, it's not a matter of getting DNA testing: apparently if the adoption authorities declare this as a fact, then so it is. I'm not sure it's wise to give adopted people information about their real parentage once they're adults, because it can create terrible pain to the blood relatives, and cast all kinds of doubts and aspersions in their mind. Ultimately you just think, "What's this woman really offering here?" Is she saying my mum had played around before she met my dad? Is that it? Is that what you're saying? This is an Irish Catholic family here. You're asking us to accept that? It's a very difficult thing to deal with.

Again, we go back to my mum and dad's wedding photo, and there's a baby being held by my Auntie Agnes, and of course it must have been *me*. Unmarried mothers had to suffer the whole Catholic guilt trip at that time—"Aurrgh, yurr're filt-y, yurr goanna be ostracized from the community!" Painful, painful stuff, and I personally don't need reminders. I'm sure you can now understand why I'm consistently involved with orphanages.

There's very little you can do to defend yourself once someone attacks you from a press angle. For me, that's like a full stop. You cease to exist. But of course, me being me, I would like to find that she was better than that, and there's every possibility that this may be the truth so, you know, time heals. Never say never, not when it comes down to things as important as another human being desperately trying to find out what it is they came from. But you're never gonna do that through *Richard and Judy*.

So, this woman pushed too many buttons at a particular point where it was very hurtful to us as a family. It's a weakness in me, I suppose, but maybe as the years drift by I might find an opening for her. But not just yet. It just came too pushy, too hard, and running to the press with it was really, really grotesque, and managed to offend all of us Lydon brothers.

I just have such a love and understanding of orphan kids. It's terrible, it's confusing, I'm well aware of that. She doesn't know who the father is; she just believes that my mother was her mother.

I was really worried that this would all be a terrible burden on my dad, at his stage in life. Through the 1990s and 2000s, I'd got really close to him. He was definitely trying to research me in terms of "Did you think we didn't love you?" It was a hard question to answer. If you think about it, that PiL song I mentioned in the first chapter, "Tie Me To The Length Of That," was probably through their eyes a rather cold estimation of their role in giving me life, with all its images of postnatal trauma. My dad always felt that I held him responsible for the meningitis I suffered, but I always felt—and this is a conversation that we actually had—that he thought I was faking the aftereffects, like my memory loss. That drove me crazy for years and years.

Dad had very curly hair, sometimes done up into a quiff. My mum was convinced he had a bit of Spanish in him, from the Spanish Armada. That's the Irish for you, going back that far! He was a very smart dresser, and always very precise about the way he'd dress up on Sundays—a very stylish fella. Most excellent ties, good suits—hard, tough, laddish, and absolutely impeccable detail—shoes polished to the point that you could see *through* them.

When he was young, whiskey was his drink of choice. One night, he almost hit my mother. I don't think I was older than five, but I flew at him, and he changed everything. He told us, "I'm your dad, I love you, and I'll never drink whiskey again." And, you know, he never, ever did. How beautiful is that?

Imagine how painful it is to go through meningitis and losing your brain, and when your memory of *that* comes back, just like a bolt in the head . . . You're sitting in the pub with your dad and you remember that, and you know it's not a lie, because you ask him and he tells you it was so. I looked at it this way: I almost could've lost my dad all the way back then; he could've been a dreadful, stereotypical Irish drinker.

All the times that I thought my dad was trying to ruin my weekend by having me out working underneath his cars, he was really actually trying to relate to me—not through words, because that wasn't really his way, but through situations. He was trying to bond. I wasn't capable of understanding that at the time. You can talk about Catholic guilt, but there's far more serious guilt trips, about parenthood and childhood and what the relationship between them really means. I look back not in anger but in sadness, and I regret totally that I wasn't as smart then as I am now.

He was still young when they had me—he didn't have all the

answers, poor fella. I'm not one to moralize or preach but I think young couples having kids at eighteen and nineteen aren't fully capable of understanding the problems that will come in the future, because they haven't fully realized themselves. But try telling the young Johnny Rotten in the Pistols anything at all! That fucker wasn't going to listen, was he?

For all the farce that *I'm a Celebrity* turned into, something else good did come out of it: I was on the radar, TV-wise. When I went in, I joked that I might get a nature program out of it, and, lo and behold, that's exactly what bloody happened. Some peoples out there watched it and obviously thought I might have something going for me there, and indeed I did. The programs I presented through 2004 and 2005 are my absolute pride and joy, work-wise, outside of music.

The first offer came in from Channel Five in Britain, to film two one-off nature specials in Africa, called *John Lydon's Shark Attack* and *John Lydon Goes Ape*. Swimming with sharks was an adventure I was really looking forward to. I'd been studying them all my life, ever since—I have to be honest!—I watched *Jaws* as a kid. Well, as a big kid—it was 1975 and, at nineteen, I was not fully developed, haha. As I've said many times before, if you're paid to do what you love to do the most, you're on to a winner, and I'm glad to say that's really happened quite a few times in my life.

Our five-week filming schedule for these two shows began in Cape Town in South Africa, where Rambo and I were to have a crash course learning to dive. We had to earn our open water PADI (Professional Association of Divers) certificates before they would let us cage-dive with Great White sharks. Which, of course, was the whole premise of the show. I have to admit, the pressure was on us: what would have happened if we'd failed our tests? It would have been a very quick show! It was a lot to take in, but we got through it, and I can truly say it was one of the proudest moments of our lives when we passed and were given the opportunity to see these fantastic creatures in the flesh.

We had diving suits measured for us, because they had to be of a certain thickness and fit perfectly, because the water is frigid. One thing just led to another: instead of the usual black garb and looking like a sea lion, me and John spotted these colors in the next room, which are usually used as trimmings. I ended up as a yellow and black bumblebee, and Rambo got Arsenal away colors—yellow and blue stripes, both with a codpiece. The South African ex-Marine who was teaching us to dive was great. I really liked him, but I'm not sure if he wanted to be near us in this gear—we did look a sight for sore eyes! Science has since caught up with us: people are now selling shark repellent wetsuits, and clashing stripes are actually regarded as an effective shark repellent.

I quickly acquired a love and affinity for being in the ocean. I'd always loved being *on top of it*, but now I loved being *under it*.

In training, we were taught about depth narcosis, also known as "the bends." On one of our ocean dives we traveled to False Bay, around an hour south of Cape Town. We were in about 80 foot of water, slouching around on the seabed, having fun, and I definitely went off on one—the closest thing would've been an acid trip. The colors suddenly became so vivid, and I just wanted to drift off into the depths and swim forever. Then came time to rise and you're supposed to rise slowly, to decompress as you go up, but I couldn't understand the ballast system, so I went up like a rocket. But for Rambo grabbing my flipper, I could have been in a serious problem when I hit the surface. While all this was going on, and I'm wondering why he was trying to grab my flipper—I thought he was just trying to annoy me—we were apparently being scouted by an enormous Great White shark. How do you miss 15 foot of teeth and muscle?

I wanted Rambo on camera with me, I should add. That was just the way it was going to be, period. I don't want things stagey-stagey, but as things naturally are. Why pretend otherwise for the cameras? But that said, I will fully admit when the camera was on me I made sure I was swimming at my best! The Channel

Five crew, I have to say, were well on board with my ideas, and had a very chirpy buzz about them. Any idea of a script was completely abandoned, and it was really a kind of a free-for-all—a series of misadventures, one fiasco following another—with the luxurious backdrop scenario of deadly dangerous huge sharks.

On the way to one of our first dives, someone had bought a newspaper, and right there on the front page was a story about a couple of poachers who had been attacked by a Great White while swimming. One was dead, the other survived. Jesus, that was an eye-opener for us.

We managed to arrange an interview with the survivor and went to see him in a shantytown in Gansbaai. When we arrived he was sitting outside one of the huts playing dominoes with his friends. It didn't look a particularly safe place, so we got the crew—with the exception of an interpreter and a cameraman—to stay in the van while me and Rambo went and spoke to the fella. But they were all very friendly to us and welcomed us in. The poacher told us how he and his friend had been out swimming for abalone fish. It was a six-hour round trip, but they were so poor they had no choice but to do it. On the way back his friend was attacked by the shark. The chap that survived had seen it coming and had been shouting at his friend to drop the fish and swim away, but it was too late, it took a chunk out of him, then came back and finished him off. He kept shouting and shouting at his mate to drop the fish, but he wouldn't do it. The survivor managed to get away but he then faced the daunting task of swimming back to the shore knowing his friend had been killed. Poor sods. It was a harrowing tale—I can't imagine what that must have been like. Horrible. He told us he'd never poach again. The whole thing certainly put us on alert, because it turned out the scene of the attack was not far from where we were set to do our Great White dive.

Over the next few days we had more practice ocean dives and I was also given special permission to swim with Ragged Tooth sharks at an aquarium. I even got to try one of those old-fashioned diver's suits with the big helmet and boots. How they ever ocean-dived in those I really don't know.

For the cage dive with the Great Whites we went to "Shark Alley" just off Gansbaai. Me and Rambo couldn't wait; this was the whole point of being here, and it was the pinnacle of my journey. Something that really appealed to me was the totally clear face mask with a microphone inside, so that they could hear me underwater. I didn't know, of course, that the damn thing wasn't working, so I was chuffed to high hell when I started singing the Morris Albert song "Feelings" to the Great Whites. Magnificent TV. I came up and asked, "Did you catch it on film?" "Oh yeah, mmmm." They wouldn't tell us until we hit shore. I was furious! It would've been the greatest thing in my life's work—singing "Feelings" to sharks on television. Yes, I am self-indulgent.

The boat captain really reminded us of the captain character from *Jaws*. He was friendly enough but he had a bit of an attitude with us and was full of himself—it was really hard to stand there and listen to what was coming out of his mouth. And his boat! It was as if it was held together by Band-Aids, the cage didn't seem securely tethered, nothing really worked on board properly. But it did the job admirably.

After we'd done the really serious stuff out at sea, we were invited back to a dinner with our entire crew, and the captain and his family and friends. The whole thing was being filmed and the captain and his friends were beginning to show off for the cameras. During dinner, he stuck his fork on my plate, and Rambo had to grab his hand. There was almost a fork battle! He ended up looking like a dribbled fool—his own daughter even got up and slapped him.

The trouble was that this was a fisherman's hut where they all go to drink. They knew we were coming and so when we walked in, it felt like we were being set up. There was a good fifty challenging fellas there and they'd all been drinking heavily, especially the captain, and so they were kind of trying to roughhouse us, but it didn't wash with us. As the evening went on, it became very unpleasant, but we dug our heels in. What made us leave? Not their nonsense. It was the first course. Can you imagine: *snails in cheese sauce*? Fuck that, let's go get something proper to eat.

South Africa had its challenges, but we loved every second of our trip there. One afternoon we had some time off from filming so we hired a helicopter to Isandlwana and Rorke's Drift, scene of two major battles in the 1879 Anglo-Zulu war made famous by the film *Zulu* and later *Zulu Dawn*. When we landed in Isandlwana it was all but deserted except for a couple of tourists. The most impressive thing about the site was the silence. The deathly hush gets you into a trancelike state, as your imagination takes over.

There was a monument dedicated to the British and Colonial forces that had lost their lives in one of the worst defeats the British Empire had suffered. And there were piles of rocks where every soldier had fallen. We climbed up to a cave where one of the last British soldiers had made his stand before he was killed. It was all very eerie.

We then boarded our helicopter to Rorke's Drift, which was about ten minutes away from Isandlwana, where a small force of British and Colonial soldiers had held off thousands of Zulu warriors. Eleven Victoria Crosses were awarded that day, the most in any single action. On our arrival we were greeted by dozens of African children—I don't think they'd ever seen a helicopter before. They were very excited to see us and thought we'd flown all the way from England. There wasn't much there except a small museum, which we visited, and the remains of the mission station and former trading post where the British forces held their line. We joined some schoolchildren on the grass and sat with them while one of the other children was describing and acting out the battle. John and John were back at school and loving it this time around. We then flew back to Cape Town to continue our filming.

However, from here, our itinerary was now to go into the jungle and mix it up with gorillas, which I dreaded, because I thought it might be the most compromising position I'd ever put myself into. I didn't know how to think or plan ahead for it.

Rather than fly from place to place, we wanted to drive, and there are no proper roads outside of Cape Town, just dangerous dirt tracks—rickety narrow passes on mountaintops. All insane and mental.

As a precursor to meeting gorillas in the wild, we were invited to the Ngamba Chimpanzee Sanctuary in the middle of Lake Victoria, near Entebbe, Uganda. It was a bit of a hassle to get there because that lake is pretty damn huge and it does kick up a good wave or two.

The Sanctuary was run at the time by Debby Cox, who was one of Jane Goodall's trainees, the world-famous chimp expert. Debby told us that humans could learn a lot from chimpanzees. What I learned was, they're bloody deadly dangerous! They are football hooligans, par excellence! They know how to throw rocks, as they did at us, and all the time they're eyeballing you. We were told that after two years old you can't go near them at all. You cannot train them, they will not have it. So maybe in that aspect we could learn a lot. But then I'm untrainable too.

The tension in their compound all stemmed from the fact that they were rescue chimps from the Congo, so they'd lost their family unit and they hadn't grown up fed by their mother's knowledge.

They were trying to stop them breeding. This one female, what a clever thing—it knew something was stopping it from having a baby, so it found a way of removing the contraceptive implant from the inside of its left arm. Every now and then they have to replace the implants, so they asked me if I would like to help carry out surgery on one of the chimps while being filmed. Then I was asked if I would carry her into the surgery, and I thought, "Ah, what a fantastic thing to do." They'd darted it, to put it to sleep—what could go wrong? I couldn't believe the weight. They are so well built and dense. I'm by no means a muscle-clad human being, and this was a real struggle. The cameraman was wetting himself, while I'm near death with this bloody great chimp. And then of course it peed down my back.

The young ones were very bumptious and joyous. They bounce all over the place, and move like lightning and love swinging off you and making eye contact with you. Those moments were deeply precious. I felt we were making some real serious connections here. One of them tried to imitate me talking to it; he'd be trying to mouth the words. I thought that was just incredible. But the hooligan element is always just around the corner.

The night before, Rambo had raised the question, "What if one of them chins me? What happens if I chin it back?" I'm in fits of laughter with that, but they will allow no such thing. They got all huffy—"No, you *can't* hit them back!" Then of course, in amongst the nippers, the next day one goes running up to Rambo, into his arms, and it's all nice and cuddly, but this other one got jealous and took a lefter at him—punched him in the eye! They'll get up on tree branches above you, and piss on you, it's like they're laughing their heads off at you. Little sods!

We were told not to think about going swimming in Lake Victoria, because of crocodiles and hippos—but worse than that were the deadly microbe diseases in the water, and also some weird snail that burrows into the sole of your foot, or your penis. Apparently, you don't feel it when it's doing this, but you know, you'll live to regret it. So what do we do with an afternoon off? We see a rowboat on the dock and go out, and the bloody thing's got a hole in it, hasn't it? We were close to shore, but all the big chimpanzees were eyeballing us. So it was snails in the willy or being gang-raped by chimpanzees!

We had nothing but two teacups to bail out the water with, because we brought a little picnic hamper. Debbie Young, or "Dobbins" as we affectionately call her, the director/producer, was in a state of mild panic. But it all worked out, because that's what you do in life—make fun of things, and get through.

Leaving the chimp sanctuary we sailed back to the mainland

and drove to Bwindi Impenetrable National Park, still in Uganda. We eventually got to the campsite for the mountain gorilla expedition close to nightfall.

The next morning, the serious hike began. We had a fantastic guide called Moses. I think the world of that man. Such fun, and all around just an excellent human being. I was like, "C'mon, where are they?" and he went, "Oh, just over there." Well, "over there" was seventeen miles away, up and down mountains, through forests and jungles—really seriously challenging to me. I'm fit in that I can howl my head off all night long, but not that kind of stuff. The only thing to save me was Rambo constantly going, "You can't stop now, the camera's on you!" There really is no business like show business. Pride is quite literally what dragged me through it.

The gorilla expert we were with, Ian Redmond, one of Dian Fossey's protégés—the woman from *Gorillas in the Mist*—was a chirpy fella, and constantly babbling on about his endless pit of knowledge. You knew this bloke was building himself up for a stupid pratfall. We got close to an old gorilla nest—they have to move every night, because their shit is full of fruit, and it attracts serious amounts of flies and siafu ants. Now, siafu ants are like warzone material: they'll eat you alive if you're stupid enough to hang around—and so the gorillas have to move.

Anyway, this fella goes into the nest, and of course—here's the killer about him—he doesn't wear boots! He has a machete, and he goes through the jungle in a big floppy hat and khaki shorts. The whole Crocodile Dundee approach. When he got into that gorilla's nest it was heaving with them ants, and he was screaming as loud as you could ever hope to imagine. They got up into his shorts too, and I think they were even having a good old go at his jewelry.

At the border from Uganda into Rwanda, where we were going to see the silverback gorillas, we were seriously held up. Inefficiency is the order of Africa, that's how it works, or doesn't—basically, they're waiting to be bribed. Until that happens, you ain't going nowhere, hour after hour. Soon this gang of little youngsters started to build into a sizeable force around us, and all they wanted, these poor little things, was *pens*. They can't go to school without a pen. We stripped our van looking for anything you could write with, and from there on in, we just bought huge boxes of Biros to give to kids wherever we'd run into them. Unlike the spoilt bastards in our country, these kids *want* to be educated. They *want* to get out of their dilemma, they *want* to improve themselves in the world—aah, it breaks your heart. There were quite a few kids with their arms half-hacked. Very hard to come to grips with that and sad they had no hope of hospital treatment.

Meeting the silverback gorillas in Rwanda's Volcanoes National Park was an incredible experience. And it *is* a meeting—you go into their company and you behave politely, because it's their house, and they were very welcoming to us. We got into their clique, and the big silverback male is sitting there, about 30 foot from us, and his big hand flopped out.

You're warned beforehand never to stare them in the eye because they'll see that as a challenge. That's all well and fine, but when you've got multicolored stripes in your hair, they're gonna stare at *you*. They're trying to say hello to you. There's always that moment where it's outside of the realm of a textbook. A lot of the time, these people that study so thoroughly actually miss nature at work.

Before we left, we threw a barbecue party for all the local kids and sent the message out to the villages. Only we didn't know that these villages had been at war with each other—that bit of information was missing in our plot. But at the time, there was a lull in hostilities between the Hutus, Twas, and the Tutsis—three tribes who used to hack at each other. Nor did we know that the black-and-white-striped penitentiary outfit that I'd chosen to wear was actually the color scheme of one of the tribes. I wore it especially, because I thought, "Ha, no one will have this." "Well, two hundred kids over there do, John!" Luckily at that point, hostilities had ceased. As soon as they saw Johnny Rotten being

Johnny Rotten, it was a gigglefest. We started sing-alongs, and all these kids started mixing.

It was magnificent and the singing was out of this world—just *embarrassingly* good. Then someone grabs you and pulls you into the middle and you're supposed to sing. "I can't! After all these years, honestly, I can't!" As if that wasn't bad enough a challenge, then it comes to dancing, and by God, they can dance. It became The Ministry of Silly Moves and Howling, and every tone out of my voice was just like shrieks of laughter. In many ways, they broke my self-control, and with magnificent results. This chanting started, and took off and echoed throughout the countryside—"Save—the—Gorilla"—a healthy message, instead of trying to eat them and sell their hands as Chinese aphrodisiacs. Such, all too often, is their fate.

The crew for *Shark Attack* and *Goes Ape* was one of the best bunches of people I've ever worked with. It was such a happy collection, no hierarchy in it, everybody chipping in, all the time ideas flowing.

Working for the Discovery Channel on my next TV nature project, however, was a very different business. *John Lydon's Megabugs* was a ten-part series shot mainly in the Southern states of America and, though I greatly enjoyed doing it, it was rather hampered by this ridiculous "wanting to stick to a script" idea, which is absolutely not my way. And I mean, you're dealing with Nature, for God's sake, you never know what's going to happen. You can vaguely have an idea, but from there on in, Nature has a way of doing what it wants to do—naturally!

If I'm to be a commentator on this kind of scenario, then I must be natural, too, but this lot would be handing me a script three minutes before shooting, and "Quiet, everyone off the set." "It's not a set, you're in a swamp!" "Can you do that walk again, but properly this time?" "Piss off!"

In their deadpan, dreary way, what was I supposed to be here? A two-bob David Attenborough? No way, stop it!

My attitude with cameras is, don't ever turn the camera off, and make sure it's got a two-hour reel in it. Do not turn it off, not for any reason, and do not point it away, do not get distracted no matter what is going on—*film it*!

A fine example of this was when we went to Florida State University, where a very crazy professor showed us some of the deadlier spiders. One particular spider was a Violin Spider, aka a Brown Recluse—its bite will make your skin rot and fall off in huge, gaping wounds that never heal. The prof hands me two little glass containers, with one spider in each, aaaaand . . . what does Johnny do? I put my head back and I put the containers on my eyes to look through the glasses like I was wearing lenses—only I tilted them wrong, and one crawls out and down my arm.

Now, the cameraman should have been filming, but what does he do, he puts the camera down to brush the spider off me. I was furious! That would have been a great bit of film, the proper thing of people watching, going, "Oh my God!"

We had another run-in over filming tarantulas in the desert. They wanted to orchestrate it, make it contrived. I absolutely would not do that. I wanted the genuine article.

We soon discovered there was a whole business going on with experts dealing in insects, so we refused to participate with that on any level. It infuriated the hell out of me and we had a major blowout with the production company, refusing to film one frame of footage until all that was stopped. Some TV shows fake these scenarios and fob them off to you as real. Sorry, but I think it's just as easy to *make* it real. Go to any patch of desert in Arizona, and it won't be hard to find a tarantula hole. The place is full of them.

I have this rigid set of values and I won't alter them, not for anybody or anything, or any amount of money. Because of this, I've found it very difficult to get a proper spot on a TV network ever since. Well, too bad, because I don't see the alternative as being acceptable. I would feel emotionally like a criminal if I went along with that.

Luckily, the science side of it had my attention to the point that I forgot we were supposed to be making programs here. I enjoyed the experts, these nutty university types. The mosquito episode, I really enjoyed. To stop things getting stale, there was that constant need to rev things up, so Rambo said, "Why don't you go in that cage?" In a huge netted area, they had something like 6,000 mosquitoes, and they were starving them to study them.

So, obviously, in I went, and Rambo said, "Why don't you drop your pants?" and so I did, and 6,500 bites later, I was cursing him to death. That evening was pure hell on earth. When we got back to the hotel, there was a hurricane coming in, and I was screaming, "Medic!" at top volume. The only thing we could find, when we walked down to a drugstore, was something which was like porridge oats in a packet. So I sat in a bath full of porridge oats for six hours with a hurricane rocking the hotel.

The next morning, I had a really bad earache—on top of all the usual moans, it was extremely painful—and a doctor finally came and told me I had a perforated eardrum. A mosquito had gone in there and perforated my eardrum and was still in there—dead by now, but in there, and he had to remove it.

My immediate thought was, "Oh my God, I'll never be able to sing again." Or—"I'll be able to sing, without the privilege of having to hear myself." I thought I was doomed—what did I do that for? How stupid. I had a few bites up my nose as well, that wasn't too much of a problem, and neither was the genitalia itching like mad. I liked the fact that it swelled up a bit because it looked well-proportioned for once. But the ear was serious pain, and it took a long while to heal itself.

Hilariously, Rambo got his comeuppance when we were on this powerboat driven by huge fans. It was midnight, it was dark, and a mullet jumped right out of the water and slapped him on the face. So he got mulletted as well as chimped!

Another funny moment was when I foolishly turned up for filming in the Florida swampland in that black-and-white-striped penitentiary outfit. It's actually the real McCoy, procured from a Southern prison by Rambo's wife, Laura. The guides there were going, "I hope there's not a sheriff about, John!"

We traveled all over the South filming this stuff. Believe me, the place is heaving with insects. Them swamps are really teeming with life, including a continual blizzard of mosquitoes. After being in that Food-R-Us-like net, it was suddenly of no consequence to me at all, because once you've had condensed mosquitoes, I found I could quite readily tolerate them in the small amounts that they appear naturally. I used to go mad trying to avoid being bitten by a mosquito, but now I couldn't give a damn. Any insect phobia is all gone.

It may surprise many readers to hear that my *Megabugs* series ended up being shown at universities in England. My brother Jimmy's son Liam went to Leeds University, and he was thrilled to ring me up and tell me they were watching it on his course. What a reward! That's a sense of achievement, that you're now involved in an education process which you always thought your songwriting did.

I very genuinely think of my music and my TV programs as complementary. There's a direct connection between "Anarchy In The UK" and *Megabugs*, because they're both equal opposites to corruption. Without meaning to be pretentious, they both work to the same basic poetic beat, a rhythm of life. There's a rhythm in how things work when you just stop and listen to the wind whistle between the trees. It's telling you something. Just open your mind and find out what that is, but don't go out there thinking you're Hiawatha. It doesn't work like that—it's not that corny or obvious. It's just, do good and let your instincts take over for a while.

14

HISTORY AND GRIEF . . . AS A GIFT

I'm not going to have the Sex Pistols rubbished. I can't afford to have anyone doing that to my life. It required an awful lot of my energy and involvement to write them songs—I don't want them to end up selling carpets or toilet cleaner. Here we are in the modern world, and record companies are getting involved in what they call "blanket agreements," where said songs can be misappropriated, and it's a constant battle. There's big money to be earned toeing the line with advertising, but to me advertising has to be used properly.

The bottom line is, you can't damage the original product. That's the term you end up having to use: the Sex Pistols are a product, a commodity, and everybody wants a piece of it. A few years back there was an opportunity of somehow connecting with a range of luxury cars. What Rambo does behind the scenes to maintain our integrity on things like this is quite phenomenal, much to the annoyance of Anita Camarata, who still represents the other three members on Sex Pistols matters.

Anita's interpretation of what is good and bad in the world is very different from mine. She's not a Sex Pistol, she never will be. She doesn't understand what we came from, what we've achieved, and who we are in the contemporary landscape. She just has a vague idea about how to sell and make money. And that can be a problem.

On a similar thread, we collectively had to deal with the Rock and Roll Hall of Fame in 2006. We were completely blindsided by our nomination, apart from the odd hint from friends in the business, that "You know, it would be good if you were in the Hall of Fame." I remember saying when the Ramones were in there before us—and, God help us, the Clash!—that they'd *already* snubbed us, and we shouldn't have anything to do with it thereafter. The Sex Pistols, third in the line of punk? Ludicrous!

Then in came the nomination, and so many bells and whistles and fire alarms went off in my head about it. You are nominated by unknowns, it's a secret ballot, but it's record-industry-sponsored, the same industry that kept both my bands, the Pistols and PiL, in debt for so many years. Why on earth would I be grateful for such a thing?

There was also a money issue about it, in that it would cost a small fortune just to get to this event. There's lots of little angles and dangles in it. We worked out that the band would be at a loss of something like $10,000, and if we wanted to bring friends or family it would be $25,000 a table, and that's just a no-no. That's how unrealistic an issue it was. It really shouldn't cost you a penny. They all want to nominate you and want a piece of the Pistols, but none of them are offering to help us out as a band.

Anita, of course, was well up for it but, right off the bat, I wouldn't do it. I didn't want the name of the Pistols to be sucked up into the industry in that way. I viewed the nomination as a finalizing of your career, a pat on the back—"Well done, now shut up and go away." Particularly in my case. Just because you're paranoid . . . Thank you, again, Poly Styrene.

I got really annoyed with the notes the Hall people were sending us upfront, all completely gleaned off bullshit websites, and they weren't really up for correcting their misinformation, which just added more fuel to it. I was wondering what on earth they were nominating us for, because they completely didn't understand our true history.

For instance, the museum guidebook they sent us claimed that when the Ramones played their first ever UK show on 4 July 1976, the Sex Pistols asked them how to form a band. Well, that very night we were actually playing a live gig in Sheffield. The Sex Pistols had already been up and at it for the best part of a year.

Through Anita, I heard that Steve Jones wanted to go but, slowly but surely, they came round to my way of thinking—apart from Glen Matlock. That's the kind of thing that can be a point of friction in a band. Some wouldn't see that as a challenge to our reputation, because indeed they *never were* following the same agenda as me. So my line was "Well, if you want to go, fine, but you ain't having me there," and that kind of put the mockers on it.

In the end, I lost my rag with the whole thing, and wrote the legendary "urine in wine" note, in which I declared us to be "outside the shitstem," and refused their invitation. Turning them down was a major coup de grâce. No one had done it up to that point. Now, every year there's one little arse out there, going, "Well, Johnny Rotten did it, so can I."

They've since put my note up as a museum piece—it's a prize part of the architecture at the Hall of Fame in Cleveland. *Uuuurgh!* It's just like the Hard Rock Hotel in Las Vegas, with its Elvis jackets and what-have-you. I hate to see living legends collecting dust behind glass cages. It's creepy to me. I love museums, but I don't want to be an actual exhibit in one. For me, history is something way back—you know, give me a couple of hundred years, but I don't want to be museumed off in my own lifetime.

I got my own back a couple of years later, when I did a TV series called *Bodog Battle of the Bands*, and we used the Hall of Fame as one of the audition venues. I've never been comfortable with competition in music, but this was not the usual talent show, with people doing karaoke of other people's songs. Here, the bands would write their own songs, and I was one of the judges. And for the first televised round, we auditioned sixteen bands in the actual museum. I viewed it as bringing life into that *dead hole.*

Oddly enough, the whole Hall of Fame fiasco didn't do us any harm internally. We were eventually all on the same page. It's an ugly feeling to be co-opted in at the backside of a thing. We hadn't played any gigs together since September '03, the touring had all sort of ground to a halt, but through the ensuing months, the idea came up to celebrate the thirtieth anniversary of *Never Mind The Bollocks* with some live events.

We actually got back together first in unexpected and frankly preposterous circumstances. "Anarchy In The UK" and "Pretty Vacant" were due to be used in the video game *Guitar Hero 3.* The plan, of course, was to use the original masters, but at the time Virgin couldn't find them. We'd already signed the deal, rightly thinking there shouldn't be a problem handing over the tapes. Lo and behold, "No!" So, whatever advance we got, we had to spend rerecording the two songs. Glen was doing some solo maneuvers, so Steve, Paul, and I went into a studio in L.A. and got Chris Thomas to fly over, and we rerecorded them in a few days. Fantastic fun.

We were already planning to play some gigs in London, so we sort of used that as a rehearsal. Initially, we thought it'd be just a night or two at Brixton Academy, but it turned into five in the end, and we added Manchester and Glasgow, for good measure.

Brixton, in November '07, worked out an absolute treat. More than a few people came, and the sound was incredible. There was a huge sense of fun in it. We even tried out something theatrical—hilariously so, looking back. Rather than just walk onstage from the wings as normal, I thought it would be a great entrance if we came in through the fire doors in the wall at the back of the stage, where the crew load in the equipment. You can usually see the buses and cars go by on Brixton Road, so it might look like we'd just got off a bus!

After all my work on TV, it was great to get the chance to play live again. It felt really good. I knew in my mind that I wasn't going to run into what people might perceive as classic Johnny

Rotten mannerisms, just standing there and sneering, and becoming a cartoon of myself. I wanted to show that the songs had another level to them—vaudeville, the evil burlesque, British music hall—a very working-classy sing-along thing, where you can say really saucy, challenging things, but with a *smile*, instead of a snarl.

We Pistols really found each other onstage and we were bang on the money—proper Pistols, all there to enjoy it. It wasn't about the money, that was the thing. There wasn't enough being made there for it just to be that. My thing is to play with the crowd and make sure a good time will be had by all, because I'm here to enjoy myself. I'm up for the cup—that must not be forgotten.

Playing to the slope there was like playing to the North Bank at Highbury. There was that great swaying in the crowd, like the terracing used to have—hands in the air, and a great deal of color. It looked fantastic to see groups of hundreds swaying one way, and another thousand swaying the other. The great swirliness of the crowd was pretty impressive. And *the roar*—quite amazing. I've never seen English audiences go off quite that way. It was a joy to behold.

If I remember back to the early Pistols days when all you could hear was someone yelling, "*Get off!*"—imagine going from that to *this*. From my point of view onstage it was absolutely magnificent, and for all the ups and downs of being in the Pistols, there's your reward. It's not the money—it's the enjoyment of having written the songs, then performing them and seeing the audience become such an important feature in the song.

So it was worth going back and working through them Pistols issues and actually using what could be construed as a negative energy in a positive way.

Unfortunately, the fallouts began when there were gaps in the touring schedule, waiting for new offers to come in. There were six months till the next batch of gigs in summer 2008, which killed the vibe. Knifing and sniping became a pastime. I don't mind insults at all, I'm very happy with them, but I won't take a lie. If somebody says I don't turn up at rehearsals and puts that out in a newspaper article, I'm going to jump down their throats. Those kinds of issues bother me.

Come the summer, we had gigs here, there, and everywhere—Vegas, all over Europe, Japan, back to Europe. We called it the Combine Harvester tour. On the "Pretty Vacant" sleeve, it used to be two buses, saying "Nowhere" and "Boredom." So why not two combine harvesters, separating the wheat from the chaff?

Early on, we headlined the Isle of Wight Festival, which was a real challenge. The promoter John Giddings made a big promise to us: "This'll be what it's all about—you've finally made it into the big time." In rehearsals the week before, we'd messed about with the idea of starting the set with a country and western version of "Pretty Vacant," which to my mind would be a delicious crowd tease—and, eventually a crowd-pleaser. But onstage that night Steve Jones and the band wouldn't go along with it properly and at the end left me there stranded, like a rodeo clown.

It was a very good gig, but it was a strange audience. I felt that we were playing to old people on Brighton pier, because everybody was in deck chairs and floppy hats. I've been told since that people loved it, but that people weren't yelling and screaming—they were stunned by us!

Other gigs were more rowdy. In Greece, the New York Dolls were on supporting us, and this mob of so-called anarchists—or shall we call them "out-and-out cunts"—ran through the crowd, all wearing scooter helmets, letting off CS gas, swinging baseball bats, and smashing the shit out of anybody in their way. "So this is anarchy for you lot, is it?" I'll say right now, I'm not an anarchist, because I've seen too many wrong moves from that sort. They're usually just spoilt middle-class kids with an attitude—like the "meat is murder" brigade. They hurt all the wrong people. Bullying and cruelty—I can't be having it.

Anyway, later on, I got slammed in the face by a missile, and got a great big cut from it. Rambo liked it—he said it was very visual. The rules are, nobody on stage except the band or our crew, so suddenly a 25-foot pole with a wet sponge on the end looms into my field of vision. I don't know where that sponge has been, it looks like it's got grease on it. Somebody shouted, "No, it's Dettol!" So I threw my arms out at the sides, like I was being crucified, and allowed myself to be dabbed, Christlike. It was making a mockery of the whole thing, in an audience participation way. The crowd got it, but the band didn't, sad to say. Some of the band were muttering, "We should fuck off, it might kick off here . . ." I was like, "I ain't going *anywhere*!"

The problem with this tour was it went on too long, to the point where we got fed up and sick of the sight of each other. The really serious good that came out of it was the conclusion in my mind: "Never again!" I'm actually always one to say "never say never," but I really genuinely feel like I just don't belong in that band anymore. I might do a one-off, but I'm certainly not going off touring with them, and I ain't writing new songs, which would be the only point of continuing any further. Any opportunity I get to write a new song, I just don't think Pistols. The Pistols are an historical accuracy, and you can't take that away. It was a truly magnificent achievement. I want it to be remembered as that. I don't want to make *Never Mind The Bollocks Part 2* because it would ruin that.

Towards the end of the tour, there was one final terrific London gig at the Hammersmith Apollo. There was a homey vibe going on that night, because the venue is just down the road from my place in Fulham. Also, it was Steve Jones' birthday the following day, and he didn't know I knew, and I got the crowd to sing "Happy Birthday Fatty" to him. That was a good moment between us. He has goodness and fun in him, but he also has that other "festering boil" side. Which I suppose we all do, because we're human. That night, he played great, he totally became the Steve Jones who leaves hairs at the back of your head standing, who hits it dead bang-on right. He can be an exceptionally good guitarist.

After that, there were a few vague promises from promoters going off into the future, but it all just fizzled out. Finally, I had a conversation with Paul Cook on the phone, and he said, "We reckon it's time to knock it on the head, John, what do you think?" And I said, "Yes, I agree." It just didn't feel right anymore. I was looking at the band and thinking, "It's still in its own time zone. As Pistols, we're not getting up to the twenty-first century," and so it became a very dull prospect for me. And that was a view shared by Paul, and presumably the others, that we didn't want to go back and do rip-offs of old stuff.

It'd be nice to think that we can be friendly outside of that band but, when we're together, we become, for some weird reason, mortal enemies. It's hard to explain, but the pressures become too much, you become too tightly knit up in each other's situations, and things get childish. What I always say about the music industry—it keeps you young—is particularly true of the Sex Pistols. It's a world of wonderful kiddiness!

I was in London over Christmas 2013, to see my brother Jimmy and the family, and I rang up Paul. He wasn't in, so I talked to his missus and his daughter, and asked for Paul to ring, but he never rang back, so that's where that is now. Everything tells you something.

I know I'm pretty damn hard work. It must've been a bit of a nightmare for Steve and Paul and Glen to cope with something a little bit different, something not "off the shelf," like I am. The problems they have with me might be absolutely firmly rooted in the fact that they had a band before me, and they think I came in and spoilt everything. I can only garner that from interviews all three of them have done over the years, implying that the Sex Pistols could've been a really great rock 'n' roll band, but for me. Well, there you go.

One way or another, I think I would've been a creative person, with or without them originally paying some attention to me. There might possibly have been better options out there, from other people that might've had similar aspirations in my direction. Who knows? That was the first option, and I dived in. There was something special between us—I can't describe what that is. You can't describe charisma. But it most definitely was there, and still is, in my mind. There's that little spark that I know is burbling around in all their little heads. Perhaps one problem is Paul and Glen still living in England—England will age you like nothing on earth.

I want us to be friends, I really do. I want us to respect each other, but I can't get them to break that barrier, that wall that's always put in between me and any one of them, individually or together. They just won't open up with me, and that's, I think, unfair. I have to learn to accept that, I suppose. For years I've endured it, and it was a good endurance, it definitely teaches you stamina and staying power. But it always comes down to the same conclusion when I rattle this in my head: I've done the best I could with these fellas, and I'll always love and respect them. And that's it. The End.

That year, 2008, really wasn't a good year for me. In the early months, I lost my dad. He died suddenly. Apparently, he'd had an argument with the woman he was living with, Mary Irwin, and her son. Dad slipped, cracked his head, had a bit of a heart attack, and—dead. The only silver lining was that, at the autopsy, the pathologist said, "I know it's terrible, but it's kind of good he did die quickly like this, because he was riddled with cancers, and otherwise he would've died a very slow, painful death." Like, "Pffff, that's good news, is it?" Yep, apparently so. It was blindingly painful.

Mary Irwin, his girlfriend, was related somehow to his cousins—keep it in the family, I suppose, the Lydon way. It was nearly thirty years since my mother's death, and I never had any resentment for his girlfriends. I expected him to be a human, but I never liked this particular woman because I thought she had a very bad nature. She was very pushy, and mean-spirited. The problems began when she said, "I'm your mammy now!" You know?! "I'm a grown man, I've done a lot. What do you take me for, a fucking idiot?" I think she was hoping *she* was the true love of his life, and not just some old biddy you bunk up with when you're old. That's what Irish people do, and I'm sure it's the same everywhere else—when you're old, don't die alone!

My dad never said much to me, but he guided me well, and subversively. As I said before, later in life, we got very close. Two years before he died, I declared in an interview that he was one of my best friends, so it was all the more shocking the way he died and why he died and how much I missed him. It tore my heart apart. I didn't think it would do that, but it really did.

When he died, I flew to London immediately and went directly to my brother Jimmy's. I was so overtired I fell asleep on the couch there. I felt terrible about that, because I knew Cathy, Jimmy's wife, loved her couch, and I hadn't washed in a couple of days. So then I went home to our house in Fulham, and I put a chair in the middle of the room, and I just tried to talk to my dad. "Hello Dad, blah blah blah." Whatever it is you do.

I never feel they come back, the people you love. They don't come back to you—they're gone. That's such a hard thing to deal with. It's the same with your enemies: when they've gone, you miss them. You can't honestly be a human being and say you don't.

I was telling you, "I see things." It's never about specific people, it's about energies, feelings that you pick up. But with my dad, I knew he was gone, his energy was no longer around, and that was the loneliest, *loneliest* thing I'd felt, ever since the death of my mum. Just sitting there on a chair in the front room. I deliberately placed it right in the middle of the room—almost dramatic, I suppose. I was going to put on records and play music, I'd set the system, but I couldn't get around to it. So I sat there in silence for ages—I found out later it was something like eighteen hours. My brother Jimmy came and picked me up. He went, "I know what you're doing, let us in!" And he was dead right.

At Dad's funeral, I was borderline passing out with tears, which I never did with my mum. I was expected to give something of a speech. I couldn't, I just couldn't. Words fail you. I walked up when I felt like it—I got really bored of the priest yabbering away—and I leant into the coffin, and I kissed my dad's dead body on the cheek. I looked down and went, "That's me dad!" and broke apart. I missed him so bad.

The saving grace was that the press didn't attend. Maybe there was some kind of respect there, as a lot of the journalists who would normally be well up for sticking a knife in me were really good and left me be. Maybe there is a kindness in there, and they know there's a line not to be crossed, because I've seen them rubbish other people's funerals. The prospect of a journalist running up with a camera going, "So, what do you feel like now your dad's dead?" didn't bear thinking about. It has become *that* grotesque.

Afterwards, we held a wake, to celebrate Dad's death, in a North London pub not far from where I used to live in Benwell Road. It was a great turnout for him. It was a community thing—that's why so many people came to pay their respects. We're all interrelated; it's an astounding affirmation of community in the most painful way possible. So many people cared. It was a real gathering of the clans, working-class style, taking place in this pub that's a well-known gangster hangout!

I was standing next to some top lads who turned up for my dad—proper Arsenal. They loved him because of the manor, and what we are, and the community spirit therein. And where does the trouble begin? From the Galway second-arse cousins. They were disgraceful and disgusting. Here I am trying to celebrate the death of my dad, and one of the daughters of a cousin stands in front of me, raises her dress and does a clippety-cloppety dance, and asks me, "Look, Oi can daance! Can ye get me oan *X Factor*?" The answer was "No!" and her response was, "Yer a cunt!"

That's how they behaved with us. How ugly is that? There's some serious sickness in people. It's like, *we* are meant to be the Irish abroad, and we're getting Irish from Ireland behaving not very Irish at all. I felt like I had to run auditions at my own father's funeral. She wasn't the only one. There was a couple of them that stood in front of me and had the audacity to sing "Danny Boy," a song I don't particularly fucking like.

Jimmy, at the time, was recovering from cancer—we didn't know if he was going to fully recover. Ouch. Double ouch. The loss of my mum was a hard one, barely into my early twenties, but the loss of my dad left me feeling for quite a bit that I had no point or purpose. I don't know how I would've got through it, but for Nora reminding me that she'd gone through this too.

I feel really sad talking about this, and I know I'm boring you. People can fuck off if they don't want to read it, because—this—is—life. All these fucks can run around with their punk agendas, but they don't understand what humanity is. My idea of punk *is* humanity, it's not vacuous nonsense like, "Are you wearing the latest outfit? Cool, dude!" Everything I do is always about my community, my friends, my family—and my family's gone, but for my brothers. I want you to understand what life can be like. Thank you for listening.

The truth, I've found, is far more interesting than the tittle-tattle they fill history books with. Nothing is as easily explained away as the powers-that-be would like it. The American Civil War was not really at all about freeing the slaves. That's nonsense. No war is ever fought over moral issues. It's always about economics. You only have to look at history.

It's fascinating finding things like the Irish used to be called "black." They were viewed as black Americans, too, in them times. Black was an all-encompassing term for lesser mortals.

My association with Bodog extended beyond just their *Battle of the Bands*. Around that time I was working on a series for them called *Johnny Rotten Loves America*, which never came to fruition. The idea was to explore bits of American history that are little-known, or swept under the carpet. They wanted us to track down "buffalo soldiers"—black U.S. cavalrymen, who are often written out of history—but Rambo suggested black Confederate soldiers instead, because no one believes that they even existed.

We tracked down a retired schoolteacher called Nelson Winbush, whose African-American granddad fought in the Civil War against the North. I found him to be one of the most absorbing fellas I've ever had a chat with. He still remembers his grandfather, and was in attendance at the funeral, where there was a Confederate flag draped over his coffin. Nelson got his flag out, and pictures of his grandfather. He showed us the pension book his grandfather got from his Southern state. They didn't usually give pensions out, because the place was ravaged in the war, and there was no money, but he still got a pension for his services to the Confederacy.

I was fairly gobsmacked not knowing any of this. American history isn't so easily explained then, is it? There's a great sense of intrigue. I'm naturally nosy and want to find out what's really going on here. We began toying with the concept of putting together a separate program on the true history of the Civil War, absolutely from a black perspective, but then the Bodog thing dismantled itself very oddly and sadly the show never got made. I'd love to go back one day and revisit it.

I was receiving other offers for TV work, but they were impossible to take on, because they'd demand that I sign long-term contracts, and not work anywhere else, and not choose my own issues in the programs, and basically be led by the nose with a financial contract as the carrot, and that was very uninteresting. They'd have rights to me, and basically I do what I'm told. And that's *impossible* for me! I can do one-offs on subjects I really like, or maybe just one series, but I'm not going to be anybody's puppet for years on any TV network, and basically do any old crap. No, no, no, I'm not quite ridden to the knackers' yard that way.

Presumably because I'd been in *I'm a Celebrity* I got an offer to do a thing called *Celebrity Circus*. I dislike circuses intensely—that's an A1 route to cruelty to animals. I also have an absolute hatred for zoos—I see them as prisons for wildlife. It was like they hadn't paid any attention to the content of the nature programs I'd already done. You get really angry with them, but at the same time you have to sit down and explain to them why this is horribly wrong. And you know at the end of the day they're still not listening. You'll get some ridiculous remark, like, "Well, we can put more money in." Sorry, no, I mean what I say.

And it's not like I could go and raise money myself for a TV production, because I'm obviously going to put anything like that into making music.

That's when the advert for Country Life butter came along, for British TV. I understood all the pitfalls in it, and hummed and hawed about taking it on—it just seemed so nutty. It was a case of "What? They're prepared to put their faith in *me* to help sell their butter?" From the first meeting, the respect coming from the ad company, and from Country Life themselves, was almost overwhelming. They were taking a real risk with me, and were going to give me pretty much a free hand to be myself, without much tedious scripting. They were really upright and correct and professional with me, unlike anything I'd experienced from a record label. There was no dishonesty, or bribing you, or forcing you into situations you didn't like. No trickery in the contract wording. So refreshing.

Then I began to see it from their point of view, and saw the fun in it. It began to seem so perfect, so utterly, mindblowingly right—the most anarchistic thing I've ever been presented with—a butter campaign! Wow, what a challenge! How was anyone going to cope with that? But then, after the initial shock, you look at it and think, "What's going on in this? Well, I do *eat* butter, I do *like* butter, and you can't make a good curry *without* butter—or without ghee, which is clarified butter. Ever tried eating baked beans on toast without butter?" Hmm. By God, these boys had got me.

This wasn't the same thing as using an old Pistols song to sell cack, which would ruin that song forever and erode its seriousness in many fans' minds. I was very happily buying into the line that we were going to promote British farming. Indeed, that's exactly what we did, we bolstered British industry! The ads were about buying British, and I thoroughly enjoyed larking about in the fields in a country gent's tweeds—in fact, I found those to be much more practical for protecting yourself against freezing, drizzly British conditions than any nylon skiing gear. The rapport with the people involved became mind-alteringly open, in what could easily have been a corporate debacle. I think we turned it into something really impressive. Fortunately for the British farming industry, it did them the world of good too.

The ads went so big, so quick. It almost felt like I'd lost the reins, it just became so enormous. Of course there was the negative hatred, and again this thing of "You've sold out." I've had to face that nonsense all my life in making music. There will always be the naysayers out there but, at the same time, you've got to say, Mr. Rotten managed to put up British butter sales by 85 percent. "So there is an audience out there!" People are aware of me, and I am respected.

Then, lo and behold, I found myself in the middle of a butter war with New Zealand Anchor, putting out all manner of insults over the internet. Even Ireland's Kerrygold had a dig at me: "If you claim to be Irish, you should've backed Kerrygold!" Well, you know, none of them asked me. Suddenly, I'm a valuable commodity!

You take on something risky like that, and then one thing leads to another. Suddenly a huge opportunity in your creativity opens, and all of a sudden people are paying attention again. I was still at a stalemate with record companies, and any time I tried to get anything off the ground I'd always run into a financial barrier that you had no way of overcoming.

So every penny I earned from those ads went straight into reactivating Public Image Ltd. There was enough there—not an enormous amount, but a bulk lump sum—that I could put up to get a band together and into a rehearsal framework. From there, it worked out that we could actually survive on touring, and get enough together to record, and make an album our way without having to have a record label. In the end, it worked out fine: we're now our own label, PiL Official Ltd, and we have our own publishing company. All that freedom, thanks to those ads.

WHO CENSORS THE CENSOR? #5

PASSIVE RESISTANCE

People I admire: Christiane Amanpour, the CNN journalist. I love her because she stops the corruption of history in the making. Whenever she's on TV, I find her riveting. David Attenborough, because of his passion for nature, naturally. Gandhi is my ultimate hero. If you're gonna have one, have one with absolutely no weapons other than wisdom.

For me, it's the effort that counts, not results. Never mind winning cups and leagues, strugglers are worth more emotionally than achievers. I really admire strugglers, I have empathy for them, people who are trying to make a change rather than sitting on the laurels of victory. That value is in all of these people. They put great trains of thought into your head and get your mind spinning, and that's wonderful.

I've got an open mind but a closed heart to politicians. And I have a definite closed door on all religions, especially new-age head rubbish. I can't suffer that. All their psychic leanings drive me crazy because it's such a waste of good energy. They're just advertising agencies, selling us the same cack regurgitated. This adoration of a higher power that makes all the decisions for you. That's ball-cutting stuff, and I like my testicles very much.

Anything that properly gets my brain twirling, negatives included—*especially* negatives, sometimes—is all to the power and benefit to me ultimately. Know thine enemy: the more you get to know him, the more you realize he's really your friend. You start having empathy—that word, again—and therefore the bitterness of us-and-them is dissolved, and you find ultimately you can have common ground.

Of course there are philosophies out there that I could never contend with, racism being one of them. It's absurd: we all come from the same bottleneck way back when. You judge a person by their deeds, and nothing else really accounts for very much. The bottom line is, is that a good person, or a bad person? Is that a liar and a cheat and a fraud, or someone genuine?

Being brought up in a very mixed neighborhood I never had to contend with "Oh look at them, they're all different." The thing was that we were all very different, but all very much the same. We all had the same problems, all had the same kind of schools to go to, living in the same kind of housing. What's the point in squabbling amongst each other over that? For me, the focus would be on—putting it very nicely—the bastards what put us there, whoever's responsible for our serious lack of opportunities. Because it's not us, we're industrious, thank you. We want to get ahead in the world.

Of course, you'd always get the "Send 'em back" brigade, and it's still going on, isn't it? And send who back? You tell me: who can really account for being 100 percent British, and what does that really mean? It's an island that has been very open to all manner of race, creed, and color, for pretty much as long as it has been populated. The English language itself is an adaptation of European languages, mixed with a bit of this, that, and the other. If you're going to go all the way back, is it Albion we're talking here? Just the Saxons? What about the Angles before them? The Celts? And who was there before the Celts? Where did all these folk come from?

The way the world is, people just move about. The blood that runs in my veins runs in every other human being on this planet. The same. One blood, that's what we are, we're one species. There's no variants that make us not be able to mix and match. We're not like, say, chimpanzees and gorillas. As long as we can have sex with each other, we're the same. And if you want to call that mongrel, as in all dogs are the same really—well, that's what we are. It's that infinite variety which as a species will sustain us, constantly refreshing the gene pool. There is no other way. Very laughingly, you can look at the inbred nonsense of royal families, and you can see it in them—they are all kind of half-headedly dopey, aren't they? Particularly the menfolk.

But then our lot have quite a bit of German in there, via Greece. And Russia. Hello, Habsburgs, how are you doing? Thank you for being so non-English. Racial purity doesn't make sense once you do any kind of study. Class purity makes even less sense. All that highlights is the greedy who don't want to share their portion of the pie—a completely clear-cut nonsense.

I understand my folk here much more than I do the attitudes of the landed gentry. There's nothing really but a curse of education that separates working class from middle class. There's an attitude in the education principles that teach a sense of superiority and inferiority. There's not much difference other than that, there really isn't. You can't say that the middle class have all the money, not from what I've seen. Who's creating these gaps between us and feeding us these false agendas? That's where I'm looking.

But I don't call it immigration, I call it migration. As a species it's very healthy for us to get up and move around the planet. Sometimes certain groups of people have to do that for economic reasons. Nobody's doing it just to be spiteful. Everybody loves the idea of a homeland. I used to, but I've kind of got the bigger picture now. It's a home planet to me.

I'm as thrilled and feel as at home in Shanghai as I do anywhere else on earth. I love the vibrancy there and I felt proud to be a human being watching China develop. There's many bad things in China—believe me, I'm not a fool about it—but I can see that the get-up-and-go they're creating there is very interesting. That's something where Britain has lagged behind. There's no gusto anymore. There seems to be a lot of laziness, idleness. I can't bear to hear anyone say, "There's nothing I can do." Of course there is. That's bloody nonsense.

We were in China with PiL for about a week in March 2013. Oh, their eagerness to learn! The government are starving them of information of what goes on outside their boundaries, so when you're there it's thrilling, trying to communicate with them. They're very talkative and friendly and open, and they just want to *know*—what are the bits that are missing in the picture for them? Slowly but surely you're putting into their minds the idea that they're being manipulated and that's surely a good thing. That opens them up, gets them to start thinking for themselves as individuals, when they realize what censorship has denied them. And how are my antics in the West not to be mentioned in the East? That's a puzzlement to me.

The government officials who have to approve your visa

analyze every word in every song in your set list. Surprise surprise, they decided that, yes, we were okay. The trouble was of course that a lot of our records and my other work was completely unknown there, so what we were trying to do really was push through our latest album *This Is PiL*, and get people to understand that there were other strings to our bow. So the set involved a full PiL catalogue. It was quite a surprise to them. But when they heard PiL, they definitely got it. The textures, the tones, the progression in it, and the openness and the joy and the pain in the music definitely scored big points.

We had just about no sleep with the excitement of it all. That's how I am. We even rehearsed there because that's where we were starting the tour. Beijing, I have to say, was *not* an eye-opener: the pollution was so bad you daren't open your eyes. It's catastrophic to expect people to live in that environment. Here I am in Los Angeles, where people moan about the pollution all the time and rightly so, but hello, it ain't in the zero pointage in comparison. Beijing is so way off the map, it's incomparable.

Onstage, you literally can't get enough air in your lungs to perform the songs properly. It's like slowly choking. Very frightening too, when you can't get the air in, particularly in the gig because it was rammed to the rafters in Beijing—even less oxygen than usual, and on top of that everybody smoking, including myself!

What you've got to do there, and in Russia, is you've got to not come down hard on the bootleg side, because that's the only way they really have of discovering you, through illegal trade. Although there's gangsters there somehow profiting off that, that's the only way they can get the information. You have to grin and grit your teeth. You're being ripped off, but at the same time you're passing on information, which is far more valuable. If ever I did any of this just for the money—well, I wouldn't be in China, would I?

I suppose places like China are the new frontier for Western music, and their innocence towards us was equally shared, because I was as naive about them as they were about me. That's where the talking can begin. You're both equally puzzled but you're intrigued and fascinated, and it leads to excellent situations in dressing rooms. Normally, I run away from those kind of scenarios, but when it's in new and unexplored territories, I'm up for that. All night long. Because that absorption of information gives you the energy to do the next gig, way more than sleep—being able to walk out of these things and know you've done something good, and you've learned so much.

I'm thrilled that other bands were turned down by the Chinese authorities. They were all the ones that joined in on those student union complaints about "Free Tibet," or whatever. Well, we should be freeing ourselves first. And I don't view Britain particularly as a free society. It still faces censorship, and if that don't work, then you'll face mockery through the press. For so long there's been a media culture there that's been swaying people into believing that music has no effect anymore, so why bother? I say, look to the owners of those publications. Need I say any more? All of those papers and TV channels that push that agenda of, why bother? That's because they've got their pile, haven't they? So it's back to trying to keep us stupid, it's back to religion in another guise. Never give up. There's nothing to give up for.

I believe in changing things, but not at any cost. I agreed with the Poll Tax riots in 1989, for instance, but I didn't understand the rioters then attacking a McDonald's. What the hell was that? The riot was about specific things: "Let's go and solve those specific problems with the Poll Tax." How does a cheeseburger come into this?

The Tottenham riots in 2011 were equally foolish. The point and purpose of it were lost to the mob, which you might assume to be out of control. But there are manipulators in there that run private agendas. In any crowd situation, be very wary who you follow so willingly. Don't be jumping behind the banner of the loudest mouth in the crowd. Make sure you know who that fella is and that you agree with his agenda. That's just common sense to me, but that's something you learn being brought up with football. We're not easily led to go charging down a street by the first arsehole that says, "This way, chaps!" You've got to have earned the right to my support.

So I'm dead against that kind of senseless rioting, particularly when it ends up with innocent people being murdered. There was one situation in 2011: why on earth were those idiots wrecking up this woman's little hairdresser's shop? The viciousness had been hoodwinked into other things, and none of them solve any problems. I'm telling you: I don't believe that violence solves anything, it just opens the door to the lowest common denominator to manipulate the mob and the end result is always stupidity. Always. Whatever cause you had to be demonstrating there in the first place is at that point gone, lost forever.

So they just ended up looting plasma-screen TVs. Once again, we're back to the power of advertising. That's the ugliness of it: the message through advertising is "Everyone must have one of these." So, everybody went out to get one. They can't earn it by fair means, so it becomes by hook or by crook. An invite to thieves, and nobody relates it to what started the escalation.

Police shooting people on the streets—my God, that was

a big, big issue that kicked off the Tottenham riots. It says a lot about Britain that the rioting was widespread because everybody in the country realized at that point that the police force was a headless chicken, and no one was in charge of anything and they didn't have the means to stop it. Which is useful information to people like me, but then it was thrown out the window with "I want Adidas and Samsung!" Pah! The masses!

Some of the racial tensions I notice when I'm back in London these days are so unjustifiable. Your poor Polish chap that gets off the plane, he's there to work and to make money. His attitude to me is proper working class. I don't see him as stealing anyone's job. He's not the enemy, it's the government that's created that agenda and deliberately wants to set you up against these folk. Where actually they're the same as us, they just want to do the best they can for their families.

I don't want this to be taken the wrong way, but Rambo and I, we have this laugh, to see how many cornices on the outside of Georgian buildings have been knocked off by Polish workers. Stripped down of ornamentation! Ornamentation is definitely on the way out!

Beyond that, I bemoan the modern architecture in London. Some pieces I like but generally speaking I find it coldly indifferent and soul-destroying. The ugliness of steel piping on the outside, and temples of glass—it's so impersonal. I can't find a message in that kind of architecture that's in any way friendly. In many ways, too, the old Georgian architecture was an imperialistic look *down* on us, but there was a beauty in it. It was at least something to aspire to artistically. There was effort in the stonework and attention to detail, which is always riveting. There's somehow an aspirational quality to it.

But oh God, that one that looks like a coffee percolator! Is it the Gherkin? Oh God, no! And that new glass "Shard" thing that's sticking up, scraping the sky. It looks so antisocial. To me it just looks like it's tearing and ripping the sky. It's a very evil piece of work. I want modern, and I want update, and I want new achievements, but I don't want them to be at the sacrifice of the people that have to live in these environments. Pay attention to what our needs are and make it comfortable for us to live in, and make us proud to live there. And that's not what's happening. Modern architecture has somehow disassociated itself completely, as indeed modern art has. It just seems art for art's sake. We're completely uninvolved.

Maybe they just build them as a tax write-off. They don't occur to me as being proud monuments of a nation's achievements.

In terms of the policing and surveillance on the streets below, it's all about protecting the very wealthy. You'll probably get less for murder than you will for damaging their property. That's telling you lots about where society has ended up. That kills creativity and when people can't be creative and contribute, they use that extreme talent to other means. If one of them be crime, that's how it ends up. If all roads are closed, you plow through the field.

That sense of neighborhood seems to be gone. I can't speak for the young, but I can speak *with* them. Times have moved on and you've had a very coldhearted Conservative government—a coalition in name, but it's just two cunts for the price of one. Before that, it was an even more distant Labour government with Blair, that all combined has led to some serious problems. This Britain is slowly dissolving and unraveling and it's not great to see.

In many ways I look back at it and I think the Sex Pistols were a way, way early messenger of doom. Yet we were offering hope, because once you realize these problems and who's creating them, you at least have a chance to change it. And then we're back to censorship. It's still there, as bad as it ever was, if not worse now. They learned from the likes of the Pistols, and people of similar attributes, how to close us down. The media is a great tool for that. Them inquiries into phone-hacking and all of that—you find out the government's in collusion with it, you find out the police are in collusion with it. It's quite a bizarre truth but one that really has to be looked at.

I don't know what answers are going to come out of the investigations into this, and the trials. I saw Rebekah Brooks, that former editor of the *News of the World*, on TV and the Americans were laughing about it. She said in court she didn't know that what she was doing was illegal and indeed she was found not guilty at the end of the trial. But I mean, come on, then, what chance do the rest of us stand? She was at the time going out with that Ross Kemp from *EastEnders*. My, oh my, what a wicked web we weave!

Then there was the revelation about how she'd been *lent* a police horse for a couple of years and that David Cameron had ended up riding it. Oh my gosh. I wonder if that horse was ever used to dispel rioters. It's intriguing: the tentacles of corruption, how high they go! Corruption in any country is from the top downwards, not the bottom up. So don't be nicking my class because they're flogging a few bits of gear at the end of the street. Have a word about those that have the money to import it, don't set us up as the mugs. Or take the easy route: just answer the *Daily Mail*'s rallying call, and blame it on the oiks. In this respect we're the oiks, *all* of us.

15

DEEPER WATER

I'd always wanted to reactivate PiL, but exactly who should be in the band wasn't completely clear to me. When I think back to PiL beginnings, I've still got Jah Wobble in my mind as being there in heart and soul. It's not a puzzlement to me, it's just my memories are fond of this fella. Not so much with Keith Levene, obviously. I can't help that. I understood fully why Wobble couldn't work with him. But I also knew that Keith's attention to detail was exactly why he couldn't work with Wobble. The two couldn't be in the same room together, and it wasn't easy having to pick one member over another.

I simply didn't want Keith in my life again. It's completely clear: he's a cunt. He's very talented, but he doesn't like himself, and therefore the rest of the world must suffer. You cannot deny the power and beauty of the guitaring in "Poptones." It's utterly wonderful, the juxtaposition of up and down. It's flora! It's tapestry! He's a cunt that can play good—incredibly good—but he's still a cunt.

So I called Wobble, and gave him the option of "Shall we work together again?" Everything was fine on the phone, sort of. But then came the question of money, and his manager had extra-special ideas about Wobble's big bad self, and how important he was to the whole thing, and he should get more than everybody else. I'm sorry, that ain't PiL—goodbye! If you're going to play them kind of games after all those years—well, where's payback? All that investment that I put into this, from the beginning—that don't count? And you still want to be paid *more*?

Wobble was my mate and he always will be, regardless of the rows or inconsequential tittle-tattles that have gone on in between. Those aren't going to change anything. He knew me before the band, and he should know me after. Hand on heart, I couldn't work with him again, but that doesn't mean that our friendship should stop. That's his decision and good luck to him on it. But for me, people that I've ever hung out with and respected as friends, stay that way, regardless of the errors of their ways. You have to forgive your friends. That's what friends are for.

Still, it was impossibly, ludicrously funny for me to hear, in 2012, that Levene and Wobble—worst enemies!—had buried the hatchet, and joined back up again, and were declaring that I can't sing. So what do they do about that? They go and get an *X-Factor*–style Johnny Rotten tribute act, to literally imitate me, in their new mock-PiL band. Exactly the thing that both of them said they couldn't bear about me, they get from a mimic! They've made a mockery of themselves in that. No progress in that at all.

The people I really wanted in PiL, deep down, were Lu Edmonds and Bruce Smith, from mid-late '80s PiL, as my guitarist and drummer respectively. I rang up Bruce, and I hadn't spoken to him for maybe twenty years. He immediately went, "Hello, John!" and I burst out laughing. It's odd, because when I spoke to Wobble, I didn't recognize his voice on the phone—he sounded like he was an infomercial trying to get all the words in very quickly. It felt wrong. But with Bruce, we were right there, as mates, instantly. I knew from that very moment that it was the correct thing to do.

After his tinnitus problems, Lu had given up trying to convert

Western computers to gamelan, and gone acoustic. He'd become a kind of cultural ambassador in former Soviet republics, and places like Kurdistan. He'd travel to different communities, bringing with him instruments indigenous to the region, and teach the locals their own lost culture. He's the most wonderful, generous, and creative person. I begged him to work with me, and his whole wall of rejection was "Oh John, please don't spend your own money!" Aw, what a sweetheart. How could I not want to work with a man who cared that deeply? He didn't want to see me lose, and viewed himself as a risk or a liability. I pleaded with him to see it differently and he did, and now here we are—PiL, back again.

Then, bass-player-hunting we went, and through our then tour manager, Bill Barclay, we found Scott Firth. His CV ranged from the Spice Girls to Stevie Winwood. Wow, that's an open mind! Or at least it's someone who knows that sometimes you have to work with chavs and slashers. You knew that he had a work ethic. I talked to him on the phone, and I really liked him. It was clear right from the outset—he loves his wife, loves his kids, he's got that area of his life sorted, and it's all centered around that. Fantastic, I'm listening to a stable-minded human being!

I turned up late to the new PiL's first rehearsal, and I'm really glad I did. When I walked in, they'd gotten into "Albatross," off *Metal Box*, which I didn't even put forward as a thing we should rehearse, but they were rehearsing it. It was a dead straight line from the studio door to the mic. I asked the crew if the mic was on, walked right up, turned to face the drummer, and—bingo!—just dropped straight into it, and I knew instantly that we were the band. It really was a most excellent moment. It was much better than I ever remember it before—more solid, much more deliberate, pointed and purposeful. That's what Scott's done, he's added that clarity to the bass. In the early days it was always distracted and fractional, because the bloke was learning.

As soon as we got out there playing, in the run-up to Christmas '09, we'd got the patterns tightened somewhat, and that allowed all manner of flexibility for the vocals. The flexibility in the past was all in the instrumentation, which was flopping all over the place, but that's how it worked. This was a much better way and just sounds tougher, and the words land better. I'm given more space, and I can sing better because of that.

It felt just magical to be playing *those* songs again. They're the story of my life, almost. Stuff like "Death Disco," about my mum. I'd have tears in my eyes most nights. On "Public Image" Lu would play a Turkish saz, wired up like an electric guitar. He'd have a vast array of instruments up there, because he was very bored with regular guitar. They were just getting him to where sonically he feels the most honest to his heart and soul.

We did what you might call "more challenging numbers," like "USLS1," to give people's heads another space to explore, and to show that I'm not a one-trick pony, by any means. It's not all a juggernaut trundling off down the highway relentlessly. I'm a bit of a tourist, when I travel, and I stop off at all the right sites. We really pushed some of these songs to the limit, like "Flowers Of Romance," where Scott added upright bass, and "Religion," which we made into an evil crescendo, with extra-heavy bass, and my sermons about the Pope, and pedophilia in the Catholic Church.

We were playing for two hours or more a night, closing with a riotous "Rise" and "Open Up." The only problem with those first shows was that my voice got sore towards the end—who doesn't in December, in England? What a clever sod I was to do my reactivation tour then!

The following year we went on and played all over—America, Japan, every nook and cranny of Europe—and I'm glad to say everyone was pleased to see us. At Shepherd's Bush Empire in July, we went off after the main set, and there was a really large fella being given oxygen and resuscitation, because he'd had a heart attack during the gig. There he was, being oxygen-ed up, and he pushes the mask off and goes, "Johnny, can you give me your autograph please!" I went over, and the paramedics thought I was trying to kill him or something. "Go away, John!"

I got into a controversy by refusing to pull out of a festival in

Tel Aviv. All these hippie groups, and Elvis Costello, were claiming that by playing in Israel, it was like I was supporting apartheid. I've been accused of many things in my life, but that one topped it. "Don't go to Israel, Johnny, it's a fascist state!" But how can you call a Jew a fascist after World War Two? Huh?! Such ignorance.

Before we went to Israel a group of protestors turned up at a few shows. What a strange bunch. The threat of violence was zero. They were probably demonstrating in the '60s and now they were in their sixties. They reminded us of the Sally Army. There were people on stilts and dressed as fruit. It was like a circus carnival with the most *Blue Peter* sticky-back plastic placards you'd ever seen. In Liverpool they organized themselves into a church choir and were singing outside the venue: "Oh no John, no John, no John, no . . ." Absolutely hilarious. Nice people though.

It wasn't the Israeli government sponsoring us, it was a promoter. We were going there to play for *people*, not politicians. In my book, it's an act of cowardice to deny a population access to something that could really oil the wheels of change in that part of the world. That's how it's done, one-on-one communication. It's surely better if people are there listening to the ideas from me, from my mouth, rather than the only thing they hear is that you don't want to play there. It doesn't matter about the reasons—you've just negated them as human beings. It's not ever a population's fault or responsibility, it's the politics that dictate to them.

The irony was, we were doing "Four Enclosed Walls" at the time, with its "Aaaa-aaaall-llaaaah" call-to-prayer refrain. The Israeli government certainly weren't sponsoring *that*. In fact, I'd love to go to Palestine and play for people, but I'm still waiting for the invite.

We conducted two solid years of touring in the most brilliant and happy environment I've ever known. Musically and in all ways, our DNA seems to sync so well together. This, to me, was finally PiL fully realized—a cohesive unit of people that really get on well with each other, and don't let tiffles cause fraction. None of these people hold resentments, and neither do I. It eats up too much energy. Get it out, shout it out, it's gone. It's what I've always wanted it to be, in every band I've had. You won't achieve it by bullying or snarkiness or backbiting. Or wage demands. You achieve them out of respect, and by God, we really do respect and love each other.

In the summer of 2011 I came back to England to start work on the first PiL album in almost twenty years. I'd done my homework, but God had other plans. I'd racked up quite a few song ideas on my own, but they all got burned in a horrible house fire at my place in Fulham. The only really important thing here was that Nora didn't die in it. That would've been the end of my life. If it'd been the end of hers, quite clearly it would've been the end of mine. I'd not have been able to cope with that.

The fire was caused by an electrical fault in the tumble dryer. We knew there was a problem with it, but we didn't think it would set the house on fire. It completely destroyed the kitchen, and to this day, we barely have a kitchen, because some three months prior, we'd canceled the insurance policy, thinking it was a waste of money. Life bites you on the butt. And we really don't have the time or inclination to rebuild the kitchen. We've found that we can live quite happily with a two-ring electric cooker and a mini fridge. The only trouble is, we might want a sink put in, because we're using a little vanity bowl in the downstairs toilet, and anytime anyone comes over, they're looking at dishes and can't wash their hands.

There was a certain tragedy in losing all that work I'd amassed on various CDs and in notebooks—they were literally all lying there on the kitchen table—but it seemed pissy in comparison to the narrowly averted death of a human being, particularly one who is so precious to me. In a way, though, it really was for the best, because it was quite literally, for all of us, start from scratch. Start all over again. We found that environment incredibly refreshing. You get handed these bombs, these grenades, and they blow up in your face, but, from the splinters, you can pick up new ideas. So everything works out for the better if you're open-minded enough to adapt.

The whole disaster actually gave me the confidence again to realize that I know how to write a song, and I didn't need the safety blanket of a bunch of old ideas. I had to start afresh, and the pure joy of knowing that Nora wasn't burnt to death gave me the kick, the energy, and the joy to say to myself, "This album's going to be about *life*!"

We'd booked a studio converted from an old farmhouse in the Cotswolds, called Wincraft. It belongs to Steve Winwood—him again! We didn't go there for the sheep and the bushes, but by the same token, we didn't want to be in the inner city, with all that traffic-jam stuff. It's a distraction and it tires you out. There, we could meander in and out when we felt like it. It was a matter of the four of us committing to our future, being 100 percent dedicated. Nora stayed in London. She would've loved to have been in the Cotswolds, but it would've been wrong. It would've meant that my activity after hours would've been all revolving around her and me, and therefore I wouldn't be 100 percent committed to the project.

Everything for the album that became *This Is PiL* was written and recorded on the spot. There was no ego involved when we put songs together; it was like experimenting all at once, and hovering around a basic principle of a song, but expanding it into all the wondrous things that music can offer. The total joy of freedom. That's what our favorite one of the lot, "Deeper Water," is all about. It's about smiling in the face of adversity—"Face the storm/I will not drown." It's like our band's anthem.

It was inspired by a little pattern that Lu was playing on guitar, and a whole heap of conversations that we'd had with each other on the tour bus. Obviously, I've been getting into going out on the ocean in boats, but also, Lu is celebrated for having admirals in his family tree. When he was very young, his family bought a yacht and they all sailed around the world.

From that, I quite literally wrote the lyrics down, and we went straight into it, in a very live format, and in front of a load of British journalists who Rambo—and our wonderful press agent and radio-plugger Adam Cotton—had brought down to check out PiL in the recording studio. So the pressure was unbelievable, and yet we thrived in the fear of falling flat on our arses, and "Deeper Water" was born. Fantastic! It was a day where we knew just how good it can get. It's a song all of us feel very deeply emotional about.

We'd never try to duplicate that moment, however. We now know we have that potential in us, so we don't force it. Anything is possible. The more you pile the pressure on, the better the work. And the further inside yourself you go, the more outwardly you can project.

The whole album was a live production, with some twiddly bits done afterwards. Some of it was even influenced by our surroundings. "Terra-Gate" started out as something Lu had said to me: we were watching the lambs and the sheep in the neighboring fields being separated one afternoon, and you could hear the mother sheep *baaaa*-ing, like, "Where's me young 'un?" It was really upsetting, and Lu explained that it was because of the shedding gate. They separate them, and off to the market go the nippers, or the mothers that have had their quota of young. Hence, the terror gate. Us as a population, we're basically treated this way too—we're divided, we're separated, and taught not to like each other.

"Human" came about one beautiful summer's day out there in the Cotswolds, drinking cider. I just thought, "I've got to grasp this mood," because it reminded me of hot days when I was young—how we'd mess about on the bombsites, and what fun it was when you were carefree, and how it isn't like that anymore. I'm not five years old anymore, and I'm aware of so many problems around me.

The five-year-olds in big cities in Britain nowadays also don't seem to have the same sense of freedom we had back in the early '60s. Then, there were hardly any cars on the streets, and there were just piles of kids running around everywhere. Now people are frightened to go shopping because of gangs of youths gathering on the corner. But what else have these kids got to do? This institutionalized, "caring" government, and surveillance cameras on every corner—it's all created a really, really tense situation. There's no jobs, no prospects, no sense of community. There's just division, derision, and chaos. Not healthy.

I suppose it's all dealing with what I see when I come back here. I loathe and despise the bastardization of pubs, for instance, which used to be our community centers, into *swine bars* and gastropubs. Those can be fine too, but there's too much of it, and it's all so cold and indifferent. So, all that explains the line, "I think England's died." And there's a real longing for the place I remember from childhood: "I miss the roses."

Some of the other songs are about the London I love. "One Drop" has Finsbury Park as the backdrop, and growing up, and your childhood experiences, and how you see yourself fitting in—how coming from something has colored your perspective, for life—for better and worse. As it says, "We come from chaos, you cannot change us."

The song is also about getting on the good foot, and finding the groove in life. You don't have to be running around causing commotions all the time. It can be amusing occasionally, but if you really want to solve life's bigger problems, you have to understand the flow of nature, and learn to go with it, and not be constantly swimming upstream against the current.

I will always believe in the multiculturalism I grew up amongst. "Lollipop Opera," to me, is the soundtrack of my youth. It's a juxtaposition of all the musics that surrounded me growing up in Finsbury Park—everything from Jamaican and African, to Greek and Turkish. All those influences, sounds, and noises—the chaos of it all, yet the fun in it.

And then there's one about Reggie, my mate from Finsbury Park. My parents were Irish, his were Jamaican, and it never bothered any of us. It really was a very mixed bag, and you didn't feel like you were in any way special or different for hanging out together. Reginald views Finsbury Park as the Garden of Eden, which is hilarious. That's why he deserves a song, because of his lifestyle and the way he experiences life. He won't let the bastards grind him down. He's always a bag of entertainment. He is absolutely to me the epitome of what makes Britain great. I'd better not say anymore—he doesn't want much of a public eye on himself.

This Is PiL is full of heart and soul. I think it's the most serious body of work I've ever been a part of, with absolutely serious players. And every single one of us feels like we've been chiseled by the alleged shitstem.

Even though we are our own label we use outside distribution to make sure the records are in the shops and available online. It's no use making a record if no one can buy it. Cargo, our UK distributors, did a fantastic job on *This Is PiL*. The distributors in Europe, Japan, and Australia were good too, but for whatever reason, America was an uphill struggle. Hardly anyone could find it, even in independent or chain stores. Which was bitterly disappointing. The distributors we ended up with—who shall remain nameless—really seemed to drop the ball. But we live and learn. Having our own label has been a real learning curve; it's been very hard work, but ultimately very rewarding.

Who knows what the public want, but it's about time in my life that I declared what it is *I* want. Public Image is a hole I chuck money into. I have no expensive hobbies or habits—except PiL. I know the songs really truly mean something. They always have, but they're ever so much more positive now. We're looking now for answers, not just tearing the arse off issues.

Is there really a generation of people out there still listening and paying attention to music? Well, I know there is, the amount of young kids that come to modern PiL gigs, and they're all so eager to run off and start their own bands. That's the reward—not the money, not the chart position—because it means I've got good records to buy for at least the next five, six, ten years. Fantastic.

I still buy records. If I'm going to eat a chicken, I want the breast and the thigh also. That's my approach to music. I will always want that variety. I'm a consistent purchaser. I do not go to record companies for freebies. In fact, I resent freebies. I like the fact that I'm paying out of my own wallet, and I like to know that that money is going back to the band one way or another. That, however low the percentage is, I'm helping them. I like to feel a part of that.

I don't believe in illegal free downloads. These are creative people who have done marvelous things, and if you want to possess and own that slice of their achievement, please put some money in their pocket. Because otherwise they ain't going to be able to afford the next one. Every time you listen to something amazing in music, it inspires you to do something amazing in your own life. That's the payoff of your investment, pure and simple.

What's happened via the internet, causing small and creative record companies to fold, is such a catastrophe. That's why if you look at the Grammys these days, it's forever Taylor Swift and Jay-Z, and it always is going to be. They've hogged the top line. That was exactly the position when we first started with the Pistols. All the top spaces were filled, and the industry didn't want anything to challenge that dictatorship.

Beyoncé, Rihanna, Jay-Z: they're all Las Vegas—big production and lights and fireworks. It's all distractions. It's not really about very much else. Dancers. Showbiz. I've seen them live—Nora likes all that stuff—and the songs become very empty and pompous in a live performance. It's actually assumed and accepted that nobody sings live. I see no culture, there's no learning in it for me. They leave me so cold.

Music in the '70s was thrilling because there were so many differences, and I mean *extreme* differences, not like modern times now. There's this propensity to mélange everything into one prevailing sound. Modern music just sounds like shifty blends—more like elevator music—and I think it's to the detriment of music in the future. A very ugly pudding comes out of it.

Most of it is just metronomes on a computer, beyond personality, humanity, or involvement. It's all part of this attitude, "Music can't change anything, why bother?" It's all part and parcel of the same agenda, to take away the power of music from us. I'm firmly entrenched in the belief that a bloody well-written lyric to a bloody fine tune gets a bloody good response.

In the face of all that, I've played some of my favorite ever gigs in latter years—the two nights at Heaven in London 2012 were astounding, particularly the second. We had a great crowd at Glastonbury 2013, too, and we definitely rocked it. People who see us know they're getting the top class.

PiL, you see, is the full cultures. Lu's coming in there with everything he really truly understands, from the Muslim world and the former Eastern Bloc. My Irish background is giving me a great sense of open vocal-ness. I might as well be an Arab with a shamrock. What was I buying when I was young? Greek and Turkish folk music, and reggae, and chuck Alice Cooper in there for good measure! It's not like you have to go on a great learning curve. It's intuitive, and it comes out of your childhood. I can't do "Do-Re-Mi," but I can give you an "Aaaa-aaaa-llaaaah!" like you never heard, and yet it'll still be all right.

So, as the song says, if you're in a storm at sea, go out to the deeper water, it's easier to ride it out. Closer to shore, that's where the danger is. That's where you get swamped, and there's Johnny Rotten, going ever and ever deeper and deeper into the wonderful world of music.

PiL, as it is today, is my pride and joy, the culmination of my life so far. But it amazes me when I see these all-time top albums lists, and *Never Mind The Bollocks* is hovering around Number Ten or Fifteen, or even Twenty. What?! Oh, for fools! Not that I care about the positioning, or chart systems, but come on, that album changed everybody's lives, one way or another—for or against! It altered your perceptions. It had—*phwoooooaar!*—seismic repercussions on British culture, then into the outer world.

It's affected the Royal Family to the point where you can't really discuss them without the Sex Pistols cropping up in the back of your head. It's hard to deny, and very hard to avoid. In many ways, it's an unloving, but deeply personal and well-felt relationship we have with the Windsors. We're bound together. I almost have a certain affection for them. They're born into a gilded cage they can't escape from, and that's a terrible thing to have to endure. God bless 'em.

As for being an anarchist? Well, there's so many variants on the word "anarchy," it's ludicrous. In America, there's a whole bunch of organizations that are all "anarchist this, that or the other." But they all end with "dot org," which more or less tells you they're government-sponsored. And they're all feeding fabulous theories and lies about "Whoa, let's dismantle society and start again!" Well, that's not quite what I'm into, matey. I don't want the power of the bully with the biggest gun to take over, and that's what would happen.

What I want to see is a line of sense and sensibility creeping into things, where people start to think what it really is that they do, and realize that every action can have an equal and opposite reaction. Start thinking of yourself as an individual, and then you'll start respecting others as individuals. You have to learn to love yourself before you can learn to love anybody else. I don't have a moral agenda: what I have is a set of values, and I value each and every single one of us on this planet.

I won't use words like "legacy"—I'm still alive!—but we have to look after what we created in the Pistols. We don't run a museum here, we run a house of accuracy, and we don't want that ever to be allowed to get out of hand and be misinterpreted. It's happening all the time.

Awards don't interest me. Funnily enough, the other day I heard that the council in Tunbridge Wells are discussing whether to put up a memorial plaque for Sid Vicious. I don't think I ever knew this, but apparently he spent some of his early years there.

If he's watching out there somewhere, I imagine he'd take that with a great sense of fun, because that's absurd by any stretch. It's also perfect, and beautiful! Oh, I'd love to be at the unveiling. I hope it's a serious ceremony, with old council biddies in fur coats and tiaras. A suit-and-tie affair, hahaha! It'll be like the opening of a supermarket!

The thing about us is, the shriller and more angst-ridden the opposition gets against something like that, the more you drive and push for it. Suddenly it's like, why not? Automatically, people like me and Sid have got many issues about things like that, but the more they say, "No," the more we say, "Yeeeaaaah!" And vice versa? Dead right. If these things are not actively appreciating your life's work and your efforts—I mean, I know we're talking Sid here, but still!—what use are they? You can't fluff over that.

But certainly a blue plaque is more than I ever got. There was a suggestion of one years ago in Finsbury Park but the council utterly and completely refused. And I'm very pleased too.

In October 2013, though, I happily went to the Dorchester Hotel in London to accept the "Icon" award from BMI (Broadcast Music Inc.), who help collect royalties on behalf of songwriters and publishers. It was their highest award for songwriting and it was really the only time I've ever received anything from the music industry where I felt like the people actually understood me.

Right up until getting into the ballroom where the black-tie ceremony was held, I was suspicious. Awards go to people that toe the line and do things in a predictable, set-patterned way. A chap like me isn't like that. I'm incapable of toeing the line because I don't have the patience to understand what the line is supposed to be.

I turned up with some of the family and a bunch of top lads from my old manor, and we had a terrific night. Being honored in a room full of fellow songwriters really meant something. I even got to sing "EMI," with one letter changed, to BMI's bossman Del Bryant. It was a big night for him because he was retiring. He was a great fella and came over and sat next to me and Nora and had a chat, which he didn't have to do.

So what do you agree to, and what do you not? The previous summer, the Pistols were invited to appear at the London 2012 Olympics closing ceremony. Their idea was we'd be plonked on the back of a truck going around the outer circle of the Olympic stadium. They wanted to deck the truck out like the Marquee club and have us on parade waving at the crowd while doing "Pretty Vacant." It was naff panto. Watering us down to a sideshow.

To be honest, I wasn't sure about the Olympics. As a whole I felt it was a waste of money that could have been better spent on the NHS itself. I didn't want to do it live with the Pistols anyway, I didn't want to do anything with them, because I was firmly into doing the *This Is PiL* album and touring it, and I didn't want to be seen as "Oh look, he's going back to that!" I wanted that chapter closed.

I only got seriously interested when the director of the opening ceremony, Danny Boyle, took a personal interest in us. I don't think the closing ceremony people who originally approached us understood what we were about at all. Danny really wanted to include us and just would not give up. He was a Pistols fan and knew the cultural importance of what we had achieved, so we eventually had conversations with him. I liked him, and he told me the whole layout of celebrating the National Health Service. I was all for that—"Of course you can use us in a video scenario, that would be delicious, actually!"

Our stipulation was they *had* to include "God Save The Queen." They were pushing for "Pretty Vacant" only. We dug our heels in for both. They came back with the idea of only using the guitar riff of "God Save The Queen." Our argument was it meant nothing without the words. Quite frankly it doesn't. Then they only wanted to use two words: "God Saves . . ." They were getting squeaky on us as they knew the Queen herself would be there. But it *had* to be "God Save The Queen." All or nothing. It took a while but we won that battle. Perhaps we should have also gone for "Bodies" like Rambo had suggested back at the initial meeting with the closing ceremony people. He later told me Anita Camarata had been kicking him under the table as he said it, haha.

In the end we were the first band used on the opening segment—there was a short sharp blast of "God Save The Queen" including those very words as the camera panned up the Thames. A nod and a wink to our boat trip. It shows you how much things had changed. The BBC, who were broadcasting the ceremony, had denied "God Save The Queen" even existed in 1977; now they were showing it on the opening of the Olympics. "God Save The Queen" was followed by a tiny snippet of "Under The House" by PiL. It was Danny's idea to use it—he had told us it was one of his favorite ever songs. I admit I barely heard it, but was happy to know it was there. Somewhere.

"Pretty Vacant," which we agreed on for the ceremony, was pretty damn near excellent as a song idea, because the entire Royal Family was sitting there staring at this video—Johnny Rotten calling you "vay-cunt!" "Nice to see you!" And there is some possible plausibility in my belief that they might actually have enjoyed it. Because that's what we British are—we're kind of nuts. It could actually be taken as "That's very witty."

So, in this instance, one minute thirty seconds of full-on video screening was a much better way to see me. In America it was actually edited out. NBC, who broadcast it, put commercials all over our part.

I watched the ceremony in a hotel room in Poland. It was very difficult because we needed to get the cable switched on in the hotel—huge palaver. Very good watching that in bed—I felt quite chuffed. I liked it because of the politics of it, and there was a Dickensian element in it, and it was celebrating a full expanse and range of what British culture is. It's not all joyous ballet dancers and classical orchestras that make Britain Britain.

The NHS was clearly being celebrated as a major British achievement, and indeed it is. It is an astounding creation. It's just it should be available to the British, instead of anyone who gets off a plane or a train, which is what's bankrupting it. It caters for all nations, let's put it that way, and therefore there's not enough money to cater for the people who actually support it.

Please, please, don't go mistaking my views as similar to those of that twat, Nigel Farrage, leader of the UK Independence Party though. It's the connotations he places on that, that bring up the ugly scenario of racism and nationalism, which is not what I'm saying at all. It's about, I don't go to Korea and demand a heart operation for free. It's opportunists who do this stuff, I'm afraid, and this is what politicians are, they jump on it and swing it to their agenda. Farrage just holds animosity for all things outside—*Ausländer*, as the Germans would call them. It's pretty damn fucking disgraceful.

What I'm saying is, please do take care of British citizens first. And emphasize the word "citizens." If you've been welcomed in, and you're in, that's it. You're British. Hello, how do you do?

When I was young, there'd be a lot of people from all over Europe who would come to England just on a tourist visa and be claiming dole money, and it was never questioned. It's the same kind of scenario now. It's never been properly clearly defined that it's a no-no. It's just abuse and theft. It's been going on a long time and nobody seems prepared to deal with it, because of the interpretation, quite wrongly, that it would be "racist." It's a minefield.

Who'd be Prime Minister, you say. Well, maybe I would! I'd make this seem very clear, because the situation *is* clear. It's because there are so many taboos in what you can and can't say that it becomes an irresolvable quagmire, because nobody's actually saying what is really going on.

In July 2012 BBC TV had me on *Question Time*. To be very honest, I turned up frantic, mid-tour. When I'm touring I find it hard to concentrate on anything else. It's quite a switch from live gigs to live TV. Your mind needs to refocus. I was excited about doing it but also very wary.

I'd watched the host, David Dimbleby, ever since I was a kid. I was thrilled at the prospect of meeting him, but I was also cautious, as he's a smart bunny, and I thought he might have it in for me. But he turned out to be such a great fella—we had a good chat after the show. If you wanted to find someone whose political point of view was closest to mine, it would probably be someone like David. He struck me as a person who can sort the wheat from the chaff.

I have met a few exceptions over the years, but in general I don't like politicians. I know that most of them are in it for the bump-up, which eventually leads to business management, or corporate seating. They're not my kind of fellas. They're what keep the country down.

I was on the show with the Labour MP Alan Johnson and the Tory MP Louise Mensch. I wasn't sure about Alan at first. He was wearing too sharp a suit, and he came across as a bit smarty-pantsy. But I can be that way too at times, haha. I spoke to him and his wife after, away from the cameras, and we really got on well. The Tory woman was harmless enough, I suppose, but I was not convinced by her drug references. Isn't it funny that when politicians are caught taking drugs it always ends with them having a bad experience, and saying how much they regretted taking it in the first place.

Once the show started, I don't think I stepped out of line or talked nonsense. I told it like it is, about educating young people about drugs, and the evil of top-flight bankers—the unmentioned dictators of a country's economic policies.

My only squabble was with a social worker type in the front row, who was trying to lecture me about the damage drugs can do to young people. But the point I was making was that those kids need access to as much unbiased information as possible. They must learn for themselves.

Overall, it was a really very positive thing to do. The next day when we left the hotel, myself and Rambo, and our press agent and friend Adam Cotton, God help him, decided to do a pub crawl all the way back to London—and the show was filmed in Derby, by the way! There'd been a major rainstorm where Derby was flooded. It must've been the water that influenced us. And let's face it, *Question Time* was a tough thing to walk into as Johnny Bloody Rotten. I needed a break! Every pub we stopped in, people had seen *Question Time.* It was quite surprising how popular we'd seemingly made the show. People watched because they knew I was on, and they wanted to hear what I had to say. People of all walks of life and all class systems—*all* were favorable. Not one negative from anyone in the general population.

Right from the minute I adhered rigidly to the nickname Johnny Rotten, I painted a target on my back. I've always known that, and no matter how many different items of clothing I go through, or hairstyles, the target remains.

I came out of the paddock full steam, and it created a lot of flurry around me, and a lot of resentment. Who is this upstart? And I still get that vibe from people, they still view me as being negative to their cozy idea of what a musical reality is. And so, by all means, I'm going to get the hammerings. Bloody hell, I'm an anvil: you can hammer it all day long, you ain't going to dent it.

And in fact, I rather like the attention of resentment and jealousy. When people take it that far, they're almost rewarding you for your own efforts. It's complimentary!

To this day, there's still people that send hate mail. Many, many more send favorable stuff, but there's always that odd on—"I want to kill you"—that you get in the post. I get that, still. A lot, from the same kind of people. You have to keep in mind that these people can't help themselves, and at least you're entertaining them, or giving them a purpose in life, even though it be your destruction and downfall.

One way or the other, I've got to look at it like, "Well, at least I'm a means to an end." There are a lot of things to hate in the world. I don't think I should be one of them. There's far better targets, but I'm more than happy to accept it.

Sometimes it can be just too bonkers. There have been women that have come up with fantasies. They send real serious hate mail, and leave things on your front door. Just unacceptable. You hear about the stalkers around Madonna—but I get that too. And it happens quite a bit. There's always one of that kind out for you in every different country, and you have to protect yourself from it.

At the present time, some of my cases are ongoing, and some have been quiet for a mo. One particular girl used to get very, very insane. She sent letters declaring that she was the real German heiress, and Nora was the fake and had to be replaced, and then we'd fall madly in love, and everything would be all right from there on in. Eh? What?!

She actually ended up working on a music TV show. I talked to her, and she'd managed to break out of it. Her job required so much energy and effort that she had no more time for that, and saw it as foolish. So hopefully that's what she actually believes. It's nutty that it can unfold that way, but once she'd found a career for herself, that was the answer to it.

It's the problem of the modern human being that we've got more time on our hands than we can handle, so it goes into the wonderful world of craziness. And I hope that no one misconstrues that as meaning that we should all be back in slave camps, where there'd be no time for that kind of behavior! Once we were plebs, but now we're left to our own devices, and some of us come up with the wrong agenda. It's a bit like, if you have a late-night club scene in a town, you're gonna get crime. Are you able to take the crime, because the late-night scene is so good and inspiring and artistically proper?

So, as I've said, I don't answer my own hotel room door. I need a bloody witness at all points. Your open-to-the-public thing is jeopardized by this one lunatic, who, if they really go for it at some point, will try and put a bullet in you—if that be the tool of choice. John Lennon was an example of that. It happens. People attach to what they see as celebrity figures, and it becomes very dangerous. Many people I've talked to—actors, even playwrights once—they're all telling me they get it. Somebody just decides, "I've got to kill you, because you let me down, you're not making the music you should. If I was you—blah blah blah!"

I have a great sense of foreboding, fear, and empathy for Adam Ant, for instance, because I know he's had this a lot, and I know what it's like. It gets very, very dangerous. It seems to come and go, then raise its ugly head again for no good reason. You have to be very aware that everything you say or do is being misjudged by one of those very few people out there with that propensity towards psychopathic behavior. I don't know what makes them so lonely and so bitter and twisted, that they spend their whole lives, firstly adoring and loving you, and then spin it, for God knows what reason, to wanting to hate and kill you.

I love to go out signing autographs and talking with the fans at gigs, but I have to be aware that in that element there is that danger, that lurking propensity towards ultraviolence. They don't even understand what it is they're doing themselves. They feel completely justified. Psychotic! I don't have any disciplines really to steer that away, other than the way I am. Look what Adam had to do—he had to leave Los Angeles. Christ! And he ain't no wanker!

At the other extreme, I've been getting untold grief over the years from the so-called authorities, particularly when trying to clear immigration.

Coming into America, they'd pull me over every single time, even though I was an "alien resident" green card holder. I was viewed unfavorably. I'm not making a special case of myself, but I did get extra-frustrated with delays. I can't sleep on planes. I'm uncomfortable. I ache. It doesn't matter what I do or take, it doesn't work. All of those things that are supposed to help you relax and sleep just annoy me. They only make me extremely tired and bad-tempered when I'm facing an immigration official.

That's when they want to ask me foolish questions, and things can go wrong. You have to bite your tongue, and I'm the kind of bloke that can't do that. They'd take me into a cubicle, and it'd be like, "Yeah, you live here, but why haven't you applied for full citizenship?" I later found out that this was partly because of my old amphetamine conviction in 1977 which still showed up despite me holding an American green card.

They've made my life hell in U.S. immigration. Every single time, I was made to sit there for two hours, with the other undesirables, most of whom couldn't speak English. The immigration officer would go, "Well, it's because the English won't drop the file on your conviction." Law can be tediously cruel.

So, enough was enough, and I applied for U.S. citizenship. It was then they told me about the open file, and yet they still allowed me to go through the procedure, even though I'd delayed it so long. It wasn't an easy decision. In my heart and soul, I feel like, "Have I walked away from something here?" Yes, I have—I walked away from unnecessary abuse. I'm the same person, but I've just now got bigger guns. And I'm a pacifist and all that, but you've got to stand up for yourself. If that's your only alternative then you've got to take it. Sorry, but my days of anal searching are gone. I won't allow myself to be physically and mentally abused any longer.

Then, there would sometimes be problems at the other end, arriving in London. The worst time was when my father died, and I was absolutely distraught. I got on the first plane out, and of course got pulled up at Heathrow. I'm in bits at the airport, and they're absolute cunts to me. Just being wicked.

After that happened I got to the point where it's like, "If you don't want to let me back into England, I don't give a fuck. I'll turn around and go back." There's nothing in that society that stood up and defended my rights. Obviously, all this goes back to "Anarchy In The UK" and "God Save The Queen"—*of course it does*! What a fool I'd be to say otherwise. Even though those songs are now how everybody thinks—it's "shoot the messenger," really. When I go back and see everybody taking a pop at royalty on English TV, I think, "Where were you when it counted?" Bunch of second-rate cowardly comedians. Now they do it in this personal-attack way. I've never tried to do anything in a personal-attack way. I'll say it again: it's the institution, not the individuals involved.

And so my personal freedom has continuously been assaulted and abused because of everything that I ever did and said.

England is absolutely dedicated to humiliating you into old age. It kills and stifles creativity. You're supposed to be dead at forty, and the rest of your life you're supposed to rot in misery. It's always "Why bother, it's all been done before," or "Act your age"—those are *such* passion-killer statements. It's a fear of change, a fear of anything new and exciting. Thank God for California. Johnny Rotten definitely gives this place accolades. They go bungee-jumping here at eighty-five, they're diving off mountains—it's full-on activity, keeping the brain alive. England just doesn't seem to want that.

It took me a long time to become an American. I finally completed the process in late 2013. Now I feel like I was born an American. The rights and freedoms of all, and the belief that all of us are born equal—these aspects are in the American ideology. It's not perfect in its working relationship with its population, but it's a hell of a lot better than the shit I've had to endure all my life under a British government.

I've never had that problem with America. It's always liked me. Bizarre as the governments here are, and how they don't like anyone to be anti-American or whatever, I can say what I want. It's appreciated as being a vital part of the Constitution, so in many ways I've achieved incredible success being accepted in America. This is the country that fought against the corruptions of monarchy and imperialism and the British Empire of that day, and the trouble that created for everybody around the world. Not that America is completely innocent itself in that respect, but what a fantastic place to live.

Nora and I once drove all the way from New York to L.A., just the two of us staying in the motels. It took about a week, just exploring that terrain. It was a complete adventure. I love not only the impressiveness of the cities here—the difference between

New York and L.A. and New Orleans—but also the landscape. Driving across the country, how it changes from state to state. Every thousand miles is a completely different climate. It's utterly fascinating to go from pine forests wrapped around San Francisco up north there, into the deserts of Utah and Nevada. *Phwoar!* And the swampiness of the Southeast—*wowzers*!

My God, this country's nuts! The natural disasters here—they don't happen on a small scale. It's not like in England—"Oh my God, look at the flooding"—here, everything is industrially bigger. Earthquakes, fires—they're all seriously major. From an English perspective, when I was young, I'd think, "Oh, those Americans are always showing off and making a big scene out of it." Well, you need to! It's such a big place geologically, that mad things happen.

Mentally and physically it's improved my health, improved my outlook on life, and it's taken away the opportunity of despondency. In many ways, I miss the culture in Britain, and in many ways I don't need to miss it, because it's *in me.* It's firmly there, the good side.

I know many English people who've not been able to make it work, living here, and have eventually gone back. There's a very big English community here around Santa Monica, and their problem is, they're trying to be English in a non-English situation. Big mistake. When in Rome, do as the Romans do. It doesn't mean you're going to lose your identity. You've gone to live abroad but you're refusing to acknowledge newer, better, or different ideas as even a possibility, and so you hoodwink yourself into your own prison and failure.

Nora and I don't hang out with expats. We have a very small circle of friends—one or two, say. Apart from that, it's what me and Nora want to do. We do everything together. I don't really like or want a large collection of acquaintances.

Here's the laugh of laughs, though: not far from our house in Malibu is Herb Alpert—the Latin-jazz trumpeter who was also the "A" in A&M Records! There's some difference in the size of our respective properties, let me tell you. He has half a mountain. But I know it more than bugs him that I live here. Talking with the neighbors, they've told me so. Well, that's your comeuppance, you fuck.

London has tended to be work, over the last decade, but for the last couple of Christmases, we've been back to see my brother Jimmy, whose cancer is now in remission. Last year we had a beautiful Christmas party at his house, and we love it. His kids come over, Katie and Liam, with their significant others, and it's a house of really happy people.

And so this is the way we are, just quiet—in our loud way. Quiet for me, I've got to be fair, is fairly *un-quiet.* I can see that now, but that's only because I'm in moments of contemplation, putting this book together.

When I'm out in the public eye, that's full-on John! The only things I'm ever secretive about are my domestic scenarios, which I don't think anybody in the public has any right to. I've always gone out of my way to keep my family out of the gossip magazine nonsenses: the *Hello!*'s of this world. You have to have a reality to go home to, not a TV crew.

The foregoing pages have, I think, let a little light in on my life. If that isn't met with all due respect, then we'll just close the shutters again. We're not bad people, us lot. Generally speaking, we're very good-natured and mean nothing but joy to the world. As far as I'm aware, I've brought nothing but joy to the wonderful world of anything I've ever touched.

When the offer to appear as King Herod in a U.S. touring production of *Jesus Christ Superstar* came in, my initial reaction was "Stop it, Rambo, you're teasing me again." Then, shock-horror, I thought, "Oh no, this git, this instigator, really is stirring my pot, like he did with *I'm a Celebrity.*" Right off the bat, I obviously said there was no way on earth it could ever work, but he goes, "I think it could," and Nora said the same. It was their common sense that got me around to thinking, "You know what, I could do this." With all the scripting and stage directions, it was essentially forcing myself to take orders. Very seriously: the *Final Challenge*!

So I'm doing a musical and what better one to do than *Jesus Christ Superstar*? Ah, the hate mail. The naysayers had already made their minds up anyway. As Forrest Gump said, "Life is like a box of chocolates," and them lot are the delicious soft centers. It is a reward to be chastised by the ignorant. You know, it's all right, there's nothing wrong with rock 'n' roll musical theater. I liked *Quadrophenia* very much—well, maybe the first hour.

The shock value didn't matter tuppence to me, really. It was about what I'd get out of it as a human being. My learning journey through life. I just love the commitment, and the challenge, and the grate on my nerves.

I didn't know any of the other cast personally beforehand, but I didn't underrate them at all. It was a really mixed bag. Brandon Boyd from Incubus was going to be Judas. Michelle Williams, the girl from Destiny's Child, was playing Mary Magdalene, JC Chasez of *NSYNC would be Pontius Pilate, and Ben Forster had already played Jesus in the UK production. And John Rotten Lydon would be King Herod. I'm here to sing with the King of the Jews—who could ask for anything more?

In April 2014, there was a press launch event in New York. On the day before, I had a costume fitting and then a bunch of us were invited out for a meal including myself, Rambo, Ben, JC, Brandon, and the promoter Michael Cohl. I met Tim Rice and Andrew Lloyd Webber for the first time backstage right before the press conference.

At the event itself, there was a bit of an atmosphere. When you meet people for the first time and nobody knows what they're doing, certain types of personalities can take offense where no harm was meant. It was six of one, half a dozen of another.

It was all jolly hockey sticks, and all very theatrical with all the Tim Rice and Andrew Lloyd Webber people. The whole cast was nervous. They reinforced that thing I always suspected: we wallow in the fear, we can't avoid it. In fact, we go looking for it, and then can hardly handle it, it's so tense. We're addicted to self-inflicted torture!

It felt like a band-on-tour vibe. Everybody was really excited. It was going to be six to eight shows a week—proper hard work. I was only going to be onstage to do my one song, but that was almost an extra pressure—I better get that one thing right.

From there, I dived right into it. Rehearsals started in New Orleans in June at a wonderful old converted fire station. It was the perfect place after all the hustle and bustle of New York. I left my ego at the door. I went into a genre, and a type of singing, that I had nothing but negative feelings about, and I found it to be a very generous and rewarding world where people really do share.

I was careful *not* to check out previous performances of "Herod's Song." Rik Mayall RIP had done it in one UK stage version. He used a lot of early Rottenisms, apparently, so that way I could've ended up parodying myself, by default, without even intending to.

In rehearsals, the female singing instructor said, "Don't worry, John, we're not going to bother you with the 'do-re-mi's.' " I've been singing for nearly forty years, and yet I was really worried about hitting a bum note in practice, before I even sang the song. And I couldn't go into my usual trip of not talking and sitting there shaking with nerves, because that would've been unfair to everybody around me. We were doing it in fairly tight quarters.

At the core of it, I wanted to understand stagecraft, and not from a cynical outsider's point of view. It was a bloody great opportunity to just *fiddle about* in a proper way. I had good chats with Laurence Connor, the director, who I found very helpful and great to work with, and the bloke playing Jesus, Ben Forster, came down to help me—he was fantastically friendly. Everyone was willing you to do your best. I only ever thought that happened with panto. There was a great sense of "bond," and "ensemble," which to me had always just conjured visions of pianos falling down staircases.

It stops you being shy, which is always gonna be a problem of mine, but it also stops you being an egomaniac, which is also a problem of mine. And one, sadly enough, is a direct response to the other. With both of those things left to the side, it was a delicious learning curve.

Within two days I found myself bouncing around, learning dance moves, and being very happy doing it—very happy adhering to somebody else's song. And being able to shapeshift in and around that, and being given the space to do so. They were saying, "Well, now you can add your flavor, if you want, John." Fantastic generosity, sharing, and camaraderie.

They wanted me to play King Herod a bit like a slimy game show host, but with a Rotten twist. The clothes were an important part of that. I *had* to be involved in the design. I *love* my clothes. I needed to feel I was wearing the outfit and it wasn't wearing me. I didn't want anything that had gone before in previous productions. So I worked closely with the production and the design teams to come up with something special. I probably drove them mad, but it had to be right. I'd originally toyed with the idea of doing it as a slick '50s Teddy Boy but in the end we decided on a variation of a nineteenth-century Southern gentleman kind of look. Do you want a bit of detail? Johnny will give you attention to detail! Tangerine, purple, and mauve paisley brocade jacket—with black satin lapel, collar, and cuffs—embellished with Swarovski Crystal sequins. White butterfly-collar shirt with a thick purple Ascot knotted cravat-type tie, and an orange-gold waistcoat. Black trousers and patent leather Prada brogues. We had the suit made up in New Orleans and when I saw it on my arrival it was exactly how I had pictured it. When I put it on I fell completely into the role.

I practically ran to the rehearsals each day. Everybody was so ripe and ready for it, and then—*bang!* The promoter, Michael Cohl, decides to pull the whole tour.

I was in my room doing phone interviews, and I was angry

because I wanted to get to rehearsals—I wanted to do some work around the old piano with the vocal coach and the piano player, in advance of actually fully rehearsing. I'd got dressed early—I wanted to go down early. I went into Rambo's room, and he was in there in a meeting with two of the tour's management team. They were all very quiet, and I bounced in—"Hello everybody!"—and Rambo stood up and went, "I've got some bad news—it's off."

I thought he was joking. No, he wasn't. My face hit the floor. It was really, really sad. Rambo says he almost saw a tear in my eye. They filled me in, and it was a dizzy moment in the head. My first thought was "I've got to go and meet the crew and cast, to see how they feel." They were all gathering in the hotel lobby, and we sat around the bar and just talked. It was very, very, very sad. But then we started to break the ice with bad, dark-humor jokes, and then eventually—"Well, let's go out and have lunch."

Eventually, the production company took us all out to an afternoon lunch/dinner, the last supper as we called it, and that went on to "Well, let's go clubbing," and quite frankly you couldn't pick a better place than New Orleans in the entire world for that. We really had a laugh—this huge mob of theatricals, which I was more than happy to be one of. It was like one of those dance movies, where everybody gets up and dances out on the street—that's kind of what we did. We danced away to the clubs, and danced from one club to another.

It was such a tragedy, because to my mind this was going to be a most excellent tour for me. I didn't have the responsibility of "leadership" here, I was just "one of." It was a very good feeling. I knew what my role was, I knew what my part was, I knew how I had to do that, and not let everybody down by showing off and being a ridiculous whatever—but then all of that was taken away from all of us. Amazingly catastrophic.

At the same time, I met some really excellent people. My God, my heart goes out to them, because everybody had committed, we'd all given up lots of work, and lots of other things we could've been getting on with, and the check's just withdrawn from you, and you're facing financial ruin, in some cases. Some of them had rented out their apartments and had nowhere to go.

This is what happens in life. It would have been a sensational production. It would have worked. I can happily say it wasn't any of *our* faults.

For the moment, of course, that now leaves the way clear for a new PiL album, which we're hoping to start around the time this book lands in the shops. It's a pity that it comes in on the tail of a letdown, but I won't be using that as a backdrop. Disappointment to me is a constant. All that work and learning won't be wasted. I'm going to pull all the positives out.

I'm also looking back at acting in a different way now. It's given me a confidence to accept those challenges and to work inside other people's scripts. I think I have enough going on inside my own head now, that I can take those challenges on without challenging the challenge.

I love my life. I manage to get myself in all manner of pickles, usually of my own making. I can't forever and a day be out there—duh!—left to my own devices, because I'm used to Johnny Rotten, but I don't want Johnny Rotten to be just Johnny Rottenisms in another field. Because that's not how this is going to work. In the same way as when I did shows like *Shark Attack*, *Goes Ape*, or *I'm a Celebrity*, I had to leave my ego outside, and get on with what I was doing, and just be myself, and actually learn to like myself.

That's all been very good research, and I found out a great deal about myself. I've got to get over this self-doubt thing. It's still there. It always will be, that fear of letting people down. It's a hard anvil I've tied around my neck. But it's what gives me the energy to go on and go forth. Like, if there's more wild and crazy offers there in the future coming out of this, then I'd be more than happy to take 'em on.

It's that endless challenge of "What's next?" You know: "Okay, done that . . . what's next?"

Actually, what *is* next?

Come ooooooon, what's next???

THE FINAL NOTE . . .

One last thing I want to see in this life is for me personally to hit the highest note possible. And if that requires that I burst every blood vessel in my body, and my brain pops out of my ears, that would be for me the most wonderful way to die. I don't necessarily mean on a stage, in a theater or an auditorium; I could do that on my deathbed. But to finally reach that note that connects with God . . . spelled backwards is still "dog."

But that's serious, right? I know I'm wrapping that around a cocoon of "Wow, is that boy off on something!," but I've given myself this job of singer, communicator—translator, really—and that's what I'm still searching for. That final note. I can feel it in me. It's about finally reaching something that's, aaah, beyond human, beyond your alleged capabilities, and you don't know what that is, but that's what you're striving for.

I think that's what everybody strives for. What was that Andy Williams song: "To dreeeeeeam the impossible dream." As an idea, it's on the lunatic fringe, I understand that, but I don't know anyone who isn't a lunatic, frankly. We're all striving for something we can't possibly get to. For me, it's hitting that note that only a dog whistle can reach, or the lowest bass note that'll make your bowels drop out. None of us have those capabilities, but if you can actually feel that you've reached there, that's the perfect time to die.

And this is what it'll say on my gravestone: "That last note was *unlistenable*". . .

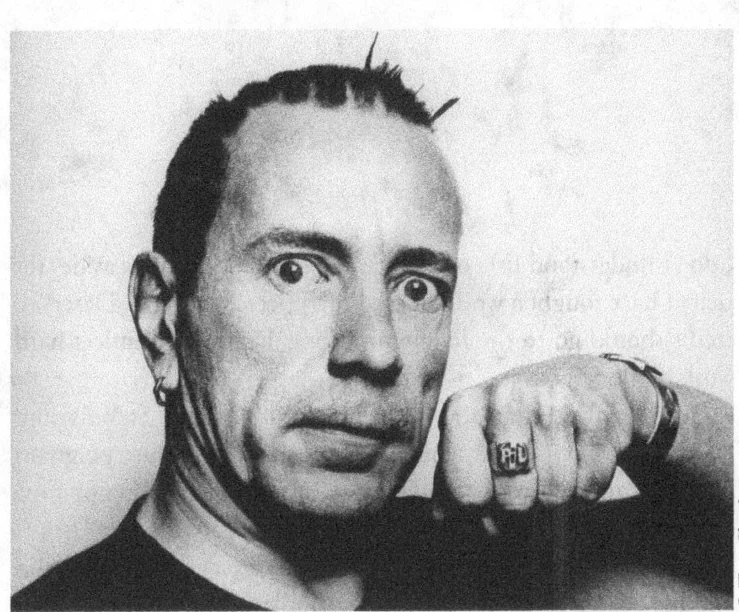

© Tom Sheehan

ACKNOWLEDGMENTS

I don't understand lists of credits, never have, but titles aside, for years I have fought a world of credit-hoppers and fakers. Deserved credit should go to the following people for their relentless hard work on this book.

Additional editing and research: John Rambo Stevens and Scotty Murphy. Thank you for your creative genius, poignant opinions, and generosity. You have learned, shared, and taught me a true sense of art and reality. Thank you.

I would also like to thank Adam Cotton's brilliant genius and his ability of always taking the simplest route to get the best results. This is the true definition of a press agent. Thank you.

I want to thank all three of you for your ongoing loyalty and trust. Thank you.

I also want to say thanks to "the Eastender" and all at Simon & Schuster for giving me the opportunity to tell my story. Thanks also to Agent Pocklington. Thank you.

I'd also like to thank in no particular order . . .

All the managers I've ever worked with because of the different aspects they helped bring to my career. Giving me insight I may not have otherwise considered. Thank you.

My accountant, Harish Shah. Thank you, Harry, for your loyalty and dignity. Thank you.

All the lawyers I've needed over the years, and a special thanks to Alexis Grower for your understanding of the artist. Thanks also, Pierre Alexander. Thank you.

I want to thank *everyone* at Virgin. The whole lot of them. People need to understand I never had a problem with any of the people at record companies—it has only ever been with the faceless accountants. I have worked with some very creative people with a real appreciation of the artist: Simon Draper, Ken Berry, Keith Burton, Jane Venton, Sian Davies, Sue Winter, Kaz Utsunomiya, Sarah Watson, and Paul Bromby to name but a few. Thank you. Also special mention to Michael Alago and Howard Thompson for their support at Elektra Records and making the record industry entertaining. Thank you. I'd also like to thank everyone at Universal, who now own Virgin and EMI: there are some real old-school creative people there like Johnny Chandler who really *care* about music. Thank you. Sorry if I missed anyone at any of the record companies, but let's be honest, there was a lot of you . . .

When I look back the most original photographers whose pictures have lasted the test of time are Ray Stevenson and Joe Stevens. Both had a great eye and their photographs still relate to me personally. Thank you.

I want to thank all my bandmates over the years. All of you. Yes, every one of you. I especially want to thank my present PiL bandmates Lu Edmonds, Bruce Smith, Scott Firth, and all our crew. This is PiL. May the integrity continue . . . Peace, love, and respect from Bunty Bratislava. Thank you.

Hello to Murray Mitchell, old PiL guitar tech. Thank you.

I'd like to give an enormous amount of respect to all Japanese fashion designers, especially Issey Miyake, Yohji Yamamoto, and Comme des Garçons. Thank you.

I don't have a huge social scene, that's not what me and Nora do. I have billions of acquaintances, the ladder is long, thank you, but the friends I have are for life. Thank you.

John Gray, thank you. Paul Young, thank you. Reggie Williams, thank you. All the lads from Finsbury Park, there's-no-place-like-home-is-where-the-heart-is, thank you.

I owe my love of *Real Racing 3* to Tony Rackley in Los Angeles. He tried to get me interested in cars. He did, just not in the way he thought, haha. Honk Fart. Thank you.

Scotty Murphy, nothing but integrity for over fifteen years. Thank you.

Dave Jackson, absolutely the most wonderful stage designer. A fantastic bloke to work with and a very close friend. Thank you.

John Stevens. Rambo. Without whom none of this would be possible. When we met everything changed. We managed to take our friendship into a business situation and make it work in an exceptional way. Thank you. Mr. John and Mr. John. Thank you. My best mate ever. THE END. Thank you.

Andrew Perry, thank you for working with me and you have a list of people who made you what you are today. It's wonderful for me to see a journalist break the mold . . . Thank you.

Andrew Perry thanks the following: John Lydon, for giving me this life-changing opportunity, and for all the inspirational music that has electrified and educated me since I was twelve; Rambo, for getting me on the team and sorting everything; Rambo and Scotty Murphy, for their indefatigable attention to detail in approving the text—research is a 24/7 pursuit, I learn; and Adam Cotton, for letting me be his first interviewer in '07, and being a diamond.

Also: Kevin Pocklington, my agent at Jenny Brown Associates, for fishing me out of journo world while also being a PiL trainspotter; and Kerri Sharp, my editor at Simon & Schuster, for getting involved, letting me get on with it, and talking Torquay. Up the Gulls! 'Nuff respect to Jonathan Wingfield, whose baffling publication of a 35,000-word proto-biography of Lydon was pivotal in my achieving long-distance authorship.

This is my first formal opportunity to thank everyone at *Mojo*, *Q*, the *Daily Telegraph*, and *eMusic*, without whom I could easily have been filing bastard insurance contracts all these years. The more I think about it, that's a BIG thanks.

To mi breddren—Declan, Andy (old punk!), Leon, the Diskos of Pembrokeshire, Sophie, Rhys, Super Anna, Glad and Mark, Rockin' Harry, Sherry, Toddy, Amber, Padfield (what, no sandwich?), Sompey Paul, Hannah and Paul, Chilli and PV, TSOOL (RIP, sniff), and all the lively peeps I couldn't possibly list—you're spiritually in there somewhere.

My part in this venture is dedicated to my family: my late father, who opposed punk as vociferously as Bernard Brook-Partridge, yet deep down was more Rotten than he ever realized; my mum, just turned eighty, who's a shining light to us all; my brother, Jimi, who in his dual capacity as my attorney, still owes me two hits of ether, and Sheillah, Bumble, and Abs; the "outlaws," Gill, Ange, Ian, and Samuel; and finally, my girls—Lisa, my babe, and Rose and Georgia—Daddy loves you all, and he's coming back dressed as a tigger . . .

INDEX

Note: The initials JL in subentries refer to John Lydon; PC = Paul Cook; SJ = Steve Jones; KL = Keith Levene; MM = Malcolm McLaren; GM = Glen Matlock; SV = Sid Vicious; JW = Jah Wobble.